The Politics of Carbon Markets

The carbon markets are in the middle of a fundamental crisis – a crisis marked by collapsing prices, fleeing actors, and ever increasing greenhouse gas levels. Yet carbon trading remains at the heart of global attempts to respond to climate change. Not only this, but markets continue to proliferate – particularly in the Global South.

The Politics of Carbon Markets helps to make sense of this paradox and brings two urgently needed insights to the analysis of carbon markets. First, the markets must be understood in relation to the politics involved in their development, maintenance and opposition. Second, this politics is multiform and pervasive. Implementation of new techniques and measuring tools, policy development and contestation, and the structuring context of institutional settings and macro-social forces all involve a variety of political actors and create new forms of political agency. The contributions study the total extent of the carbon markets, from their prehistory to their contemporary expansion and wider impacts.

This wide-ranging political perspective on the carbon markets is invaluable to those studying and interested in ecological markets, climate change governance and environmental politics.

Benjamin Stephan is a postdoctoral researcher at the Centre for Globalisation and Governance, Hamburg University, Germany.

Richard Lane is a PhD researcher at the Centre for Global Political Economy, University of Sussex, UK.

Routledge Studies in Environmental Policy

Land and Resource Scarcity
Capitalism, struggle and well-being in a world
without fossil fuels
*Edited by Andreas Exner, Peter Fleissner, Lukas Kranzl
and Werner Zittel*

Nuclear Energy Safety and International Cooperation
Closing the world's most dangerous reactors
Spencer Barrett Meredith, III

The Politics of Carbon Markets
Edited by Benjamin Stephan and Richard Lane

The Politics of Carbon Markets

Edited by Benjamin Stephan
and Richard Lane

Routledge
Taylor & Francis Group

LONDON AND NEW YORK

First published 2015
by Routledge

2 Park Square, Milton Park, Abingdon, Oxon OX14 4RN
711 Third Avenue, New York, NY 10017, USA

Routledge is an imprint of the Taylor & Francis Group, an informa business

First issued in paperback 2016

British Library Cataloguing-in-Publication Data
A catalogue record for this book is available from the British Library

Library of Congress Cataloging in Publication Data
The politics of carbon markets / edited by Benjamin Stephan and Richard
Lane.
pages cm. — (Routledge studies in environmental policy)
Includes bibliographical references and index.
1. Emissions trading—Political aspects. 2. Carbon offsetting—Political
aspects. 3. Carbon dioxide mitigation—Political aspects. 4. Air—
Pollution—Political aspects. 5. Environmental policy. I. Stephan,
Benjamin. II. Lane, Richard (Research student, University of Sussex)
HC79.A4.P646 2015
363.738'746—dc23
2014010958

ISBN: 978-0-415-70713-8 (hbk)
ISBN: 978-1-138-20515-4 (pbk)

Typeset in Bembo
by Swales & Willis Ltd, Exeter, Devon, UK

'*The Politics of Carbon Markets* offers an enlivening antidote to carbon market fatigue. The assembled articles avoid moribund debates between proponents and opponents of carbon trading by anchoring carbon markets in the broader concerns of environmental politics; and not a politics of governments and policy instruments, but a postfoundational "lively politics," which interrogates questions of nature, culture and power. Benjamin Stephan and Richard Lane's edited volume is a must read for those seeking to understand the past, present, and future politics of carbon.'

Simone Pulver, University of California at Santa Barbara, USA

'Can carbon markets be part of the solution to global climate change, including in the Global South? Can they make development more sustainable as well as reduce emissions? This edited volume takes a historical perspective on the evolution of carbon markets, mainly in the North. With close analysis of the political construction of markets, the volume offers food for thought for those who want to understand how it might evolve in developing countries in future.'

Harald Winkler, University of Cape Town, South Africa

'Carbon markets have long been understood to be exacerbating the very climate crisis they were supposed to help address. What is less understood is why they stagger on regardless. This timely volume greatly illuminates the deep politics of these so-called "zombies" and of the zombie scholarship that sometimes accompanies them. In so doing, it lends welcome empirical substance to the study of that more encompassing zombie phenomenon known as neoliberalism.'

Larry Lohmann, The Corner House, UK

Contents

Illustrations

Figures

Tables

Contributors

Philippe Descheneau is a doctoral candidate and part-time professor at the School of Political Studies of University of Ottawa. His research concentrates on the Clean Development Mechanism and the development of carbon markets from the perspective of international political economy and sociology of markets. He has research interests in the role of money, finance and resistance in environmental markets.

Anita Engels is Professor of Sociology at the Centre for Globalisation and Governance, Hamburg University. Her fields of research are carbon markets, corporate greening and climate change and society. She is also vice speaker of the Cluster of Excellence 'Integrated Climate System Analysis and Prediction (CliSAP)' (DFG EXC 177).

Richard Lane is a DPhil candidate in the Department of International Relations, and researcher at the Centre for Global Political Economy, University of Sussex. His research is focused on the historical interrelation of economic and environmental governance – particularly the development of pollution trading schemes at the dawning of the neoliberal era.

Markus Lederer is a Professor of Political Science with a focus on International Relations and International Governance at the University of Münster in Germany. His research interests cover climate politics, in particular carbon markets, REDD+ and the role of developing/emerging economies in carbon governance.

Heather Lovell is a Reader in the School of Geosciences at the University of Edinburgh. She has held positions at Durham and Oxford Universities, the University of Technology (Sydney), and in the UK Parliament. Her research is on the social aspects of climate change and energy and focuses on two main topics: carbon markets, and low-energy housing.

Donald MacKenzie is a Professor of Sociology at the University of Edinburgh. His current research is on the sociology of finance, focusing on automated trading, the use of mathematical models, and the evaluation and trading of bonds. He has worked on topics ranging from the sociology of nuclear weapons to the meaning of proof in the context of computer systems.

Elah Matt holds a PhD in European Union climate change and transport policy from the University of East Anglia. Her research interests include environmental policy instruments, climate change policy, and eco-innovations. Currently, she is a Post-Doctoral Mimshak fellow and a visiting fellow at the Tyndall Centre for Climate Change Research.

Kathleen McAfee is currently Associate Professor in International Relations at San Francisco State University, after a career in international development. Her current research focuses on 'selling nature to save it' through market-based responses to climate crisis carbon-offset trading, payments for ecosystem services, and REDD. She has been a UN consultant, participant-observer in international social movements.

Chris Methmann holds a DPhil in Political Science from the University of Hamburg. His research centres on climate politics, climate security and climate-induced migration and has been published in the *European Journal for International Relations*, *Security Dialogue*, *Millennium* and *International Political Sociology*. He works as a campaigner for the German advocacy group Campact.

Peter Newell is Professor of International Relations at the University of Sussex. He researches the political economy of climate change and low carbon energy transitions. He is co-author of the books *Climate Capitalism* and *Governing Climate Change*, co-editor of *The New Carbon Economy* and author of *Climate for Change* and *Globalization and the Environment: capitalism, ecology and power*.

Chukwumerije Okereke is Reader (Associate Professor) in Environment and Development at the Department of Geography and Environmental Science, University of Reading, UK. He is also a Senior Visiting Fellow at the Environmental Change Institute, University of Oxford. He is a Lead Author for the IPCC (Intergovernmental Panel on Climate Change).

Matthew Paterson is Professor of Political Science at the University of Ottawa. His research focuses on the political economy of global environmental change. His most recent book is *Climate Capitalism* (with Peter Newell). Most recently, he has been working on carbon market politics and on transnational climate change governance. He is also a Lead Author for the IPCC.

Tianbao Qin is a Professor of Law and Vice Dean of the School of Law, Wuhan University, China. He is a Legislative Expert for China's Ministry of Environmental Protection and the Secretary-General of the Chinese Society of Environment and Resources Law (CSERL). Currently, his research focuses on climate change, energy security law and policy.

Arno Simons is currently completing his doctoral thesis on the discursive construction of emissions trading in documentary communication, at the Technische Universität Berlin. As a member of the Innovation in Governance Research Group he also works on the historical development of other environmental markets, such as markets for biodiversity or habitat credits.

Clive L. Spash is a Professor in the Socio-Economics Department, Wirtschafts Universität Vienna, where he holds the Chair of Public Policy and Governance. He is an economist who writes, researches and teaches on public policy with an emphasis on economic and environmental interactions. He is also Editor-in-Chief of *Environmental Values* and has over 100 publications in books and journals.

Benjamin Stephan is a researcher at the Centre for Globalisation and Governance, Hamburg University and holds a DPhil in Political Science. His research is concerned with transnational climate governance – particularly carbon markets and avoiding deforestation (REDD+) – as well as energy and transportation policy. His most recent book is *Interpretive Approaches to Global Climate Governance* co-edited with Chris Methmann and Delf Rothe.

Eva Sternfeld is the Director of the Centre for Cultural Studies of Science and Technology in China at the Technical University of Berlin. She is a sinologist and geographer working on environmental issues, resource management and urban development in China and holds a PhD from the Institute for East Asian Studies, Free University Berlin.

Johannes Stripple is Associate Professor at the Department of Political Science, Lund University, Sweden. He investigates climate change policy and emerging practices of a carbon constrained world. Johannes has edited (with Harriet Bulkeley) *Governing the Climate: new approaches to rationality, power and politics* (CUP 2014).

Jan-Peter Voß is Junior Professor of Sociology of Politics at the Department of Sociology and head of the Innovation in Governance Research Group, at the Technische Universität Berlin. His research is interested in the relation between politics, knowledge and innovation combining approaches from policy and governance studies with those from science and technology studies.

Acknowledgements

This book would have never seen the light of day without the support of a number of people. We would like to thank Kerstin Walz for help with compiling the manuscript and ensuring that the formatting is consistent and complete. We would also like to thank Louisa Earls and Helen Bell at Routledge for their work and the assistance they have given us while putting this book together.

Of course, this book would not have been possible without a dedicated group of contributors who provided such interesting and thought provoking chapters. And in contributing to an internal review process they also helped to ensure the quality of the chapters. We would also like to thank Steffen Böhm and Rafael Kruter Flores for their help and contributions during the review process.

Richard Lane would also like to thank Peter Newell and Samuel Knafo for their guidance and Sahil Dutta, Matthieu Hughes and Steffan Wyn-Jones for their unstinting support. Most importantly I wish to thank Niamh Kelly, who, as always, makes all this possible.

Benjamin Stephan would like to thank Anita Engels and Matthew Paterson for their advice and support. A big thank you goes to Matilda and Jule for being such lovely distractions and of course to Jenny, who helped to create the space to work on this book.

1 Zombie markets or zombie analyses?

Revivifying the politics of carbon markets

Richard Lane and Benjamin Stephan

Introduction

On 10 May 2013, atmospheric carbon dioxide levels, as measured at the Mauna Loa observatory in Hawaii, reached a symbolic new high of 400 parts per million – a level not seen on earth for the last 3 million years (Gillis 2013). On the same day, the Financial Times Stock Exchange (FTSE) 100 index in London also closed on a symbolic high, in this case, the figure of 6,625. This number breached the psychologically important 6,600 level not seen since the pre-financial crisis days of October 2007, and appeared to herald, at least in the UK, what economic commentators and politicians fondly call 'the green shoots of recovery'. This specific conjunction of record-breaking emissions and financial market figures is coincidental, however the relationship between the objects they partly represent – the environment and the economy – is anything but. Within academic and popular analysis, predominant economic, legal and political science perspectives maintain that by pricing greenhouse gas (GHG) emissions, carbon markets foster economic growth while mitigating GHG production. That is, they mediate between the environment and the economy. The 'new carbon economy' (Boyd *et al.* 2011) – comprised of international UN-controlled cap-and-trade and offsetting mechanisms as well as regional, national and subnational markets – is therefore considered as embodying at least the possibility of decarbonising the global economy (Newell and Paterson 2011). Carbon markets directly entail the generation of growth sectors – specifically financial instruments but also carbon offsetting projects. As such, the idea of carbon markets can be seen to resonate well with a longer standing ecological modernisation perspective on environmental governance (Bailey *et al.* 2011).

Sixteen years after the signing into force of the Kyoto Protocol, the confluence of the Mauna Loa and FTSE 100 highs casts a shadow over the possibility of the carbon markets disentangling these figures. In 2014, the impacts of climate change, long apparent to the Alliance of Small Island States (AOSIS) and other low-lying countries in the Global South seemed finally to reach even the UK and US. Record-breaking extreme weather events resulted in Ed Miliband, the Leader of the UK's opposition, declaring that the UK is 'sleepwalking into a national security crisis on climate' (Helm and Doward 2014) on the same day (16 February 2014) as US Secretary of State John Kerry described climate change during a speech in Jakarta, Indonesia as 'perhaps the world's most fearsome weapon of mass destruction' (Gordon and Davenport 2014; Helm and Doward 2014). This air of shrill panic however, when viewed in the longer history of global environmental politics, seems like simply the latest shock in a kind of long drawn-out horror story of one environmental crisis after another, stretching back to the emblematic birth of the contemporary environmental movement with the publication of Rachel Carson's *Silent Spring* in 1962.

If a certain sense of horror is the tone with which to approach climate change, then the critical wing of the discipline of political economy has recently furnished us with an apparently apt metaphor for the development of the carbon markets: the zombie. Zombie capitalism (Harman 2009), zombie economics (Mirowski 2013), and zombie neoliberalism (Peck 2010) are terms that have rapidly multiplied in both academic analysis and popular commentary since the 2007 financial meltdown.[1] This notion of zombie neoliberalism: empirically defunct yet still dangerously powerful; without a lively legitimacy yet proliferating and voracious, is an apposite metaphor to describe the ongoing development of the carbon markets. Indeed, the activist and scholar Oscar Reyes brought the lurching figure of the zombie explicitly to the study of carbon markets in 2011. Although Reyes was content to refer to 'carbon market zombies' in the title of this study he otherwise left this analogy largely unexplored. He simply noted that, in spite of the apparent evidence from the UN's Clean Development Mechanism (CDM) that offsetting has resulted in increases rather than decreases in GHG emissions, 'the carbon market zombies stumble on' (Reyes 2011: 129).

Carbon market zombies do not appear limited to the CDM, however. Despite being staggered by multiple crises, carbon markets continue to be at the core of the global response to climate change; and indeed can be seen lurching into ever new domains. One month before the Mauna Loa and FTSE 100 highs, the world's largest carbon market, the EU Emissions Trading System (EU ETS) was struck yet another apparent death blow.[2] In the wake of persistent over allocation and the EU parliament's initial vote against a 'Backloading' plan to temporarily remove 900 million tonnes of carbon allowances from the persistently over-allocated market, on 16 April, the price of EUAs fell 50 per cent in only ten minutes, from an already moribund €5 to €2.63 (Fioramonti 2014: 91; see also Methmann and Stephan, this volume). In May 2013, the World Bank also abandoned its yearly State and Trends of the Carbon Market reports where it calculated the volume of the market (see for example World Bank 2012) in favour of a publication referred to as 'Mapping the Carbon Pricing Initiative' which no longer includes these estimates (World Bank 2013). It is not difficult to imagine why the Bank – intimately involved as it has been through the provision of funding and expertise – would shy away from a seemingly clear advertisement of the apparent failure of the carbon markets.[3] These developments came after the EU ETS lost a third of its value in 2012 (from US \$148 billion to approximately US\$ 100 billion), and the

December 2012 sale of 5.58 million permits by the EU for €6.45 million – a cost too low to incentivise meaningful investment in low carbon technologies among the regulated industries (Fioramonti 2014: 91). Similarly the CDM, which was estimated at US$33 billion in 2007, has seen a 90 per cent year on year decrease in the price of Certified Emissions Reductions (CERs) to 2012 to about 40 cents per offset. This is roughly 10 cents less than 'what analysts say it costs developers in fees to get issued with credits and well below costs involved in investing in carbon-cutting equipment' (Fioramonti 2014: 93–94).

In spite of these issues, the EU continues to be firmly wedded to its ETS, arguing that climate policies need to work 'in a way that will not hamper economic growth in Europe but which leaves companies maximum flexibility to cut emissions at least cost' (Hedegaard 2011). Elsewhere around the globe, carbon markets are moving forward in a variety of national and regional forms: California and Kazakhstan started their emissions trading systems in January 2013 followed by a series of pilot schemes in China later that year (see Engels, Qin and Sternfeld, this volume). Additional trading systems are being considered in Chile, Turkey and several Brazilian states (see Lederer, this volume); and there are many advocates for an integration of REDD+[4] – a global mechanism to reduce emissions from deforestation – into the carbon market (see Lovell and MacKenzie as well as McAfee, this volume). Carbon markets appear to have attained the zombie status of being essentially unkillable.

Although not without rhetorical potency (or indeed its critics – see Newell, this volume), it is not the goal of this chapter to summarise the ongoing development of the carbon markets as a kind of zombie apocalypse.[5] We argue instead that the zombie analogy should be brought to bear on the existing academic literature on carbon markets. The zombie was originally introduced to indicate the unkillable nature of apparently defunct neoliberal governance and economic theory, but an equally important aspect of the zombie is that it acts, but without agency. The living dead do not just stumble on, they stumble *mindlessly* on. It is our contention that the current literature on carbon markets, while continually proliferating, is marked in general by a failure to adequately account for the politics inherently involved in the development, ongoing maintenance and contestation of the carbon markets. Analyses of carbon markets are full of (implicit) zombies: actors that lack political agency or accounts that draw on restricted notions of politics and the political.

Building on the special issue *The Politics of Carbon Markets* in the journal *Environmental Politics* (see Stephan and Paterson 2012; see also Simons *et al.* (2014) for further work building on the discussion in this special issue), this edited volume aims to revivify discussion on carbon markets. It investigates the market system, and its insertion into and influence on, climate and environmental governance within the global political economy. It does this by analysing the routines, institutions, techniques and technologies established, or refuted, through practices of social and material negotiation. The special issue from which this current volume developed was comprised of contributions that took a variety of perspectives on what politics involves, what it means to act politically, and how political actors, forms and objects are constructed. This edited volume develops this work by broadening it with new empirical results and additional theoretical insights. First, however, we turn our attention to the existing literature on carbon markets. Where are the zombies? To what extent, and with what effect, do dead accounts of politics roam the academic landscape?

Looking for politics in the literature on carbon markets

There is an extensive and ever swelling body of literature on carbon markets. Initially dominated by economists and legal scholars, more recently political scientists, sociologists and geographers have contributed to our understanding of carbon markets. We have surveyed the existing literature focusing on the politics of carbon markets. It can be split into three broad categories. The first strand of literature engages with the *mechanisms of markets*. It encompasses the economics, law and technopolitics literature that is interested in how different elements of the market affect its performance. Second, there is a broad interest – predominantly from political scientists – in the *process of policy* that led to the current network of carbon markets. Scholars are particularly interested in the diffusion and transfer processes and the role certain actors have played in them. And last, there is a heterogeneous *critique of carbon*. This category comprises poststructuralist works critiquing carbon markets and their governance effects by problematising and deconstructing them, contributions that assess the legitimacy and justice implications of carbon markets based on particular normative assumptions and radical political economy critiques that advocate the abandonment of carbon markets. These accounts generally draw on a relatively narrow understanding of politics and are thus only able to reflect the political aspects of carbon markets in a very limited manner. Drawing on a postfoundational understanding of the political (Marchart 2007) we will sketch out an alternative, more encompassing perspective in response to this literature review.

The mechanisms of the markets

The central concern within dominant economics and legal literatures, as well as a considerable amount of social science contributions, is with questions of 'optimal design' (Stephan and Paterson 2012: 547).[6] These perspectives start from the presumption of carbon trading as cost-efficient and, therefore, the most viable solution to the collective action problem of climate change.[7] They are interested in choosing a market design that optimises effectiveness and efficiency of carbon markets. For example, how can certain design characteristics enable or disable the achievement of goals such as the creation of a certain price level for carbon or the minimisation of transaction costs? Examples of political science engagements that fall into this category are the articles by Green (2008) and Lund (2010) who assess the effects that the delegation of authority to private actors has on the effectiveness of the CDM. In focusing on the ways in which the effectiveness of carbon markets are impacted upon, through their insertion into, and development of, a variety of governance structures, the politics of the markets themselves drops out of the analysis. Emissions trading and offsetting are taken for granted as 'naturally' the optimal tools for regulating GHG emissions given the 'laws of the social world' that prescribe these as efficient and effective and proscribe other non-market forms of governance (Lane 2012).

A contrasting approach draws on the various technopolitical and performative notions of economics from Actor Network Theory (ANT) and the broader field of Science and Technology studies (STS).[8] Analyses here, similarly to the economic and political science perspectives recounted above, focus closely on the mechanisms of the markets. Accounts have been developed on how the material, technological and technical infrastructure of the markets has been assembled and how this enables economic theory to be 'performed'. Michel Callon (2009), for example, argued that the development of carbon markets should be considered as a form of 'in vivo' experimentation by economists in the creation of markets in new environmental commodities. Moreover, he claimed that before the carbon markets actually existed, they were practised at various times, in different places and forms, beginning in the US with the first experiments in pollution trading in California in 1976. Donald MacKenzie's (2009a, 2009b) analyses of the construction of emissions markets have focused in part on the technopolitics of the seemingly merely technical market components of the 'ratchet' and Global Warming Potential (GWP)[9] (see also Paterson and Stripple as well as Methmann and Stephan, this volume).

This currently small literature responds in part to economic and political science accounts that are concerned predominantly with how markets can be optimised. Here, the ways in which carbon markets are materially and technically constructed are brought to the fore. However, the question of whether this results in explicitly depoliticised accounts of the development of markets remains. This general critique of the programme of economic performativity of Callon, MacKenzie and others has been made from various quarters (Butler 2010; Fine 2005; Mirowski and Nik-Khah 2007). Anders Blok (2010, 2011), for example, has argued that Callon's presumption of the monetisation of commensurable carbon equivalencies is too straightforward and results in a failure to analyse the specific politics of environmental quantification and monetisation. Furthermore, Callon's formulation that carbon markets have been practised beforehand tends to shade into an understanding of the 'market' as a form of single substance that can be moved from place to place, time to time. It thereby occludes to some extent the more interesting question of how these different market implementations have been constructed as mere variants of a basic emissions trading form (Lane 2012).

These economic, political science and technopolitical/performative literatures focus closely on the mechanism of the carbon markets, and in doing so push politics out to the margin of their analysis, or fail to grasp the politically contested and constructed nature of the markets altogether. The economics and political science perspectives concerned with the optimal design of policy reduce politics to either a neoclassical economics account of rational actors pursuing their self-interest in contexts of poorly designed property rights, or as a kind of external constraint or necessary concession limiting the effective and efficient functioning of pure markets. From a technopolitical approach the emphasis on the material and technical performativity of economics results in an elision of the broader institutional and structural context within which the markets develop. In both cases, the role and importance of the actors – whether institutional or human – that actually bring about, lobby for, bear these policies to fruition and then contest them, is radically undercut.

The process of policy

A second set of approaches focuses on the processes involved in the development, diffusion and assembling of carbon markets. This literature is interested in the struggles for, and confrontations over, carbon markets and can, therefore, be seen as ostensibly having a direct focus on the politics of carbon markets. Significant attention has been given to how the idea of emissions trading has spread and how it became adopted in different regions. The EU ETS is the most prominent case in this context. But similar work has been undertaken on regional GHG trading systems in the US (Selin and VanDeever 2011) as well as the cap-and-trade system in New Zealand (Bullock 2012; Hood 2010). Furthermore, there is literature on the establishment of the CDM (e.g. Boyd 2009; Pulver *et al.* 2010; Shin 2010). There are, however, only a few contributions that compare developments across different markets and thus are able to identify broader patterns (e.g. Betsill and Hoffmann 2011; Paterson 2012).[10]

The EU ETS has been a focus of interest because it constitutes the first and also by far the largest mandatory emissions trading system for GHGs. The literature pays a lot of attention to the ways that policy transfer and learning mechanisms enabled the EU to adopt emissions trading (Christiansen and Wettestad 2003; Damro and Mendez 2003; Voß 2007). This is complemented by taking into account intergovernmental bargaining between EU member states and the EU's unique multilevel structure (Skjærseth and Wettestad 2008, 2009). There has also been great interest in the policy entrepreneurs (Skjærseth and Wettestad 2010; Wettestad 2005) and large policy networks (Braun 2009) that enabled the diffusion and adoption of emissions trading from its home in the US to an initially resistant EU. Particular attention has been given to the role of business actors and their motivations to engage and promote carbon markets (e.g. Meckling 2011; Pinske 2007; Pinske and Kolk 2007). Meckling, for example, analyses the role of business coalitions favouring carbon markets on the inclusion of emissions trading in the Kyoto Protocol, the adoption of emissions trading by the EU and on the reintroduction of emissions trading into the US context. He highlights that the initial business coalition consisted of energy producers or energy-intensive companies who pushed carbon markets as 'a hedging strategy that would prevent policy alternatives' and that an 'anti-taxation agenda' could be found in all of his cases (Meckling 2011: 44).

Overall, these literatures share an interest in the process of policy making that led to today's carbon markets. However, they vary greatly with regard to the degree of political agency that is made visible. Many of the contributions focusing on the involved actors clearly reference political intervention, negotiation and contestation over policies and their contents, how they are brought to bear on new issue areas, in new locations and with new institutional contexts. However, there is frequently a very narrow conception of politics and the political in operation here. Many of these accounts can be considered as effective rejoinders to approaches that focus solely on the mechanism of the markets, but in bringing an analytical focus to the actors and their activities, the insights of the first set of literature outlined above often drop out of perspective. The role of economic theory, and its relation to real world practice, the material technologies, conceptual techniques, accounting tools and practices that were the focus of the previous literature seem to play little or no role in these views on the politics of the carbon markets. Moreover, broader institutional forces and the structural context of a capitalist and specifically neoliberal global political economy mostly escape analysis here. Global businesses, extractive industries or simply capital are the silent animators of political agents, and a third broad literature can be seen as responding, in part, to this concern.

The critique of carbon

The proliferation and dominant role played by carbon markets in the global climate polity has also evoked a variety of critiques, in very different forms. Poststructuralist and political ecology scholars have deconstructed fundamental concepts of carbon markets. Other scholars have assessed the development and performance of carbon markets on the basis of their legitimacy and with respect to moral and ethical questions. A third group of contributions are based in or draw from political economy perspectives, and variously critique market implementation, call into question the very possibility of environmental effectiveness of carbon markets, and raise concerns about so-called carbon colonialism.

Poststructuralists' engagement with carbon markets is dominated by Foucauldian governmentality studies (e.g. Bäckstrand and Lövbrand 2006; Methmann 2013; Wolf 2013). Other contributions focus on how the carbon markets or different aspects of it have come into existence (Boyd 2010; Stephan 2013; see also Paterson and Stripple, this volume) while others scrutinise the governing effects these markets have both on the state and the individual level (Lövbrand and Stripple 2011; Paterson and Stripple 2010). There are also links here to the technopolitics literature introduced earlier, with some authors drawing on both strands of work (Lovell 2013; Lovell and Liverman 2010). A critique of the carbon markets has also been undertaken from a political ecology perspective – an approach originating from geography. Accounts making use of this notion have investigated how the materialities and political structures – developed at a variety of local and global levels – have interacted and been transformed through the development of carbon offset markets (Bumpus 2011; Lovell *et al.* 2009; Newell and Bumpus 2012).

Discussions about legitimacy have dealt with the input and output legitimacy of carbon markets (Lederer 2011) and are often linked to the broader debates on environmental effectiveness and justice. Lövbrand *et al.* (2009), for example, look at both ends to assess the legitimacy of the CDM on the basis of CDM projects in Chile, China and Mexico. Skjærseth (2010) analyses to what extent the EU ETS reforms prior to its third phase had an influence on its legitimacy. Furthermore, the legitimacy of carbon markets is often discussed in the context of the role of non-state actors and their ability to shape state governance (e.g. Paterson 2010). There have also been discussions of carbon trading from philosophical quarters on the basis of moral or ethical grounds. These interventions disaggregate moral and ethical concerns and claims regarding the markets on the basis of differing principles of justice (Caney 2010 and a response from Aldred 2012; Page 2011).

The literature that engages with carbon markets from a political economy perspective situates the development of the carbon markets within the broader context of a capitalist and often specifically neoliberal global socio-economic system. In investigating how GHGs and their production have been made into a commodity, they ask how carbon markets affect the distribution of mitigation efforts within societies or between the Global North and the Global South. Three forms of Marxist political economy analysis can be demarcated here. First are accounts that critique the specific development of carbon trading, but draw from an account of capital and capitalism that is not intrinsically linked to any specific mode of production or material and energetic means of value creation (e.g. Bitter 2011; Brunnengräber 2007; Newell and Paterson 2010, 2011). That is, while these accounts often indicate negative environmental, justice and legitimacy questions raised by the actual development of the markets, they are open to, or at least agnostic regarding the potential of the contemporary capitalist global system to decarbonise; and the place of carbon markets in this decarbonisation. The role and power of finance as a specific fraction of capital has often been stressed in market creation here, often drawing from a regime of accumulation or neo-Gramscian perspective (Levy and Newell 2005; Matthews and Paterson 2005; Paterson 2001; Stephan 2011). The second form of political economy accounts can be seen in contradistinction to the first. Analyses here draw from an eco-Marxist scholarly tradition oriented around a notion of 'the second contradiction of capitalism' (O'Connor 1994). Accounts in this vein maintain that as capitalist and specifically neoliberal forms of environmental governance, carbon markets are inherently incapable of leading to 'real' emissions reductions beyond a status quo scenario (e.g. Lohmann 2006, 2011; Smith 2007).

A third political economy approach can be seen as a continuation of an older eco-colonial debate within the broader study of environmental governance (e.g. Agarwal and Narain 1991). Bachram (2004: 11) argued that the markets enable responsibility for carbon intensive lifestyles in richer nations to be pushed onto the poor, with the South acting as a carbon dump. Over and above this imposition of costs on the Global South, however, is the claim that, particularly in regard to forestry projects, the necessity of intensively managing these projects to ensure that carbon is sequestered for the lengths of time claimed involves a clear neo-colonial imperative. Accounts highlighting this carbon colonialism are often grounded in case studies that detail indefensible claims by project proponents, restrictions for the local community due to a project, or even human rights violations that have occurred in its context (for a broad collection of case studies see Böhm and Dabhi 2009).

This extensive literature on the critique of carbon, while crucial, can be seen to elide an adequate focus on the politics of carbon markets on two counts. First, there is often little focus on the actual political actors responsible for bringing the carbon markets into being. This is particularly true for the governmentality literature which, in general, provides excellent insights into how certain regimes function but sheds little light, for example, on why carbon markets and not carbon taxes have become a formative institution in the climate polity (Stephan *et al.* 2014). Second, while the political actors are highlighted more clearly in the political economy critiques to carbon markets, they are in most instances reliant upon reified and monolithic conceptions of macro-social actors or structures. Carbon markets are either the projects of large corporate entities driven simply, directly and straightforwardly by the inherent logic of capital, or the causes of the market developments recede back into a necessarily structuralist context of neoliberalism.

The absent politics of carbon market analysis

What this brief review of the literature on carbon markets indicates is that in each of the areas identified – the mechanism of markets, the process of policy and the critique of carbon – there are only partial, halting and staggering accounts of the politics at play. Moreover, each of these three areas can be considered mutually exclusive with respect to the particular facets or elements of politics that are highlighted, explored, analysed and critiqued. In this way, the overall literature on the carbon markets can be considered as generating a series of zombie accounts, where undead market agents act, but they lack an adequate accounting of political agency. These agents seem to simply shuffle forward, animated by the drives of a seemingly naturalised economic rationality, or an equally naturalised neoliberal organisation of the socio-political world. In response to an academic literature dominated by accounts that have frequently failed to represent the actors they study as politically lively, we propose that a similar creature be brought to bear upon the longer and ongoing history of the carbon markets – that of the re-animated monster of Doctor Frankenstein.

The particulars of Mary Shelley's famous 'ghost story' lend themselves well to an analogical retelling of the broader history of the project of carbon trading, a history that is common to both carbon market advocates and critics alike. First, the early emissions trading experiments in the US from the mid-1970s onwards were stitched together by the still new Environmental Protection Agency from the dead, addled, regulatory flesh of the 1970 Clean Air Act. Second, this crude form was, if not brought to life as such, animated by being bathed in the galvanic fluid of economic theory. This re-animation of the corpus of emissions trading regulations in the form of cap-and-trade markets was begun with the release of the Project 88 report. This was continued through the implementation in 1994/5 of the Sulphur dioxide markets under Title IV of the 1990 Clean Air Act amendments, and completed during the UN negotiations on climate change at Kyoto in 1997, resulting in the three (market based) flexible mechanisms of the Kyoto Protocol: Emissions trading, Joint Implementation, CDM. Third, these seemingly monstrous markets were rejected almost immediately by their horrified maker, with the US's refusal to ratify the Kyoto Protocol. Finally, and of course most prophetically worrying, the story of Frankenstein ends with death in the Arctic wastes – although given the record low ice coverage recently recorded, it is probably safe to say that in the future, the Arctic will look considerably different to the frozen, pristine landscape imagined by Shelley. More importantly here, however, is that Frankenstein acts under his own volition, and it is this analogy that is useful, when attempting to highlight the lively political character of carbon trading.

The lively politics of carbon markets

What do we mean by lively politics here? We draw in the first instance, from what Marchart (2007) refers to as a postfoundational understanding of the political. This notion of the political is shared by a number of theorists (e.g. Laclau and Mouffe 2001; Žižek 2000) whose work can broadly be characterised as poststructural (for an overview and an application to environmental politics see Kenis and Lievens 2014). While politics, as the activity that takes place in and around parliament and government offices, can be political – particularly when it involves struggles concerning fundamental questions such as freedom, justice or equity – a lot of it is oriented around technocratic, postpolitical management. Instead, the political is understood here as the sphere entered whenever our social relations are challenged and their contingent character becomes visible. Controversies about appropriate norms, according to which we should organise our society, or parts thereof, take place in the realm of the political. These struggles and controversies evolve after dominant meaning structures become dislocated and leave a void that needs to be filled (Laclau and Mouffe 2001: 142). Events such as natural disasters which cannot be accommodated within the structures of the dominant discourse can have this dislocative effect. Alternatively, dislocation is caused by societal groups actively challenging the dominant order. New norms and forms of acting in and understanding the world can then develop and be contested by various different groups within society. Importantly, norms increasingly lose their political character over time as they are routinised, become naturalised and are increasingly taken for granted.

Drawing on this notion of the political a number of scholars have criticised the current discourse on climate change as post-political (Methmann 2011, 2013; Swyngedouw 2010; see also Kenis and Lievens 2014). Politicians repeatedly insist that the potentially apocalyptic impacts of climate change render responses to it as beyond politics. Careful technocratic management is presented instead as the only legitimate response. In line with this critique, the establishment of carbon markets and the commodification of GHGs and carbon sinks constitutes a profoundly post-political approach. Carbon markets are presented as both economically and environmentally effective. In doing this, debates around the fossil fuel dependence of our social-production (e.g. Lohmann 2011; Mitchell 2011; O'Connor 1994) are displaced by technocratic discussions about the design details of emissions trading systems or offset mechanisms. This adds to what Blühdorn (2011) has called the ecological paradox: namely, that we are very well aware about the dawning ecological disaster but there is not any substantial action to prevent it. We are, rather, keeping ourselves busy by discussing how to streamline the CDM project approval process or how exactly baselines should be constructed for REDD+ instead of fundamentally calling into question whether the goals and the current mode of climate governance are adequate (see also Methmann 2011).

We do not disagree with this assessment. However, we argue that there is more to the story of carbon markets than just the depoliticisation of climate politics. Drawing upon a postfoundational understanding of the political we can also flag the political moments and agencies involved during the development, implementation and operation of carbon markets (see also Stephan 2012). Carbon markets, once they are implemented, might seem to be purely technocratic in nature but they do not lose their contingent character. There is always the possibility of dislocative events that might challenge the taken-for-grantedness of carbon markets. These dislocations might be external to carbon markets, for example the financial crisis creating a general – albeit apparently short lived (see e.g. Mirowski 2013) – scepticism regarding markets and financial institutions; or the ongoing inability to agree on a successor to the Kyoto Protocol resulting in a collapse of demand for credits traded under the Kyoto markets.

In addition to these external events, dislocations can also be seen to arise internally to the markets. As Barry (2002) has argued, scientific and technical procedures such as those forming the basis for carbon markets do not only have depoliticising effects. Instead, they should also be seen as providing the basis for new objects to be opened up and new sites of disagreement to be developed (see, for example, McAfee, Spash and Newell, this volume). Also, the calculations, technologies and narratives that generate and compose the carbon markets remain, themselves, fragile. They do not always produce clear and definite results, technical procedures fail, calculations are insecure and readily contested (see also Methmann and Stephan, this volume).

While only a small number of the contributions to this volume explicitly situate themselves within it, the majority of chapters fit well within this overarching perspective. A postfoundational understanding of the political allows us to theorize the politics in the creation, maintenance and contestation of carbon markets in terms of depoliticising and repoliticising narratives, institutions, techniques and technologies. Depoliticisation is never entirely complete, it cannot wholly close out the possibility of further debate and contestation of what is taken to be merely technical, objective or natural. However, these more-or-less technical, objective and natural facets of carbon trading also result in the development of entirely new and previously unforeseen and unimagined political possibilities, with the expansion of the political into new areas of contestation.

The politics of carbon markets

To explore the politics of carbon markets this edited volume has assembled twelve contributions that assess different aspects of carbon markets from a variety of different theoretical angles, and with a diverse set of analytic foci. Of course, outside of either an oppressive carceral environment or Foucault's notebooks, a truly panoptic gaze is impossible for any particular analysis, and the chapters of this edited volume are as much subject to constraints on the breadth of political coverage as are other contributions to the overall literature. As a collection, however, this edited volume does aspire to something more. The book is split into three sections: *The politics of carbon before carbon*; *The politics of carbon*; and *The politics of carbon after carbon*. Rather than organise the contributions by empirical focus, or theoretical approach, we have made use of a distinct historical perspective putting the chapters into conversation. Through this, we highlight the constant play of politics, and the shifting, reconstituting sphere of the political in the carbon markets.

Contributions to each of the three sections of the book investigate the ways and means through which politics is involved in the carbon markets. In *The politics of carbon before carbon*, the chapters focus on the prehistory of carbon trading. They highlight political contestation and the development of new political agencies prior to the development of the markets, how this politics framed and contextualised the carbon markets and, indeed, made their development possible. The markets are themselves continuously politically reconstituted and contested, and in *The politics of carbon* this ongoing dialectic of maintenance and contestation, of challenge, expansion and seeming retrenchment is explored. Finally, the presence of the markets has ramifications and broader implications for both climate change governance and environmental politics in general. These are explored in *The politics of carbon after carbon*, where chapters focus on the influence and impact of the existence of the markets, and of carbon as a tradeable commodity.

The politics of carbon before carbon

Richard Lane begins this volume by going back to the prehistory of emissions trading in general. Drawing on the work of Timothy Mitchell (2011) and developing an ANT-inflected analysis, Lane investigates the history of the 1975 growth ban in the US. This ban is a crucial, yet so far unexplored event in the development of early pollution trading and helped put in play the context that informed the development of carbon markets. The growth ban is taken to provide the signal case proving the laws of economics: namely the inefficiency of the 1970 Clean Air Act – later reinscribed as the archetype of so-called command-and-control regulations. It was in response to this ban that EPA regulators developed the first emissions markets. However, the growth ban was not a natural result of socio-economic laws. It was made; and in order to understand this making, Lane begins with his analysis just after the Second World War and focuses on the technical changes implemented to the economy and the environment as objects of politics by the nascent discipline of environmental economics.

Arno Simons and Jan-Peter Voß continue the analysis of the prehistory of carbon markets. Beginning with the theoretical roots of emissions trading in economic theory in the 1960s they investigate how a 'policy instrument constituency' developed around emissions trading as an approach to environmental regulation. They draw on STS studies to reconstruct how the conceptual basis for carbon markets was developed, and how this was fostered by and through specific policy actors. They are particularly interested in the 'centres of policy calculation' where the development of carbon markets was orchestrated and from where they subsequently diffused. Through their notion of 'policy instrument constituency', they are able to account for the actors who have contributed, picked up on, or promoted these ideas, and in doing so they are able to show that both dimensions – ideas and actors – are closely linked. They argue that the development and diffusion of the carbon trading idea is connected to the increasing growth of a carbon trading constituency that can be seen as related to these earlier developments.

Heather Lovell and Donald MacKenzie analyse the history and politics of expertise around allometric equations. This is a technique to measure and calculate a tree's biomass and carbon content and thus enables us to make estimates about emissions reductions from a forest. This is crucial for REDD+, for example, which is designed to make payments for measured, reported and verified (MRV'ed) emissions reductions. Lovell and MacKenzie argue allometric equations are 'immutable mobiles' (Latour 1987) transported from a forestry context geared at timber production, where they were used to calculate the timber volume of a forest, to the climate policy world to calculate emissions reductions. Here these equations act as a market device that assures emissions reductions from forests are the same – no matter where they occur. However, allometric equations are not the only MRV technique. This 'historic' approach to measurement competes with newer remote sensing techniques, and Lovell and MacKenzie's chapter helps to illuminate the politics around this competition.

Matthew Paterson and Johannes Stripple close out the politics of carbon before carbon section with a chapter that scrutinises what they call the five moments in the commodification of carbon: the invention, proliferation, verification and differentiation into 'boutique' and 'Walmart carbon'. First, their chapter details how the notion of the tCO_2e (the tonne of carbon dioxide equivalent) came into existence. Second, this is combined with a closer look at how these market units and standards have been diffused. Third, they are interested in how commensurability – the fact that a tonne is a tonne is a tonne – can be assured. The notion of virtuous carbon that Paterson and Stripple introduce emphasises the interconnectedness of virtuality and virtue in the context of carbon markets – carbon is, on the one hand, highly abstracted and virtualised, and, at the same time, there is a strong ethical imperative to engage and participate which de-emphasises ethical contestation. The concept of virtuous carbon provides us with a better understanding of how carbon markets contribute to a depoliticised engagement with climate change.

The politics of carbon

In Chapter 6, Elah Matt and Chukwumerije Okereke analyse the coming into being of both the EU ETS and the CDM from a neo-Gramscian perspective. Drawing on Gramsci's notion of passive revolution, Matt and Okereke show that the adoption of emission trading and carbon offsetting policies can be understood as an attempt by the 'climate-accommodating carboniferous block' to stabilise its position. In the light of emergent findings on anthropogenic climate change it had increasingly been challenged by environmental NGOs. By promoting carbon markets, this block not only managed to co-opt many critics, but it also helped to promote the policy mechanism that supposedly has the smallest impact on its core business. This neo-Gramscian assessment links both a focus on particular actors and their role in the development and implementation process with an account of the broader economic and social structures.

In Chapter 7 Markus Lederer picks up on the paradox that was outlined at the beginning of this introduction: carbon markets are proliferating in the Global South despite being in severe crisis (at least in the Global North). He helps us to understand this paradox by shedding light on the reasons behind carbon market adoption in the Global South. He starts by sketching out the specific characteristics of these Southern versions of carbon markets. In addition to a functional role that carbon markets play in the cases he has assessed, Lederer highlights two sets of actors that have a key role in establishing them: first, there is an influential constituency that favours carbon markets in the Global South. It is somewhat different to the constituency that has existed in earlier carbon markets as the World Bank and companies from the South feature much more prominently. Second, the Southern states themselves play an important role. Drawing on the idea of the 'developmental state' Lederer shows that contemporary carbon markets in the South are a state-led industrial policy.

The chapter by Anita Engels, Tianbao Qin and Eva Sternfeld continues on from Lederer's chapter by providing a detailed overview of current developments in China with its pilot emissions trading systems at the subnational level. In its 12th Five Year Plan China announced the creation of twelve emissions trading systems at the provincial and city levels. The espoused goal of these systems is to gain experience with trading mechanisms and then upscale these pilots into a national system. The case of China is an interesting one as these carbon markets are being created in an economic, social and political context that is radically different from that in which prior systems have developed. Furthermore, the broader Chinese economic context is transforming from a system of centralised planning to a market economy. The introduction of carbon markets is not only affected by these changes but also contributes to them. Engels, Qin and Sternfeld provide us with a detailed description of the emissions trading pilots and help us to understand the position of these instruments pilots in the larger Chinese context.

Following two empirically oriented chapters, Philippe Descheneau brings us back to a deconstructive mode with a contribution that focuses on one particular moment in the commodification of carbon – its monetisation. Drawing from the STS literature as well as economic sociology and the political economy of finance, Descheneau considers the possibilities and the extent to which carbon can be considered as a form of money. He argues that not only is making money from carbon possible via the markets that have developed, but that this process has also enabled the construction of carbon as a *form of money*. The importance of the social underpinnings of the commodification of carbon is highlighted and Descheneau investigates the seemingly merely technical moments in the creation of carbon markets. The development of registries, exchanges platforms, and forms of linkage between markets – making different types of carbon fungible – are related to both broader questions of the social meaning of money, and to the means through which ecological services are valued.

The politics of carbon after carbon

Clive Spash begins this last section with a contribution that is unique in a collection of this kind. Based on his personal experience as a scientist within Australia's Commonwealth Scientific Industrial Research Organisation (CSIRO), his chapter gives an account of the impact of carbon markets and the development of a powerful constituency around them on the politicisation of knowledge production and scientific activity. Spash details his treatment at the hands of CSIRO management and parliamentarians during and after the publication of an article critical of emissions trading (Spash 2010). At the time, the Rudd administration was in the process of debating a possible introduction of emissions trading. Spash's article, which was critical of the implementation of emissions trading, was subsequently suppressed and the author gagged. Later, the academic quality of the paper was questioned in parliament in spite of it being accepted for publication by a respected academic journal. Spash analyses this experience as a concrete case of politicisation – driven by the very existence of carbon as a commodity – of knowledge production at the science–policy interface.

Peter Newell investigates the CDM Policy Dialogue – a stakeholder consultation process as a response to the increasingly vociferous critique of the CDM from both carbon market participants and critics. As Newell points out, seemingly failing in its two core goals (generating low-cost emissions reductions and providing sustainable development), alongside collapsing demand for CERs has led to a severe legitimacy crisis of the CDM. Newell reads the CDM Policy Dialogue as in part an effort to contain this crisis and reinstate the legitimacy of the CDM. In this way the politicisation of a seemingly settled issue is brought to light, alongside subsequent attempts to once again depoliticise it. While Newell shows that much more room was given to market participants and their concerns about administrative burdens and transaction costs, he maintains that the involved civil society organisations were not merely 'taken for a ride'. Instead, he argues that this case demonstrates the role resistance has in constituting the carbon market (Paterson 2009) and the broader dynamics of climate capitalism (Newell and Paterson 2010).

In Chapter 12, Kathleen McAfee puts carbon markets in a broader context by discussing them in relation to the green economy debate. She draws from Swyngedouw's (2010) argument regarding the post-political treatment of climate change and shows that the discourse of green economy follows a similar pattern. However, during the Rio+20 summit in 2012 the dominant green economy narrative was substantially challenged, and hence repoliticised. McAfee starts by problematising the economic ideas and theories that underwrite the green economy concept in order to elucidate their depoliticising origins. Framing REDD+ and carbon markets as a concrete application of the green economy idea, she points out a series of problems that occur in practice. This leads her to further consider the position of a number of governments and many non-governmental actors from the Global South at the Rio+20 summit as presenting the most manifest criticisms of the notion of green economy. McAfee shows how these criticisms disrupt the totalising and post-political character of the green economy narrative and carbon markets.

Chris Methmann and Benjamin Stephan's chapter provides a fitting conclusion to this edited volume as it brings it full circle. They highlight the political moments in the constitution of carbon markets, assess concrete depoliticising effects and point out possibilities for repoliticisation. Methmann and Stephan start by sketching out how carbon markets have been framed as the most suitable mitigation policy and how contestations, for example within science, need to be reconciled in order to enable carbon's commodification. They carve out the depoliticising dynamics at the micro-level, showing how mitigation activities are disentangled from their social surroundings through the focus on carbon. Social relations beyond the notion of emissions reductions are not represented and hence cannot be disputed. But Methmann and Stephan do not leave it at that. They conclude by highlighting the fragility and contingent character of carbon markets through a number of examples – over allocation and collapsing prices; carbon market crimes; the ongoing scientific discussion on the suitability of the concept of GWP – where carbon markets have been challenged.

Conclusion

The carbon markets are in the middle of a fundamental crisis. In the absence of binding GHG reduction targets, market prices have collapsed and an increasing number of actors have, if not completely abandoned, then severely reduced their engagement with carbon trading. However, it is very unlikely that in spite of the various – internal and external, contingent or necessary – problems plaguing the markets, they will be abandoned any time soon. Carbon trading is deeply ingrained in the broader institutional set-up that constitutes the current climate polity. Indeed, alongside this crisis we see ongoing and continual proliferation of the carbon markets – particularly in the Global South.

It is more important than ever to understand the politics that constitute and surround carbon markets. While there have been engagements with the politics of carbon markets before this edited volume, they often have been limited in nature. Analyses that paid attention to the technical details omitted the actors and socio-political context that stood behind the proliferation of carbon markets. Those focusing on these diffusion processes, on the other hand, did so without considering the broader structural context within which carbon markets are implemented. Broadly critical accounts and those that sought to explicitly deconstruct the taken-for-granted nature of the carbon markets failed to recognise the active and generative roles played by actors in the development of policy, and the role of what were presumed to be the merely technical aspects of markets.

This edited volume brings together contributions that together transcend this apparent mutual exclusivity of perspective, and highlight differing aspects of the politics of carbon markets. The analyses are arranged in a way that spans the extent of the carbon markets from their prehistory up to their impact on current climate governance, and this is crucial to enabling a clear grasp of the multivalent politics of the carbon markets. The book thus moves beyond the deathly narratives largely populating the existing literature. By enabling depoliticising and repoliticising practices and events to be brought into relation with each other, the development and utilisation of political agency can be observed in this edited volume in a number of forms: from the level of the material and technical, the human, the institutional and the societal. In this way we believe the volume can bring back to political life an academic literature overrun by shambling, depoliticised actors animated only by brute structural or prematurely naturalised causes.

Acknowledgements

The authors would like to thank Arno Simons, Chris Methmann, Matthew Paterson and Peter Newell for helpful comments on an earlier version of this introduction.

Notes

1   The metaphor of the zombie was used initially, as Chris Harman notes in his 2009 book *Zombie Capitalism*, by economic commentators to refer to the 'undead' and continually threatening state of exposed banks (Harman 2009: 12).

2   This was only the latest, but perhaps most fundamental issue in a series of problems the EU ETS had been facing. Previously, value added tax carousels had caused billions in damages to EU governments, and phishing and hacker attacks on companies trading accounts had resulted in the repeated closure of registries (see Methmann and Stephan, this volume; also Chan 2010; Shapiro 2010). Furthermore, free allocation of allowances according to a 'grandfathering' system resulted in significant windfall profits for the electricity industry during the first two phases of the EU ETS (Gilbertson and Reyes 2009; Sandbag 2010).

3   In the acknowledgements of the report the authors merely state that 'the report does not provide a quantitative, transaction-based analysis of the international carbon market as current market conditions invalidate any attempt and interest to undertake such analysis' (World Bank 2013: 3) – without revealing the collapse of the market, one might add.

4   REDD+ stands for Reducing Emissions from Deforestation and Degradation. It is a mechanism currently being negotiated under the United Nations Framework Convention on Climate Change to help tropical developing countries to reduce their emissions from deforestation and forest degradation.

5   In fact, a number of scholars have critically engaged precisely with horror-imbued apocalyptic narratives of climate change (see e.g. Swyngedouw 2010; Methmann and Rothe 2013). They maintain that apocalyptic narratives do not contribute to a meaningful engagement with climate change. They rather result in a post-political treatment of the issue that negates critical interrogation or democratic dispute and leads to technocratic management as the only viable response (see also Methmann 2011).

6   A thorough survey of this literature goes beyond the scope of this introduction. For a selection see Svendsen and Vesterdal (2003), Kartha *et al.* (2004), Michaelowa and Jotzo (2005), Tuerk (2009) and Ellerman *et al.* (2010).

7   This view is usually backed with reference to an apparent base of empirical knowledge on earlier emissions trading mechanisms (e.g. Ellerman *et al.* 2000; Tietenberg 2006). For a critical engagement with the cost-efficiency claim see Lane (2012).

8   The 2009 special issue of the journal *Accounting, Organisation and Society* (Callon 2009; Lohmann 2009; MacKenzie 2009a) is central and most regularly cited here. See also Lippert (2012), Lohmann (2010), Lovell and MacKenzie (2011) and MacKenzie (2009b).

9   The ratchet is a mechanism within the 1990 US Clean Air Act that was designed to automatically reduce the overall emissions permit allocation in the face of expected individual over allocations; the GWP is an index produced by the IPCC and allows comparison between different GHGs and the development of a composite commodity – carbon.

10 Drawing on a *varieties of capitalism* argument, scholars have compared the adoption of the CDM in different countries (Benecke 2009; Friberg 2009; Fuhr and Lederer 2009; Schröder 2009) or the way in which companies from different EU countries differed in facing the new regulatory challenge EU ETS (Engels 2009; Engels *et al.* 2008). These studies, however, are less interested in the development and diffusion of the mechanisms. Rather, their focus lies on the specific character of their adoption.

References

Agarwal, A. and Narain, S. (1991) *Global Warming in an Unequal World: A case of environmental colonialism*. New Delhi: Centre for Science and the Environment.

Aldred, J. (2012) The ethics of emissions trading. *New Political Economy*, 17 (3), 339–360.

Bachram, H. (2004) Climate fraud and carbon colonialism: the new trade in greenhouse gases. *Capitalism Nature Socialism*, 15 (4), 5–20.

Bailey, I., Gouldson, A. and Newell, P. (2011) Ecological modernisation and the governance of carbon: a critical analysis. *Antipode*, 43 (3), 682–703.

Barry, A. (2002) The anti-political economy. *Economy and Society*, 31 (2), 268–284.

Bäckstrand, K. and Lövbrand, E. (2006) Planting trees to mitigate climate change: contested discourses of ecological modernization, green governmentality and civic environmentalism. *Global Environmental Politics*, 6 (1), 50–75.

Benecke, G. (2009) Varieties of carbon governance: taking stock of the local carbon market in India. *The Journal of Environment and Development*, 18 (4), 346–370.

Betsill, M. and Hoffmann, M. J. (2011) The contours of 'cap and trade': the evolution of emissions trading systems for greenhouse gases. *Review of Policy Research*, 28 (1), 83–106.

Bitter, M. (2011) Contradictions of the commodity carbon: on the material and symbolic production of a market. *In:* E. Altvater and A. Brunnengräber, eds., *After Cancún*. Wiesbaden: VS Verlag für Sozialwissenschaften, 71–93.

Blok, A. (2010) Topologies of climate change: actor-network theory, relational-scalar analytics, and carbon-market overflows. *Environment and Planning D: Society and Space*, 28, 896–912.

Blok, A. (2011) Clash of the eco-sciences: carbon marketization, environmental NGOs and performativity as politics. *Economy and Society*, 40 (3), 451–476.

Blühdorn, I. (2011) The politics of unsustainability: COP15, post-ecologism, and the ecological paradox. *Organization & Environment*, 24 (1), 34–53.

Böhm, S. and Dabhi, S. eds. (2009) *Upsetting the Offset: the political economy of carbon markets*. London: MayFly Books.

Boyd, E. (2009) Governing the Clean Development Mechanism: global rhetoric versus local realities in carbon sequestration projects. *Environment and Planning A*, 41 (10), 2380–2395.

Boyd, E., Boykoff, M. and Newell, P. eds. (2011) The 'new' carbon economy: what's new? *Antipode*, 43 (3), 601–611.

Boyd, W. (2010) Ways of seeing in environmental law: how deforestation became an object of climate governance. *Environmental Law Quarterly*, 37 (3), 843–916.

Braun, M. (2009) The evolution of emissions trading in the European Union: the role of policy networks, knowledge and policy entrepreneurs. *Accounting, Organizations and Society*, 34 (3–4), 469–487.

Brunnengräber, A. (2007) The political economy of the Kyoto Protocol. *Socialist Register*, 43, 224–225.

Bullock, D. (2012) Emissions trading in New Zealand: development, challenges and design. *Environmental Politics*, 21 (4), 657–675.

Bumpus, A. G. (2011) The matter of carbon: understanding the materiality of tCO_2e in carbon offsets. *Antipode*, 43 (3), 612–638.

Butler, J. (2010) Performative agency. *Journal of Cultural Economy*, 3 (2), 147–161.

Callon, M. (2009) Civilizing markets: carbon trading between in vitro and in vivo experiments. *Accounting, Organizations and Society*, 34 (3–4), 535–548.

Caney, S. (2010) Markets, morality and climate change: what, if anything, is wrong with emissions trading? *New Political Economy*, 15 (2), 197–224.

Chan, M. (2010) *10 Ways to Game the Carbon Market*. Friends of the Earth.

Christiansen, A. C. and Wettestad, J. (2003) The EU as a frontrunner on greenhouse gas emissions trading: how did it happen and will the EU succeed? *Climate Policy*, 3 (1), 3–18.

Damro, C. and Mendez, P. L. (2003) Emissions trading at Kyoto: from EU resistance to Union innovation. *Environmental Politics*, 12 (2), 71–94.

Ellerman, A. *et al.* (2000) *Markets for Clean Air: the U.S. acid rain program*. Cambridge/New York: Cambridge University Press.

Ellerman, A. D., Convery, F. J. and Perthuis, C. de (2010) *Pricing Carbon: the European Union Emissions Trading Scheme*. Cambridge/New York: Cambridge University Press.

Engels, A. (2009) The European Emissions Trading Scheme: An exploratory study of how companies learn to account for carbon. *Accounting, Organizations and Society*, 34 (3–4), 488–498.

Engels, A., Knoll, L. and Huth, M. (2008) Preparing for the real market: national patterns of institutional learning and company behaviour in the European Emissions Trading Scheme (EU ETS). *European Environment*, 18 (5), 276–297.

Fine, B. (2005) From actor-network theory to political economy. *Capitalism Nature Socialism*, 16 (4), 91–108.

Fioramonti, L. (2014) *How numbers rule the world: the use and abuse of statistics in global politics*. Available from: http://www.contentreserve.com/TitleInfo.asp?ID={9E27C21D-43ED-4648-8FE7-D736056CBC67}&Format=50 [accessed 28 February 2014].

Friberg, L. (2009) Varieties of carbon governance: the Clean Development Mechanism in Brazil – a success story challenged. *The Journal of Environment and Development*, 18 (4), 395–424.

Fuhr, H. and Lederer, M. (2009) Varieties of carbon governance in newly industrializing countries. *The Journal of Environment & Development*, 18 (4), 327–345.

Gilbertson, T. and Reyes, O. (2009) *Carbon Trading: how it works and why it fails*. Uppsala: Dag Hammarskjold Foundation.

Gillis, J. (2013) Heat-Trapping Gas Passes Milestone, Raising Fears. *New York Times*, 10 May.

Gordon, M. R. and Davenport, C. (2014) Kerry Implores Indonesia on Climate Change Peril. *The New York Times*, 17 February, A8.

Green, J. F. (2008) Delegation and accountability in the Clean Development Mechanism: the new authority of non-state actors. *Journal of International Law and International Relations*, 4, 21.

Harman, C. (2009) *Zombie Capitalism: global crisis and the relevance of Marx*. Chicago: Haymarket Books.

Hedegard, C. (2011) Climate protection is not DEindustrialisation, but REindustrialisation: doing things smarter and more efficiently! Speech given at the Zero Emission Conference in Oslo, on 21 November 2011. Available from: http://ec.europa.eu/commission_2010-2014/hedegaard/headlines/news/2011-11-21_01_de.htm [accessed 7 March 2014].

Helm, T. and Doward, J. (2014) Climate change is an issue of national security, warns Ed Miliband. *The Observer*, 16 January 2014.

Hood, C. (2010) Free allocation in the New Zealand emissions trading scheme: a critical analysis. *Policy Quarterly*, 6 (2), 30–36.

Kartha, S., Lazarus, M. and Bosi, M. (2004) Baseline recommendations for greenhouse gas mitigation projects in the electric power sector. *Energy Policy*, 32 (4), 545–566.

Kenis, A. and Lievens, M. (2014) Searching for 'the political' in environmental politics. *Environmental Politics*, Online first DOI: 10.1080/09644016.2013.870067.

Laclau, E. and Mouffe, C. (2001) *Hegemony and Socialist Strategy: towards a radical democratic politics*. 2nd ed. London/New York: Verso.

Lane, R. (2012) The promiscuous history of market efficiency: the development of early emissions trading systems. *Environmental Politics*, 21 (4), 583–603.

Latour, B. (1987) *Science in Action: how to follow scientists and engineers through society*, Cambridge: Harvard University Press.

Lederer, M. (2011) From CDM to REDD+: what do we know for setting up effective and legitimate carbon governance? *Ecological Economics*, 70, 1900–1907.

Levy, D. and Newell, P. J. (2005) *The Business of Global Environmental Governance*. Cambridge, MA: MIT Press.

Lippert, I. (2012) Carbon classified? Unpacking heterogeneous relations inscribed into corporate carbon emissions. *ephemera*, 12 (1/2), 138–161.

Lohmann, L. (2006) *Carbon Trading: a critical conversation on climate change, privatization and power*. Uppsala: The Corner House.

Lohmann, L. (2009) Toward a different debate in environmental accounting: the cases of carbon and cost-benefit. *Accounting, Organizations and Society*, 34 (3–4), 499–534.

Lohmann, L. (2010) Uncertainty markets and carbon markets: variations on Polanyian themes. *New Political Economy*, 15 (2), 225–254.

Lohmann, L. (2011) Capital and climate change. *Development and Change*, 42 (2), 649–668.

Lövbrand, E. and Stripple, J. (2011) Making climate change governable: accounting for carbon as sinks, credits and personal budgets. *Critical Policy Studies*, 5 (2), 187–200.

Lövbrand, E., Rindefjäll, T. and Nordqvist, J. (2009) Closing the legitimacy gap in global environmental governance? Lessons from the emerging CDM market. *Global Environmental Politics*, 9 (2), 74–100.

Lovell, H. (2013) Measuring forest carbon. In: H. Bulkeley and J. Stripple, eds., *Governing the Global Climate: rationality, practice and power*. Cambridge: Cambridge University Press.

Lovell, H. and Liverman, D. (2010) Understanding carbon offset technologies. *New Political Economy*, 15 (2), 255–273.

Lovell, H. and MacKenzie, D. (2011) Accounting for carbon: the role of accounting professional organisations in governing climate change. *Antipode*, 43 (3), 704–730.

Lovell, H., Bulkeley, H. and Liverman, D. (2009) Carbon offsetting: sustaining consumption? *Environment and Planning A*, 41 (10), 2357–2379.

Lund, E. (2010) Dysfunctional delegation: why the design of the CDM's supervisory system is fundamentally flawed. *Climate Policy*, 10 (3), 277–288.

MacKenzie, D. (2009a) Making things the same: gases, emission rights and the politics of carbon markets. *Accounting, Organizations and Society*, 34 (3–4), 440–455.

MacKenzie, D. (2009b) *Material Markets: how economic agents are constructed*. Oxford: Oxford University Press.

Marchart, O. (2007) *Post-foundational Political Thought: political difference in Nancy, Lefort, Badiou and Laclau*. Edinburgh: Edinburgh University Press.

Matthews, K. and Paterson, M. (2005) Boom or bust? The economic engine behind the drive for climate change policy. *Global Change, Peace & Security*, 17 (1), 59–75.

Meckling, J. (2011) The globalization of carbon trading: transnational business coalitions in climate politics. *Global Environmental Politics*, 11 (2), 26–50.

Methmann, C. (2011) 'We are all green now': Hegemony, governmentality and fantasy in the global climate polity. Dissertation in Political Science. University of Hamburg. Available from: http://ediss.sub.uni-hamburg.de/volltexte/2014/6730/pdf/Dissertation.pdf [accessed 22 July 2014].

Methmann, C. (2013) The sky is the limit: global warming as global governmentality. *European Journal of International Relations*, 19 (1), 69–91.

Methmann, C. and Rothe, D. (2013) Apocalypse now: from exceptional rhetoric to risk technologies in global climate governance. In: C. Methmann, D. Rothe and B. Stephan, eds., *Interpretive Approaches to Global Climate Governance. (De)Constructing the Greenhouse*. London: Routledge.

Michaelowa, A. and Jotzo, F. (2005) Transaction costs, institutional rigidities and the size of the clean development mechanism. *Energy Policy*, 33 (4), 511–523.

Mirowski, P. (2013) *Never Let a Serious Crisis Go to Waste: how neoliberalism survived the financial meltdown*. London: Verso Books.

Mirowski, P. and Nik-Khah, E. (2007) Performativity, and a problem in science studies, augmented with consideration of the FCC auctions. *In:* D. MacKenzie, F. Muniesa and L. Siu, eds., *Do Economists Make Markets? On the performativity of economics*. Princeton, NJ: Princeton University Press.

Mitchell, T. (2011) *Carbon Democracy Political Power in the Age of Oil*. London: Verso Books.

Newell, P. and Bumpus, A. (2012) The global political ecology of the Clean Development Mechanism. *Global Environmental Politics*, 12 (4), 49–67.

Newell, P. and Paterson, M. (2010) *Climate Capitalism: global warming and the transformation of the global economy*. Cambridge/New York: Cambridge University Press.

Newell, P. and Paterson, M. (2011) Climate Capitalism. *In:* E. Altvater and A. Brunnengräber, eds., *After Cancun: climate governance or climate conflicts*. Wiesbaden: VS, Verlag für Sozialwissenschaften, 23–44.

O'Connor, J. (1994) The second contradiction of capitalism. *In:* T. Benton, ed., *The Greening of Marxism*. New York: Guilford Press.

Page, E. A. (2011) Cashing in on climate change: political theory and global emissions trading. *Critical Review of International Social and Political Philosophy*, 14 (2), 259–279.

Paterson, M. (2001) Climate policy as accumulation strategy: the failure of COP6 and emerging trends in climate politics. *Global Environmental Politics*, 1 (2), 10–17.

Paterson, M. (2009) Resistance makes carbon markets. *In:* S. Böhm and S. Dabhi, eds., *Upsetting the Offset: the political economy of carbon markets*. London: Mayfly Books, 244–254.

Paterson, M. (2010) Legitimation and accumulation in climate change governance. *New Political Economy*, 15 (3), 345–368.

Paterson, M. (2012) Who and what are carbon markets for? Politics and the development of climate policy. *Climate Policy*, 12 (1), 82–97.

Paterson, M. and Stripple, J. (2010) My space: governing individuals' carbon emissions. *Environment and Planning D: Society and Space*, 28, 341–62.

Peck, J. (2010) Zombie neoliberalism and the ambidextrous state. *Theoretical Criminology*, 14 (1), 104–110.

Pinkse, J. (2007) Corporate intentions to participate in emission trading. *Business Strategy and the Environment*, 16 (1), 12–25.

Pinkse, J. and Kolk, A. (2007) Multinational corporations and emissions trading: strategic responses to new institutional constraints. *European Management Journal*, 25 (6), 441–452.

Pulver, S., Hultman, N. and Guimarães, L. (2010) Carbon market participation by sugar mills in Brazil. *Climate and Development*, 2 (3), 248–262.

Reyes, O. (2011) Zombie carbon and sectoral market mechanisms. *Capitalism Nature Socialism*, 22 (4), 117–135.

Sandbag (2010) *The Carbon Rich List: the companies profiting from the EU Emissions Trading Scheme*. London: Sandbag.

Schröder, M. (2009) Varieties of carbon governance: utilizing the Clean Development Mechanism for Chinese priorities. *The Journal of Environment and Development*, 18 (4), 371–394.

Selin, H. and VanDeveer, S. (2011) US climate change politics and policymaking. *Wiley Interdisciplinary Reviews: Climate Change*, 2 (1), 121–127.

Shapiro, M. (2010) Conning the Climate. *Harper's Magazine*. February 2010, 31–39.

Shin, S. (2010) The domestic side of the clean development mechanism: the case of China. *Environmental Politics*, 19 (2), 237–254.

Simons, A., Lis, A. and Lippert, I. (2014) The political duality of scale-making in environmental markets. *Environmental Politics*. Online first, DOI: 10.1080/09644016.2014.893120.

Skjærseth, J. B. (2010) EU emissions trading: legitimacy and stringency. *Environmental Policy and Governance*, 20 (5), 295–308.

Skjærseth, J. B. and Wettestad, J. (2008) *EU emissions trading: initiation, decision-making and implementation.* Aldershot/Hampshire/Burlington: Ashgate.

Skjærseth, J. B. and Wettestad, J. (2009) The origin, evolution and consequences of the EU Emissions Trading System. *Global Environmental Politics*, 9 (2), 101–122.

Skjærseth, J. B. and Wettestad, J. (2010) Fixing the EU Emissions Trading System? Understanding the post-2012 changes. *Global Environmental Politics*, 10 (4), 101–123.

Smith, K. (2007) *The Carbon Neutral Myth: offset indulgences for your climate sins.* Amsterdam: Carbon Trade Watch.

Spash, C. L. (2010) The brave new world of carbon trading. *New Political Economy*, 15 (2), 169–195.

Stephan, B. (2011) The power in carbon: A neo-Gramscian explanation for the EU's adoption of emissions trading. *Global Transformations towards a Low Carbon Society, Working Paper Series*, 4, University of Hamburg.

Stephan, B. (2012) Bringing discourse to the market: the commodification of avoided deforestation. *Environmental Politics*, 21 (4), 621–639.

Stephan, B. (2013) How to trade not cutting down trees: a governmentality perspective on the commodification of avoided deforestation. *In:* C. P. Methmann, D. Rothe and B. Stephan, eds., *Deconstructing the Greenhouse: interpretative approaches to global climate governance.* London: Routledge.

Stephan, B. and Paterson, M. (2012) The politics of carbon markets: an introduction. *Environmental Politics*, 21 (4), 545–562.

Stephan, B., Rothe, D. and Methmann, C. P. (2014) Third side of the coin: hegemony and governmentality in global climate politics. *In:* J. Stripple and H. Bulkeley, eds., *Governing the Global Climate: rationality, practice and power.* Cambridge: Cambridge University Press.

Svendsen, G. T. and Vesterdal, M. (2003) How to design greenhouse gas trading in the EU? *Energy Policy*, 31 (14), 1531–1539.

Swyngedouw, E. (2010) Apocalypse forever? Post-political populism and the spectre of climate change. *Theory, Culture & Society*, 27 (2–3), 213–232.

Tietenberg, T. H. (2006) *Emissions Trading, an exercise in reforming pollution policy.* 2nd ed. Washington, DC: Resources for the Future.

Tuerk, A. (2009) *Linking Emissions Trading Schemes.* London: Earthscan.

Voß, J.-P. (2007) Innovation processes in governance: the development of 'emissions trading' as a new policy instrument. *Science and Public Policy*, 34 (5), 329–343.

Wettestad, J. (2005) The making of the 2003 EU Emissions Trading Directive: an ultra-quick process due to entrepreneurial proficiency? *Global Environmental Politics*, 5 (1), 1–23.

Wolf, S. (2013) *Climate Politics as Investment.* Wiesbaden: Springer.

World Bank (2013) *Mapping Carbon Pricing Initiatives.* Washington, DC: World Bank.

World Bank (2012) *State and Trends of the Carbon Market.* Washington, DC: World Bank.

Žižek, S. (2000) *The Ticklish Subject: the absent centre of political ontology.* London: Verso Books.

Part I

The politics of carbon before carbon

2 Resources for the future, resources for growth

The making of the 1975 growth ban

Richard Lane

Introduction

Two questions will likely occur to you after reading the title of this chapter. First, what is the 1975 growth ban? Second, what does an event that precedes the signing of the Kyoto Protocol by 22 years have to do with the carbon markets? This chapter will address both these questions. Moreover, it will begin to uncover why these questions arise in the first place – that is, why the growth ban became a naturalised, taken for granted and, ultimately, largely forgotten component of the longer history of emissions trading. To begin with the second question, even a brief appraisal of the expansive academic, popular and market-actor oriented literatures on carbon markets is enough to indicate that three apparent 'laws of the social world' (Lane 2012: 583–585; Latour 2005) underpin the development of carbon trading. These laws each relate to the ways in which the economy, the environment, and the market[1] are conceived and acted upon within the institutions and structures that comprise global attempts to govern and address climate change.

First, the economy must grow. While an obvious mainstream economic shibboleth, after the disruptive impacts of the 2007 financial crisis, questioning the necessity or good of economic growth is simply not possible within the mainstream environmental governance sphere either. Therefore, it is imperative, as EU Commissioner for Climate Action Connie Hedegaard maintained in 2011, to undertake action on climate change through the implementation of policies that do not impede economic growth and would enable emissions to be cut at least cost (Hedegaard 2011). Second, environmental pollution in general, and climate change in particular, are understood as market failures. Lord Stern made this point clearly in his famed 2006 report when he claimed that 'Climate change presents a unique challenge for economics: it is the greatest and widest-ranging market failure ever seen' (Stern 2006: i). That is, greenhouse gases are simply inadequately internalised externalities of production. Climate change as market failure is more openly challenged by a heterodox ecological economics tradition and also from an environmental justice perspective (see Lane and Stephan, this volume, for a brief survey of literature), and yet it remains central to climate change governance. The third law is that carbon markets, comprised of emissions trading in the form of both cap-and-trade and offsetting mechanisms, are the most economically efficient means of undertaking environmental governance. This has been a mainstay of the environmental economics tradition stretching back to the early 1980s (see e.g. Tietenberg 2006; Calel 2011). Again, this understanding has been challenged by market critics (e.g. Lohmann 2006; Lane and Stephan, this volume), yet this critique has proved remarkably ineffective in dislodging the canonical understanding of market efficiency propounded by a dominant economics tradition.

There are three core objects of the politics of carbon markets: the economy, the environment and the market; and three laws that define these objects: growth, externalities, efficiency. The construction of the growth ban in the US is intimately related to the development of each of these. On 31 May 1975 the deadline for US states to attain the National Air Ambient Quality Standards (NAAQSs) mandated by the 1970 Clean Air Act (CAA) passed. The failure of most states to meet these standards appeared to present an immanent ban on the expansion of polluting energy and manufacturing industries. This gave legislators and regulators their first taste of a problem that would come not only to frame US environmental regulations in future, but to define the very question that two decades later, the carbon markets were ostensibly developed to answer. As environmental economist Thomas Tietenberg put it in the second edition of his highly influential *Emissions Trading: an exercise in reforming pollution policy*, 'Was it possible to solve the air quality problem while allowing further economic growth?' (Tietenberg 2006: 7).

The realisation of an imminent and indeed immanent growth ban threatened by the CAA is taken to provide de facto historical evidence indicating that so-called command-and-control regulations are indeed economically inefficient (Tietenberg 2006; Pérez Henríquez 2013). It resulted in the search for 'an exception to allow greater flexibility in the administration of the Act and opportunity for growth of national industrial capability' (Cook 1988: 45). This ultimately resulted in the 1977 amendments to the CAA and a mechanism to allow the development of new polluting facilities and the expansion of existing ones in non-attainment areas – known as the offset provision. This, along with a rehabilitated 'bubble' concept, formed the basis of the first experiments with market mechanisms for pollution control in the US; mechanisms that would be developed into fully fledged emissions trading with the arrival of the sulfur trading markets under President Bush's 1990 CAA amendments, and then exported globally in the form of the carbon trading markets instituted under the Kyoto Protocol. While this account of the role of the growth ban in the history of emissions trading mechanisms is taken as brute fact, there are, in fact, significant problems with this understanding.

Fabricating the 1975 growth ban: objects, laws, translations

The growth ban did not simply provide the graphic, real-world evidence of certain immutable economic laws. While a recent text on emissions trading claims – without any supporting evidence – that economic development 'was

clearly being hampered by environmental regulation' by the 1975 deadline
(Pérez Henríquez 2013: 51), in fact, where states did not meet the 1975 deadline, it was the deadline that was deferred, not growth (Meidinger 1985: 453).
This does not mean that the growth ban was merely a post hoc construction,
however; a fiction designed to delegitimise federal environmental standards
and regulations in line with the development of a broader neoliberal approach
to socio-political governance in the US. Instead, in order to understand the
development of the 1975 growth ban and its importance in setting the context
that would frame the subsequent development of the carbon markets, it is
necessary to investigate what I have elsewhere referred to as the 'promiscuous
history of emissions trading' (Lane 2012); an approach that derives from actor-
network theory (e.g. Latour 2005) and economic performativity (e.g. Callon
1998). Following Timothy Mitchell (2011: 176), I argue that from the end
of the Second World War onwards the economy and the environment were
assembled and reassembled through new technical practices and technologies.
Over a 25-year period, a socio-technical world was ultimately (but not from
the outset intentionally) constructed that would enable the development and
stabilisation of the growth ban as part of the economic 'laws of the world'.

In his 2011 book *Carbon Democracy*, Mitchell argues that the environment
was made into an object of politics, or matter of concern, in the late 1960s and
early 1970s to rival the economy as part of, and as a means to, construct the
crisis of the 1973–1974 oil price shocks, thereby helping to install the market
as a fundamental general organising principle. In this chapter I take certain
components of Mitchell's narrative and reassemble them for different ends.
Specifically I focus on the role of economics and the economics discipline,
and the tools and technologies that they develop in the process of fabricating
the economy and the environment as objects of politics.[2] Following Barry
(2002, 2013), my interest here is in how this depoliticisation did not simply
foreclose political contestation around these objects, but their very fabrication as
objects – replete with their own essential laws – opened up new forms of politics, around new objects, and resulted in the development of new technologies
and techniques in order to manage this politics. Fundamental to this process
were a group of economists associated with 'the most important think tank
you've never heard of' (Rauch 2002), the Washington DC based Resources
For the Future (RFF). Over a 25-year period from the end of the Second World
War, RFF would, through its research, publications, workshops and influence
as a crucial knowledge provider to the US government, undertake a series of
translations (Barry 2013) with respect to the economy and the environment. The
notion of translation here is used to refer to a relation or process that, following Latour (2005), induces the local actors that comprise the regulatory content
and economic context into coexisting. This chapter traces the translations of the
economy and the environment around the laws of growth and external effects
that were necessary to enable the development of the growth ban.

In order to understand the history of the growth ban, it is first necessary to
explore briefly the history of the growth of the economy. The first section of

this chapter begins by outlining Timothy Mitchell's claim that the economy as commonly understood, is a twentieth-century invention, and stems in the first instance from the depression years of the 1930s. I then detail how RFF secured the possibility of ongoing economic growth, and thereby reinforced the economy as a central object of politics by separating growth from material constraints and Malthusian resource scarcity over the period 1952–1963. As pollution and environmental degradation became an increasing concern over the 1960s and 1970s, the second section investigates how RFF helped make the very certainty of economic growth that they had previously secured, into the cause of pervasive pollution, through the development of the ecologically inflected notion of 'the Spaceship Earth'. Finally, the third section covers how RFF divorced economic growth from material constraints for a second time through the development of the *Materials balance* approach. This brought pervasive pollution fully in line with the economic theory of externalities, and enabled the oil price shocks of 1973–1974 and the broader energy crisis to be linked to environmental regulation in such a way as to enable the laws of economics, as evidenced by the growth ban, to operate. In doing so, this undermined the opposition of the environment and the economy, and helped construct the opposition of the economy not with the environment per se, but with environmental regulation. In line with the increasing influence of the neoliberal perspectives of the Mont Pelerin society, Chicago School economics, public choice theory and the political scientist Samuel Huntington's 1975 diagnosis of an 'excess of democracy', this event would proffer the development of market-based tools – emissions trading – as a solution to both economic *and* environmental woes.

Resources for growth

Looking back from the second decade of the twenty-first century, a focus on the growth of the economy and material progress looks natural enough, but the necessity of this growth is nowhere near so apparent, and requires a brief diversion into the very modern histories of growth and the economy. In fact, prior to the enlightenment, none of the many accomplishments achieved in engineering, architecture, technology, navigation and in numerous other areas and disciplines 'was thought of as "progress" – simply as the ingenious contrivances of persons, mostly anonymous, to meet immediate needs' (Purdey 2010: 65).

The economy and growth

Growth as a specific economic concept first gained acceptance through the work of the classical economists of the eighteenth century (Purdey 2010: 68). However, for Adam Smith and his contemporaries, material progress was rarely explicitly discussed (Arndt 1978: 7). Notions of economy and notions of growth were closely intertwined here, and as Mitchell has argued in numerous places (e.g. 1998, 2005, 2008, 2011), Adam Smith's notion of economy was not related to the structure of production or the exchange of goods within an economy (Mitchell 2005: 128). Rather, the notion of economy referred only to the frugal and prudent husbanding of resources – and this was directly related to progress and growth. While a second generation of classical economists, most predominantly Thomas Malthus and David Ricardo, would be more explicitly concerned with the possibility (or not) of growth, they also did not focus on the economy as it is currently used, but rather the world of 'human settlement, agriculture, and the movements of populations, goods and wealth' (Mitchell 2005: 128).

After the marginal revolution of 1870s and the development of neoclassical economics, the explicit concern with growth in the discipline largely disappeared: 'hardly a line is to be found in the writings of any professional economists between 1870 and 1940 in support of economic growth as a policy objective' (Arndt 1978: 13). In response to the global depression of the 1930s, the theoretical work of economists such as John Maynard Keynes, and the econometric modelling of Jan Tinbergen and Ragnar Frisch, reconceptualised national wealth as the national economy – something that inheres not in the household or the settlement, or agriculture, or the neoclassical market, but as a 'general structure of economic relations' (Mitchell 1998: 85). This approach required, argues Mitchell (1998: 87), two conceptual shifts. First, a clear distinction had to be defined and maintained between what Frisch would call the 'intrinsic structure' of a mechanism and its exterior. Second, this intrinsic structure could not be thought of any longer as a single market – it must be thought of as 'the whole economic system taken in its entirety' (Mitchell 1998: 87). These conceptual innovations do not, however, simply denote the development of new forms of economic thought in the form of dynamic models and macroeconomics. Instead, they should be seen as marking the construction of a newly imagined, ontologically discrete, measurable and calculable object: the economy (Mitchell 2008).

As Mitchell notes, in Britain, Keynes' involvement extended beyond imagining the new object to be counted, and he and his students worked directly with the Treasury to design methods of estimating national income. In the US, Simon Kuznets of the National Bureau of Economic Research would systematise a method for estimating national income in 1941. One year later the US Department of Commerce would begin publishing national economic data, and in his 1944 budget speech, President Roosevelt would officially introduce what would become the embodiment of the new idea of the economy – Gross National Product (GNP). The enumeration of the GNP of an economy made it possible to represent the size, structure (and crucially) the growth of this new totality. Alongside this, the metrological tools capable of accounting for all the instances of spending and receiving money within a specified geographical space – the national income accounts – would help make this object coextensive with national boundaries; and in this way, the burgeoning reality of the new notion of the economy would rapidly lead to a re-imagination of the nation-state (Mitchell 1998: 89) as the bearer of this precious burden. This re-imagination was not itself explicitly theorised and was instead introduced as a 'commonsense construct' (Mitchell 1998: 89; citing Radice 1984) which provided the boundaries within which the new aggregate accounts of production, employment, investment, as well as synthetic averages such as interest rates, price levels and real wages could be measured.

While this commonsense construct was rapidly elevated to the level of essential freedom and a natural necessity, this did not have the immediate effect of disinterring it from its earthy, material constraints. Instead, in the post-war world of seemingly exhausted mineral and other natural resources, the apparent need for economic growth, understood as the intensification of monetary flows within spatially constrained national boundaries, would at the end of the 1940s and beginning of the 1950s, bring Malthus heaving back from the dead. In 1947, J. Frederick Dewhurst's expansive *America's Needs and Resources* was published, and this undertook the first ever wide-ranging audit of the material and economic position of the US. This influential report reinforced the view that the preceding global conflagration had burned too bright, for too long, and in the process had 'chewed up' much of the mineral resources of the US, leaving the country with a depleted supply of most natural resources (Dewhurst 1947: 675). As the grinding mechanical howl of the Second World War began to fade at the beginning of the 1950s, only to be supplanted by the cold war's keening echo, the old Malthusian concern with the scarcity of material resources began to take hold. Were there enough resources available for the necessary growth of the economy? Or were the lights in danger of going out?

Scarcity and Resources for Freedom

In response to these fears, President Truman formed the President's Materials Policy Commission (PMPC) in 1951, and tasked William S. Paley – then the Chair of the board of CBS news – with the role of Chairman of the Commission. In a letter to Paley, dated 22 January of that year, the President made clear the threat of 'the nation's materials problem' (PMPC 1952, Volume 1: iv) and that shortages of material resources could not be allowed to jeopardise either national security or economic expansion. In order to respond to this materials problem, the Commission, funded under the National Defence budget, was to undertake a detailed review of the future supply of mineral, energy and agricultural resources in the US. On 2 June 1952, the Commission transmitted its five-volume report: *Resources for Freedom: Foundations for Growth and Security*, to the office of the President. The core concern at the heart of the Paley report was with the continuing growth of the economy, substantiated early on in the report, when in a moment of inadvertent levity the commissioners bluntly stated that:

> [W]e share the belief of the American people in the principle of growth. Granting that we cannot find any absolute reason for this belief we admit that to our Western minds it seems preferable to any opposite, which to us implies stagnation and decay.
>
> (PMPC 1952: 3)

It is perhaps not surprising that an absolute reason eluded the commissioners, given the newly imagined status of the economy, and the only recent development of a concern with growth with specific respect to this object. Apparently not content with basing a belief in growth on just the preferences of the 'Western mind' however, the Paley report also states that its 25-year projections are based on the equally unfounded presumption that the historic average rate of growth of 3 per cent a year – newly established by the work of Kuznets – would continue over this period (Maass 1953: 207). Could this rate of growth be maintained, given the increasing material throughput required, and the increasing material scarcity that this implied? The Paley report broke with prior approaches, such as that undertaken by Dewhurst (1947), by focusing on economic factors such as the costs and prices of end products derived from natural resources, instead of the measurement of physical stocks themselves (Maass 1953: 206). Resource depletion was not, therefore, considered as a physical absolute, but was instead expressed through rising costs. This crucial shift from an absolute scarcity to (relative) price scarcity allowed the commissioners to abstain from the then common concern with resources running out (PMPC 1987) and shift to a new conception of mineral reserves. With specific respect to oil, the commissioners argued that:

> Public judgements of the prospects for future petroleum supplies have frequently been distorted because of popular misconceptions concerning the nature of proved reserves. Time after time the fact that proved reserves were equivalent to only about 12 to 15 years' production has come to the attention of publicists who have then sounded the alarm that the United States was about to run out of oil. *Reserves must be considered not as a total reservoir from which all future production is to be drawn, but as the basis of operations, a sort of working inventory.* Proved reserves are indeed like a reservoir, but a reservoir into which there is an inflow as well as an outflow.
>
> (PMPC 1952, Volume 3: 5; emphasis added)

The inflow into the reservoir could be maintained simply by ensuring that the cost of new discoveries does not exceed the general price level of crude oil and associated petroleum products; therefore, as Hans Landsberg put it in his introduction to the reissued report in 1987, '*At a cost* there is always more' (PMPC 1987: 85; emphasis in original). The Paley approach to scarcity, therefore, inverted the relation of cost to material scarcity. For Dewhurst, costs were high because material resources were scarce. The approach propounded by the Paley report maintained – on the basis of a conception of resource reserves as inventories with both input and output – that increasing costs actually results in more resource availability. This change in the measurement of scarcity meant that being no longer bound by physical and material constraints, as presented in the Paley report, the economy was freed to continue a trajectory of continuous growth – a trajectory that the work of the classical political economists (particularly Malthus) with their presumption of fixed resource supply, would have denied.

The death of scarcity and the growth of the economy

The Paley Commission report was, as Maas noted, 'the most original and significant contribution to the study of resources and public policy since the 1933 report of the Mississippi Valley Committee and the early reports of the National Resources Committee' (Maass 1953: 210). It had an immediate impact in the US, with the re-evaluation of resource programs by federal government agencies such as the US Geological Survey. Similarly the extractive and mineral industries, also brought their measurement of resource scarcity in line with what Maass referred to as the 'Paley approach' and shifted to price-focused reserve measurements (Bowden 1985: 221).

Following on from his report, William Paley had intended for the federal government to assume responsibility for ongoing, continuous assessment and analysis of resource problems and prospects (PMPC 1987). The incoming Eisenhower administration declined this proposal, however. This resulted in the founding in 1953 – at a conference attended by President Eisenhower – of the private nonprofit Resources for the Future (RFF). Instituted by Paley and members of a committee on resource availability and economic growth at the Ford Foundation, by 1977 RFF had been funded to the tune of USD 47.5 million by the Foundation (Magat 1979: 186–187). During the first decade of its existence a substantial element of the research undertaken by RFF was prompted by, and should be considered a continuation of, the 'Paley approach'. This culminated in the publication of highly influential empirical studies on resource use and scarcity by Potter and Christy (1962) and Barnett and Morse (1963). Barnett and Morse concluded that the predictions drawn from the Malthusian and Ricardian doctrines did not fit the data well and were obsolete. As they firmly stated: 'A limit may exist, but it can be neither defined nor specified in economic terms . . . . Nature imposes particular scarcities, not an inescapable general scarcity' (1963: 11).

From the early 1950s to the early 1960s, the apparent death of scarcity at the hands of the Paley Commission (Mitchell 2011: 177) further enabled the construction of an economic mainstream dismissive of generalised Malthusian and Ricardian resource scares and with a seemingly clear-eyed focus on the economy and its growth, as a distinct and central object of politics. Six years after the publication of the Paley report, growth was finally given the 'absolute reason' that eluded the Paley Commissioners. In 1958 Nelson and Laurence Rockefeller tasked Henry Kissinger with preparing a report on 'The challenge of the future' (Purdey 2010: 80; Dale 2011). Heading a panel comprised of economists from large corporations and key universities, Kissinger produced a report entitled: *The Key Importance of Growth to Achieve National Goals*. This identified the importance of economic growth as the solution to the pressure on national income of multiple and competing claims, and argued that it not only brings 'dignity, freedom, and purpose' but promises to expand 'the opportunities for individual fulfilment, multiply the incentives for enterprise, enable us to improve our educational system, permit us to increase our protection against economic hardship, make possible rising standards of national health and open new vistas of cultural achievement.

(Dale 2011)

This approach would perhaps reach its apotheosis in 1971 when Simon Kuznets would publish his *Economic Growth of Nations*. Thirty years after the official introduction of GNP, with this work Kuznets provided an empirical bookend to the construction of the economy. The innovative replacement of general resource scarcities with particular price-based scarcities rapidly came to represent the 'economic orthodoxy' on natural resources (Perez-Carmona 2013: 87) and helped reinforce the ontological footing of the economy represented by GNP. The economy could be safely perceived as a distinct sphere, divorced from a natural resource base, and driven by an inherent logic of continuous growth.

The growth of limits and the limits to growth

While the technical innovations undertaken by the Paley Commission and continued by RFF signalled the death knell of concerns with resource scarcity and secured the economy as an object capable of continued, and exponential, growth, one year earlier a popular book was published that would ignite a new fear, not this time with a lack of growth, but with the impacts of growth itself. In 1962, the former researcher for the US Fish and Wildlife Service, Rachel Carson, would publish her book *Silent Spring*. Originally serialised in three parts in the New Yorker, *Silent Spring* highlighted the ecologically catastrophic impact of the introduction of DDT and other chemical pesticides into contemporary agriculture and sold half a million copies in hardback alone, staying on the *New York Times* Bestseller list for 31 weeks (McCormick 1995: 55). This book, frequently credited as marking the beginning of the environmental revolution (McCormick 1995: 65) 'arguably opened the eyes of the American public' (Flippen 2000: 4) to the problems that accompanied unalloyed economic and industrial growth in post-war America.

The continued expansion in the production of oil and the associated growth in car ownership would also have a crucial series of impacts. By 1965 the ratio of adults to cars in the US stood at only 1.66:1 (Landsberg and Schurr 1968: 41) and this level of car ownership resulted in increasing air pollution, particularly in urban areas, leading to the portmanteau term 'smog' being coined to refer to the particulate clouds engulfing Los Angeles. The overseas and offshore production and transportation of oil would also result in highly visible environmental catastrophes. In 1969, a Union Oil drilling platform in the Dos Cuadras offshore oil field six miles from the California coast experienced a blowout, resulting in the largest oil spill seen in US waters at the time. The Santa Barbara

Channel oil spill received widespread media coverage, galvanised environmental action in Richard Nixon's home state of California, and resulted in the new President visiting affected beaches just one month into his first term (Hamblin 2013: 190).

Throughout the 1960s, pollution developed into an increasingly pressing and seemingly pervasive issue in the US, and within a country shaken by the ructions of anti-Vietnam war protests, an increasingly vociferous and politically potent movement crystallised around the environment. In 1968, the publication of the Apollo 8 *Earthrise* pictures would give this movement its own motif, and as the novelist Arthur C. Clarke would come to claim, these images '[f] or millions on earth . . . must have been the moment when the Earth really became a planet' (McCormick 1995: 80). The images of planet Earth – singular, self-contained and fragile in the void of space, would not only galvanise an environmental movement, but came to signify the development of a new approach to the environment, one focused on the closed ecosystem of Spaceship Earth.

The coming Spaceship Earth

Three years earlier, in July 1965, the then US ambassador to the UN, Adlai Stevenson, gave a speech to the UN economic and social council in Geneva on the problem of global urbanisation. In the speech, Stevenson made use of the metaphor of the Earth as a spaceship 'on which humanity travelled, dependent on its vulnerable supplies of air and soil' (McCormick 1989: 67). This speech was drafted by the former editor of *The Economist*, Barbara Ward, who would, in 1966, publish the book *Spaceship Earth*, based on her 1964 George B. Pegram lecture series at Brookhaven National Laboratories. It would not be until two years later, however, that the Spaceship Earth would make its most celebrated (Pearce 2002: 60), and perhaps most surprising appearance, thanks to RFF.

In March 1966, RFF held its sixth research forum. This was on the topic of 'Environmental quality in a growing economy', and was planned by a staff group led by Allen V. Kneese, the then director of the organisation's research programs in water quality and quality of the environment. As RFF economist Henry Jarrett noted in his introduction to the edited volume that resulted from the forum, the changing emphasis from resource concerns to those of the effects of environmental pollutants – understood within RFF as a transition from environmental quantity to environmental quality, resulted in the following question:

> Granting that most of the pressures upon the natural environment have been direct or indirect results of a prosperous and expanding economy, does it follow that further erosion of environmental quality must continue to be the price of further economic gains?
>
> (Jarrett 1966: xii)

Not only is economic growth queried here in relation to its environmental impacts and hazards, but by using the markers of growth developed in an earlier RFF study *Resources in America's Future* (Landsberg *et al.* 1963) – namely, population growth, automobile ownership, personal consumption expenditures, and the federal reserve board index of industrial production – Jarrett maintained that '[t]he underlying causes of these discomforts and hazards are to be seen in the same statistics that most of the time are hailed as indicators of economic growth' (Jarrett 1966: ix). Here then, the very assurances of continued growth provided through the metrics and measurements of the early 1960s RFF reports, and underwritten by the development and broad acceptance of the notion of price scarcity, starkly constitute the new problem of the environment.

The forum's most important and famous output would come in the form of a paper entitled *The Economics of the Coming Spaceship Earth*, by the future President of the American Economics Association, Kenneth E. Boulding. Here, Boulding argued that economists in particular had failed to come to grips with the transition of the Earth from an open system with an economy maintained in the midst of a throughput from inputs to outputs, to the Earth as an effectively closed system, where 'the outputs of all parts of the system are linked to the inputs of other parts. There are no inputs from outside and no outputs to the outside; indeed, there is no outside at all' (Boulding 1966: 2). Prefiguring the Apollo 8 iconography by two years by using the colourful vernacular of spacemen and cowboys, Boulding states that established measurements of economic success do not make sense in a 'spaceman economy' (Kula 1998: 130).

The 'shadow of the future spaceship' was already evident for Boulding, not through the exhaustion of resources, but rather through the issue of pollution. From Los Angeles 'running out of air' due to the prevalence of smog, to Lake Erie becoming a 'cesspool', to the issue of DDT raised by Rachel Carson (Boulding 1966: 12) – pollution was a fundamental concern in a spaceman economy. The diagnosis of pervasive pollution within a cowboy economy, driven by consumption and production as the basis of welfare, led to Boulding's prescription for a spacemen economy. Whereas in the former, 'throughput' as measured roughly by GNP was to be maximised, in the latter, the maintenance of stock forms the basis of welfare (McCormick 1995: 80), and throughput – production and consumption – is to be minimised (Boulding 1966: 8). Not only is pollution important then, but it is pervasive. The reason for this, according to Boulding, is that the economy or 'econosphere' is merely a subset of the 'world set' or total ecosphere, and while we tend to treat the former as an open system, the latter is not. Here then, we not only have the economy, and the concerns of critical economists such as Boulding, brought together with the ecological approach outlined by Howard and Eugene Odum (e.g. Odum and Odum 1959), but this is undertaken through the explicit subsumption of the economy under a global ecology.

The limits to growth

Boulding's Spaceship Earth would spur a raft of further books and studies examining the relationship between economic growth and the environment. Or as David Pearce put it in his intellectual history of environmental economics, '[g]radually, the unsustainable lifestyle issue became synonymous with the pursuit of economic growth, and the antigrowth movement was born' (Pearce 2002: 60). This predominantly academic debate (Perez-Carmona 2013: 89) would be brought to wider public attention with the publication in 1972 of the Club of Rome's famous *Limits to Growth* Report.

Using figures drawn from Kuznet's *Economic Growth of Nations*, and the US Bureau of Mines 1970 report *Mineral Facts and Problems*, and utilising the new technology of computer simulation to develop a model based on Boulding's concept of the Spaceship Earth, *The Limits to Growth* describes how the quality of life will decrease progressively as economic growth and population numbers increase. In order to save this situation, Meadows *et al.* recommended the implementation of a series of specifically anti-growth targets, including:

> 30 per cent reduction in birth rates; 50 per cent reduction in pollution generation; 75 per cent reduction in natural resource depletion; 40 per cent reduction in capital formation; 20 per cent reduction in food production, as it fuels population growth.
>
> (Kula 1998: 135)

The Limits to Growth was released to coincide with the UN's 1972 conference on the Human Environment in Stockholm and this timing ensured that the apocalyptic vision of the report achieved widespread, international coverage. By the beginning of the 1970s not only had the unforeseen side effects of economic growth, in the form of widespread pollution, become an important matter of political concern, but the certainty and logic of economic growth had come to be seen as the very engine of this concern. Within the Spaceship Earth, the finite limits first indicated by Boulding, and later by the anti-growth movement popularised by the Club of Rome, would place the environment fundamentally at odds with the economy (Mitchell 2011), and this would be evident within the US Environmental Protection Agency (EPA), Nixon's newly developed regulatory body instituted in 1970 in order to undertake the provisions of the 1970 CAA.

In May 1973 the EPA hosted a National Conference on Managing the Environment in Washington DC. In his opening address to the conference, acting administrator Robert W. Fri outlined the central concern then vexing the EPA:

> It is particularly important to examine critically the great American shibboleth known as growth. It is our own special sacred cow, and in its most exaggerated form it makes environmental management difficult if not impossible. It is the antithesis of stability . . . We may have to make

do – indeed, we must learn to want to make do – with smaller cars, with less energy, with recycling our wastes instead of throwing them in the city dump, and adjusting the size of our families to responsible norms.

(EPA 1973: 6–7)

A concern with the question of growth was shared at the conference through the circulation of a speech delivered in October 1972 by the EPA's first administrator William D. Ruckleshaus, who just one month before the conference had departed from the EPA in the wake of Nixon's political demise. Ruckleshaus' focus would similarly echo the language of closed ecological systems, spaceman economies and the limits to growth. Even the most economically oriented presentation, given by Russell Train, who was at the time the Chairman of the President's Council on Environmental Quality[3] and had been, since its founding in 1969, simply focused on the development of managerial tools such as the Environmental Impact Analyses developed under the 1969 National Environmental Policy Act (NEPA), benefit-cost analyses, and the need for public engagement. At this point, it was clear that within the EPA at least, the 1975 CAA deadline was not discussed in relation to any deleterious effects on growth. What would happen then – between this 1973 conference and 1975 – that would fundamentally change the EPA's perspective on economic growth and environmental pollution? How would administrator Fri's core concern with the harmful impact of economic growth on environmental stability be converted into an equally urgent concern with environmental regulations' harmful impact on economic growth?

The new limits to growth

Almost exactly five months after the *Managing the Environment Conference*, on 16 October 1973, OPEC states decided to raise the posted price of crude oil by 70 per cent. The next day, Arab Gulf states announced a 5 per cent reduction in the production of their oil in response to the US's obstructive role in the Arab–Israeli conflict. Moreover, production was to be reduced by 5 per cent a month until Israel removed its forces from the territories occupied in the June 1967 war. These two separate events, frequently run together as the 'OPEC embargo' or oil crisis of 1973–1974 (Mitchell 2011: 184–185), were the apotheosis of an apparent energy crisis that had been steadily developing since John Nassikas, the head of the Federal Power Commission gave a speech at the National Press Club on 10 August 1970 announcing its arrival. Enormous queues for petrol at the pumps in the US, and daily price hikes against a background of inflation and economic slowdown, would then appear to have delivered an object and indeed abject lesson on the laws of supply and demand, as well as almost immediately re-sacralising the economic growth questioned earlier that year by EPA administrator Fri.

Given these events, it would be easy to see the concern with the growth ban presented by the CAA's 1975 deadline as simply a revealed economic truth once the energy and the environmental crises were combined. It would be easy, and it would be wrong. Instead, the conversion of the CAA deadline into an all threatening growth ban would require the further translation of the environment from a political object freighted with its own eco-logic of finite means, closed systems, and limits, to one no longer at odds with the economy and its infinite growth imperative. The dissolution of the conflict between environment and economy would then enable the very real economic impacts of the energy crisis to be blamed not just on the OPEC states, but on burgeoning and burdening environmental regulations – specifically the CAA – of the early 1970s. Once again, RFF economists would be central to this process; in the first instance, by bringing pollution fully in line with the economic theory of externalities under the rubric of a *Materials balance* approach, and then by highlighting the economic impact of environmental regulations.

Materials balance and the rehabilitation of externalities

In a paper entitled 'Production, consumption, and externalities', that first appeared in the *American Economic Review* in 1969, Robert U. Ayres and Allen V. Kneese presented their *Materials balance* approach to the economic analysis of pollution. This approach was the first outcome from the new research program focusing on environmental quality, directed by Kneese at RFF.[4] This research program represented a shift to a concern with environmental quality for both Kneese and RFF, and is indicative of the development of the new discipline of environmental economics from its resource-focused base at this time (Pearce 2002: 59–60; Lane 2012: 590). The development of the *Materials balance* approach itself should be seen as a response to the 1966 RFF forum on environmental quality organised by Kneese. There he raised concerns about the lack of progress in the application of welfare economics principles to environmental concerns and, importantly, the 1969 and 1970 *Materials balance* texts take as their starting points the critiques of mainstream economic thought raised by Kenneth Boulding.

In order to revivify the economic approach to the environment, which RFF and its publication of Boulding's thesis had helped throw into malingering doubt, Ayres and Kneese sought to address a common failure of the economics literature to view production and consumption processes in relation to the 'fundamental law of conservation of mass' (Ayres and Kneese 1969: 283). They argued that:

> Almost all of standard economic theory is in reality concerned with services. Material objects are merely the vehicles which carry some of these services, and they are exchanged because of consumer preferences for the services associated with their use or because they can help to add value in the manufacturing process. Yet we persist in referring to the 'final consumption' of goods as though material objects such as fuels, materials, and finished goods somehow disappeared into the void a practice which was comparatively harmless so long as air and water were almost literally free goods.
> (Ayres and Kneese 1969: 284)

The Spaceship Earth thesis indicated the problem with this economic notion of air and water, and so Ayres and Kneese sought to develop a general equilibrium model which included a series of variables representing material inputs and outputs in order to 'view environmental pollution and its control as a materials balance problem for the entire economy' (1969: 284). Central to this task was the rehabilitation of the economic theory of externalities.

Externalities, which originated in the work of the British economist Arthur Cecil Pigou in the 1920s, refer to a failure to take into account the broader costs (or benefits) of individual action when deciding upon that action. For Boulding, this economic concept was fundamentally inadequate to deal with the large scale and harder to solve problems of pervasive pollution within the Spaceship Earth (Boulding 1966: 14). Boulding makes clear that pollution is not specific (and local) in the way that economic analysis of externalities takes it to be up to this point, but is in fact pervasive and unavoidable. Following from both this and E. J. Mishan's critique of externalities (Mishan 1965), Ayres and Kneese argued that, for Pigou, the importance of externalities was generally minimised, with these being considered as exceptional cases (Ayres and Kneese 1969: 282); and they highlighted how this notion of minor and negotiable deviations remained the prevalent perspective on externalities within the post-war welfare economics discipline (see also Pearce 2002: 59–60).

Ayres and Kneese argued that this understanding of externalities was wrong, and that these were not the 'freakish anomalies' (1969: 287) they had previously been supposed. Following the work of Kapp (1950) the *Materials balance* approach showed that when the flow of materials through an entire economy, with inputs in the form of material resources and outputs in the form of final goods and ultimately residuals (pollutants) is considered, externalities are not exceptional, but rather inherent to the economic process (Perez-Carmona 2013: 89). While the purpose of the *Materials balance* approach is focused on the development of economic theory and the introduction of the proper consideration of residuals production, at the same time the notion of the environment is itself reconstructed. Environmental pollution – understood by Boulding as a pervasive problem of unlimited economic growth within a limited or finite global ecosystem, and therefore an issue fundamentally outside of the remit of mainstream economics – is reconstructed as an issue of economic externalities.

The rehabilitation of the welfare economics concept of externalities within the *Materials balance* approach, and the synthesis of this with a general equilibrium model of the economy, had the effect of seemingly bringing pervasive pollution fully in line with the economic theory of externalities. This innovation by RFF economists enabled the economics profession to reassert its providence over issues of environmental regulation in the US, and to undermine the concept of the environment as bounded by finite limits. That is, *Materials balance* took the concerns highlighted by Boulding and the anti-growth movement and yet translated them into a form that removed, once again, economic growth from material constraints. This defused the conflict between the environment and the economy by undertaking a timely volte face on the environment and market failure. No longer is environmental pollution the clear indicator of the failure of markets and of a focus on economic growth, rather, it is due to market failure – pollution results when markets are not implemented in order to adequately price the environment.

The *Materials balance* approach spread rapidly through both the discipline of economics and the wider policy making consciousness – the latter through its family resemblance to and easy association with Garrett Hardin's (1968) highly influential *Tragedy of the Commons* argument in the journal *Nature*; and via Kneese himself, who was a well-known figure among Washington's environmental policymakers, due to his frequent appearances before Congress at environmental hearings, and from 1972 via his relationship with Senator Peter Domenici and his staff, who was a member of the Senate's environmental affairs subcommittee (Kelman 1981: 21). Further work in this vein was sponsored by RFF (e.g. Bohm and Kneese 1971), and by December of 1970, *Materials balance* was referred to by Robert Solow as 'the economist's approach to pollution' in his vice-presidential address at the annual meeting of the American Association for the Advancement of Science in Chicago (Solow 1971).

Aside from establishing economic provenance over pollution and natural resources, the *Materials balance* approach would enable the oil industry to further define the environment in such a way as to force coal, gas and particularly nuclear energy producers to internalise the environmental costs of their operations (Mitchell 2011: 192). As the CEA's annual report on the state of the economy in 1974 highlighted, environmental regulations, particularly with respect to the construction of nuclear reactors, were one of the major causes of the recent increased demand for oil (cited in Nordhaus 1974: 559). Alongside this increased oil demand, *Materials balance* would crucially enable the environment and energy to be put together in such a way as to allow the blame for higher energy prices to be shifted not just onto OPEC countries as Mitchell notes (2011: 173–199), but onto federal regulation undertaken in order to safeguard the environment and reduce pollution; specifically the 1970 CAA with its imminent 1975 deadline for the attainment of its primary standards, and thus help to directly develop the growth ban.

The new limits to growth – environmental regulation and the growth ban

In April 1971, RFF held a two-day public forum in Washington DC on 'Energy, economic growth, and the environment'. This forum was

> guided by an acute awareness of the apparent conflict that has been emerging between two societal objectives that are both of prime importance: providing energy to meet the needs of future economic growth and protecting the quality of the natural environment.
>
> (Schurr 1972: vii)

The *merely apparent* nature of this conflict would become clear early on as the issue of economic growth was broached first as a bedrock concern and central to the energy-environment debate (Schurr 1972: vii). Walter W. Heller, who had among other roles previously been a consultant, from 1965–1969, to the executive office of the President; the Chairman of the Council of Economic Advisors from 1961–1964; and tax advisor to King Hussein of Jordan; began the forum by reiterating Robert Solow's characterisation of *Materials balance* as the economist's approach to pollution. With this at the core of his presentation on economic growth, the concern over the conflict between growth and the environment was thoroughly undermined, leaving Heller to conclude that:

> Much if not most of the environmental damage associated with growth is a function of the way we grow – of the nature of our technology and the forms of production. By prohibiting ecologically deadly or dangerous activities and forcing producers to absorb the cost of using air, water, and land areas for waste disposal, growth technology and production can be redirected into environmentally more tolerable channels.
>
> (Heller 1972: 28)

Following this he states:

> Coupled with a conviction that economic growth can more than atone for its sins is a belief that its environmental vices can be diminished and its virtues magnified by greater use of the pricing system, by putting appropriate price tags on use of the public environment for private gain.
>
> (Heller 1972: 29)

The *Materials balance* approach then, and its rehabilitation of the theory of externalities with respect to environmental pollution leads here to an explicit reconciliation of 'the economy' and 'the environment'. But if these two central objects of politics are no longer at odds, if the environment no longer supplies finite limits to growth, then how can the unprecedented increase in energy prices and concurrent economic slowdown at the beginning of the 1970s be accounted for? Fortunately a response to this was provided by the second topic of the conference on both 'the effects of energy use on the quality of the natural environment and the effects of environmental restrictions on energy costs and availability' (Schurr 1972: viii). Papers by Phillip Sporn, the former president of the American Electric Power Company, and Richard Gonzalez, formerly of Humble Oil, sought to provide evidence of the detrimental impact on economic growth of environmental regulations. Sporn argued that even assuming what he referred to as a low end estimate of cost increases due to environmental regulation of 25 per cent, then the annual increase due to environmental regulations alone in the US's electric energy bill by the year 2000 would amount to USD 32 billion, over one and a half times more than the *total* US electric energy bill of USD 20 billion in 1969. As Sporn concluded:

When in 1776 Tom Paine said: "'Tis dearness only, that gives everything its value,' he was speaking of freedom. But could he also have been prophetically alluding to the activities of his countrymen two centuries later in environmental control?

(Sporn 1972: 88)

The US government would answer this rhetorical question in part through 'Project Independence' initiated by the Federal Energy Administration in March 1974 with the goal of evaluating the nation's energy problems and developing a framework for a national energy policy. Outlining the alternative energy strategy of increased domestic energy supply in its executive summary, the FEA states that 'Potential water and environmental constraints would have to be overcome' (Project Independence: Executive Summary; reprinted in Grayson 1975: 33). The linking of energy and environmental regulation had been a continual focus of the Nixon administration through the first few years of the 1970s (Mitchell 2011: 191). On 4 June 1971 Nixon gave his landmark energy message to Congress, and this was reiterated in his later energy policy statement in April of 1973. Indeed, Nixon attempted on several occasions to institutionalise the relationship of energy and environmental regulation through the development of an energy and environment department, but this was continually rebuffed by a truculent Congress, resulting in the Nixon administration implementing the EPA with a sole environmental remit. Finally then, this linkage was implemented just two months before impeachment hearings began against the President, in part through the lamination of economic theory and real-world events derived from the emergence, intersection and retranslation of growth and environment as objects of politics through the work of RFF and associated economists.

On 15 January 1975, President Ford's Energy Proposals would seek to re-establish the US's surplus capacity in total energy in order to end its vulnerability to economic disruption caused by foreign (predominantly Middle Eastern) supply. The President's interim measures, including deregulation of new natural gas, and ongoing programmes, including the massive expansion of offshore oil production in Alaska, and reinvigorated domestic coal production and use, would require the scaling back of environmental goals; and with specific respect to coal, the President stated that:

Use of our most abundant domestic resource – coal – is severely limited. We must strike a reasonable compromise on environmental concerns with coal. I am submitting clean air amendments which will allow greater coal use without sacrificing clean air-goals.

(reprinted in Grayson 1975: 24)

Echoing these proposals, on 12 May 1975, the then EPA administrator Russell Train gave the first pronouncement on a new concern for environmental regulators in a letter sent to Senator Edmund Muskie. No longer was former administrator Robert Fri's concern with the harmful impact of economic growth on the environment the sole or even central focus of the EPA, instead an urgent fear of environmental regulations' harmful impact on economic growth had taken hold:

> Under the existing CAA, if a national ambient air quality standard for any pollutant is being exceeded in an air quality control region after the attainment date, then no further construction or expansion of sources of that pollutant could be permitted in that region. This provision of law coupled with the fact that a substantial number of regions did not attain one or more of the standards on time posed a dilemma for the Committee. On the one hand, protection of the public health remains the predominant goal of the CAA and the Committee. On the other hand, a complete prohibition on new growth or expansion in non-attainment regions would pose very serious problems. The economic impact on certain urban areas of such a growth ban could be quite harmful.
>
> (Train 1975)

It is at this point then that we can see that RFFs translation of the environment from a finite object antithetical to the economy in the late 1960s and early 1970s, into one consonant with economic theory, is fundamental to the development of the growth ban. The *Materials balance* approach enabled environment and energy to be subsequently brought together in such a way that federal environmental regulation could be pilloried for its apparent negative economic impact, and the 1975 deadline for the attainment of air quality standards under the CAA reconstructed into the looming and leering ban on growth.

Conclusion

The story of the 1975 growth ban is significantly more interesting, and indeed important, than it is usually presented in the literature on emissions trading. The importance of the growth ban is not due to its apparent retelling of the parable of Paul on the road to Damascus. The scales do not fall, suddenly, from the eyes of the environmental spaceman, revealing the underlying economic logic of pollution as externality, the necessity of economic growth, and the moribund nature of environmental regulations. But neither is this merely a fiction; instead it is imperative to trace the ways in which the economy and the environment were developed and translated over a 25-year period, in order to fully grasp the ways through which a socio-technical world was assembled, enabling the 1975 growth ban to inhabit the role it is given in the history of emissions trading. RFF, through the work of its economic staff, was key here. It was their settling of the sphere of the economy as an object capable of limitless growth irrespective of material use that helped spur the development of the environment as a contrasting and specifically limited object – again in part through their own work. In response, this notion of the environment as a spaceship was then retranslated back into economic theory through the work of the *Materials balance* approach. This undermined the contestation between the economy and the environment and enabled environmental regulation to be cast as deleterious to the former, and with the CAA as undertaking a de facto ban on growth.

This process of translation did not suddenly stop in 1975 either. The growth ban was not yet quite established as the empirical evidence of the inherent conflict between economic growth and environmental protection. It was not until the 1980s that the 'regulatory unreasonableness' (Bardach and Kagan 1982) of the uniform air quality standards of the 1970 CAA was clearly determined as imposing an unacceptable cost burden to the US's energy and manufacturing industries. Even as late as the mid-1980s this understanding of the growth ban was not generally accepted. For example, one of the central figures in the economic analysis and justification of emissions trading, Thomas Tietenberg made no mention of the growth ban in the first edition of his *Emissions Trading*, published in 1985. In this edition, Tietenberg, in fact, discusses the failures of the CAA only in terms of the lack of prescribed standards being met;[5] and yet, by the 2006 second edition, the problem of the growth ban and its subsequent role had been thoroughly settled. As Simons and Voß outline in the next chapter, the translations undertaken by RFF that would come to enable the development of the growth ban were rapidly taken up and utilised by the nascent emissions trading constituency.

For Mitchell, the oil crisis would enable a new mode of government under the auspices, and utilising the machinery of, 'the market'. This is based in part on the notion that any political conflict could be grasped and governed as a matter of supply and demand. The 1975 growth ban can be seen to play a similar role with specific respect to the reconstructed object of the environment. The developing popularity of market mechanisms and their use within the EPA during the 1970s is due in part to the spread of a public choice inflected 'interest group liberalism' (Meidinger 1985: 463) through the organisation. This was based on a set of four core elements: environmental regulation cannot be simply deduced from the science; regulatory decisions are fundamentally compromises among contending interests, and not the application of correct principles; a stable framework must be created to enable competing interests to develop and implement compromises; and decisions should be bound to the interests of private parties. That is, the environment came to be considered as best regulated through what can be clearly seen is an analogue of the market.

The development of emissions trading is often given as simply a means (the best means) of response to environmental problems (Abyd Karmali, in ClimateChangeCorp 2009); but it actually has a much more intimate history with the creation of the environment as a contemporary object of politics. As Lohmann correctly claimed in 2006, the markets were made in the US; but this is not all. The environment, understood as inputs (resources) and outputs (residuals) to an economy, with pollution reconstructed as an externality driven by inadequate pricing, and where the market is the most efficient way of doing this, was also made in the US. Emissions trading, and specifically the carbon markets, are explicitly justified on the grounds that they are the most efficient means to deal with environmental issues, and this chapter has explored the history of these grounds, through an analysis of the translations in the economy and the environment – as objects of politics – that were undertaken in the process of fabricating the 1975 US growth ban. Of course, this does not mean that this economic understanding of the environment is set for all time. Looking forward to the chapter that closes this book (Methmann and Stephan, this volume), the carbon markets themselves – through their myriad failures and contradictions – present the possibility of repoliticising this currently taken-for-granted conception of the environment.

Notes

1 'The environment', 'the economy' and 'the market' given in the definite article are understood here as objects of power and politics. That is, they are ontologically discrete, measurable and calculable (see e.g. Mitchell 2008).
2 Space constraints here prevent an adequate discussion on the development of the market. However, I have addressed this in relation to the focus on the importance of efficiency elsewhere (Lane 2012).
3 Which was instituted as part of the 1969 NEPA Act.
4 This would be presented more fully in an RFF report entitled *Economics and the Environment: A materials balance approach* (with Ralph C. D'Arge from the University of California) in 1970.
5 The first references to growth ban appeared in the academic literature with the publication of books by Richard A. Liroff (1986) and Brian J. Cook (1988).

References

Arndt, H.W. (1978) *The Rise and Fall of Economic Growth: a study in contemporary thought.* Melbourne: Longman Cheshire.

Ayres, R. U. and Kneese, A. V. (1969) Production, consumption, and externalities. *The American Economic Review*, 59 (3), 282–297.

Bardach, E. and Kagan, R. A.(1982) *Going By the Book: the problem of regulatory unreasonableness.* Philadelphia: Temple University Press.

Barnett, H. J. and Morse, C. (1963) *Scarcity and Growth: the economics of natural resource availability.* Baltimore: Johns Hopkins Press.

Barry, A. (2002) The anti-political economy. *Economy and Society*, 31 (2), 268–284.

Barry, A. (2013) The translation zone: between actor-network theory and international relations. *Millennium – Journal of International Studies*, 41 (3), 413–429.

Bohm, P. and Kneese, A. V. (1971) *The Economics of Environment: papers from four nations.* London: Macmillan.

Boulding, K. E. (1966) *The Economics of the Coming Spaceship Earth: environmental quality in a growing economy.* Baltimore: Resources for the Future/John Hopkins University Press, 3–14.

Bowden, G. (1985) The social construction of validity in estimates of US crude oil reserves. *Social Studies of Science*, 15 (2), 207–240.

Calel, R. (2011) Climate change and carbon markets: a panoramic history. Available from: http://eprints.lse.ac.uk/37397/ [accessed 14 April 2013].

Callon, M. (1998) *The Laws of the Markets*. Oxford: Blackwell.

ClimateChangeCorp (2009) Is carbon trading the most cost-effective way to reduce emissions? ClimateChangeCorp. Available from: http://www.climatechangecorp.com/content.asp?ContentID=6064 [accessed 6 March 2013].

Cook, B. (1988) *Bureaucratic Politics and Regulatory Reform: the EPA and emissions trading*. New York: Greenwood Press.

Dale, G. (2011) International Socialism: The growth paradigm: a critique. Available from: http://www.isj.org.uk/index.php4?id=798&issue=134 [accessed 8 December 2013].

Dewhurst, J. F. (1947) *America's Needs and Resources: a Twentieth Century Fund survey which includes estimates for 1950 and 1960*. New York: Twentieth Century Fund.

Environmental Protection Agency, EPA (1973) *Final Conference Report for the National Conference on Managing the Environment*. Washington, DC: Environmental Protection Agency.

Flippen, J. B. (2000) *Nixon and the Environment*. Reprint, University of New Mexico Press.

Grayson, L. E. (1975) *Economics of Energy: readings on environment, resources, and markets*. Princeton: Darwin Press.

Hamblin, J. D. (2013) *Arming Mother Nature: the birth of catastrophic environmentalism*. New York: Oxford University Press.

Hardin, G. (1968) The tragedy of the commons. *Science*, 162, 1243–1248.

Hedegaard, C. (2011) Climate protection is not DEindustrialisation, but REindustrialisation: doing things smarter and more efficiently! Speech given at the Zero Emission Conference in Oslo, on 21 November 2011. Available from: ahttp://ec.europa.eu/commission_2010-2014/hedegaard/headlines/news/2011-11-21_01_de.htm [accessed 7 March 2014].

Heller, W. W. (1972) Coming to Terms with Growth and the Environment. *In:* S. H. Schurr, ed., *Energy, Economic Growth, and the Environment*. Baltimore: Johns Hopkins University Press.

Jarrett, H. (1966) *Environmental Quality in a Growing Economy*. Baltimore: Johns Hopkins Press, published for Resources for the Future.

Kapp, K. W. (1950) *The Social Costs of Private Enterprise*. Cambridge: Harvard University Press.

Kelman, S. (1981) *What Price Incentives? Economists and the environment*. Boston: Auburn House.

Kneese, A. V., Ayres, R. U. and D'Arge, R. C. (1970) *Economics and the Environment: a materials balance approach*. Baltimore: Johns Hopkins Press, published for Resources for the Future.

Kula, E. (1998) *History of Environmental Economic Thought*. London/New York: Routledge.

Kuznets, S. (1971) *Economic Growth of Nations: total output and production structure*. Cambridge: Belknap Press of Harvard University Press.

Landsberg, H. H. and Schurr, S. H. (1968) *Energy in the United States: sources, uses, and policy issues*. New York: Random House.

Landsberg, H. H., Fischman, L. L. and Fisher, J. L. (1963) *Resources in America's Future: patterns of requirements and availabilities, 1960–2000*. Baltimore: Johns Hopkins Press, published for Resources for the Future.

Lane, R. (2012) The promiscuous history of market efficiency: the development of early emissions trading systems. *Environmental Politics*, 21 (4), 583–603.

Latour, B. (2005) *Reassembling the Social: an introduction to actor-network-theory*. Oxford/New York: Oxford University Press.

Liroff, R. (1986) *Reforming Air Pollution Regulation: the toil and trouble of EPA's bubble*. Washington, DC: Conservation Foundation.

Lohmann, L. (2006) *Carbon Trading: a critical conversation on climate change, privatization and power*. Uppsala: The Corner House.

Maass, A. (1953) Book review of Resources for Freedom. *The American Political Science Review*, 27 (1), 206–210.

Magat, R. (1979) *The Ford Foundation at Work: philanthropic choices, methods, and styles*. New York: Plenum Press.

McCormick, J. (1989) *Reclaiming Paradise: the global environmental movement*. Bloomington: Indiana University Press.

McCormick, J. (1995) *The Global Environmental Movement*. 2nd ed., New York: Wiley.

Meidinger, E. (1985) On explaining the development of 'emissions trading' in U.S. air pollution regulation. *Law & Policy*, 7 (4), 447–479.

Mishan, E. J. (1965) Reflections on recent developments in the concept of external effects. *The Canadian Journal of Economics and Political Science/Revue canadienne d'Economique et de Science politique*, 31 (1), 3–34.

Mitchell, T. (1998) Fixing the economy. *Cultural Studies*, 12 (1), 82–101.

Mitchell, T. (2005) Economists and the economy in the twentieth century. *In:* G. Steinmetz, ed., *The Politics of Method in the Human Sciences: positivism and its epistemological others*. Durham: Duke University Press, 126–141.

Mitchell, T. (2008) Rethinking economy. *Geoforum*, 39 (3), 1116–1121.

Mitchell, T. (2011) *Carbon Democracy: political power in the age of oil*. London: Verso Books.

Nordhaus, W. D. (1974) The 1974 report of the President's Council of Economic Advisers: energy in the economic report. *The American Economic Review*, 64 (4), 558–565.

Odum, E. P. and Odum, H. T. (1959) *Fundamentals of Ecology*. Philadelphia/London: W.B. Saunders Co.

Pearce, D. (2002) An intellectual history of environmental economics. *Annual Review of Energy and the Environment*, 27 (1), 57–81.

Perez-Carmona, A. (2013) Growth: a discussion of the margins of economic and ecological thought. *In:* L. Meuleman, ed., *Transgovernance*. Berlin/Heidelberg: Springer, 83–161.

Pérez Henríquez, B. L. (2013) *Environmental Commodities Markets and Emissions Trading: towards a low carbon future*. Abingdon, Oxon/New York: RFF Press.

Potter, N. and Christy, F. T. (1962) *Trends in Natural Resource Commodities: statistics of prices, output, consumption, foreign trade, and employment in the United States, 1870–1957*. Baltimore: Johns Hopkins Press.

President's Materials Policy Commission, PMPC (1952) *Resources for Freedom*. Washington, DC: U.S. Government Printing Office.

President's Materials Policy Commission, PMPC (1987) *Resources for Freedom: summary of volume I of a report to the President*. Washington, DC: Resources for the Future.

Purdey, S. J. (2010) *Economic Growth, the Environment and International Relations: the growth paradigm*. London/New York: Routledge.

Radice, H. (1984) The national economy: a Keynesian myth? *Capital & Class*, 8 (1), 111–140.

Rauch, J. (2002) Ideas Change the World – and One Think Tank Quietly Did. Reason. com. Available from: http://reason.com/archives/2002/10/07/ideas-change-the-world-and-one [accessed 30 April 2014].

Schurr, S. H. (1972) *Energy, Economic Growth, and the Environment*. Baltimore: Johns Hopkins University Press.

Solow, R. M. (1971) The economist's approach to pollution and its control. *Science*, 173 (3996), 498–503.

Sporn, P. (1972) Possible impacts of environmental standards and electric power availability. *In:* S. H. Schurr, ed., *Energy, Economic Growth, and the Environment.* Baltimore: Johns Hopkins University Press.

Stern, N. (2006) *The Economics of Climate Change: The Stern Review.* Cambridge: Cambridge University Press.

Tietenberg, T. H. (1985) *Emissions Trading: an exercise in reforming pollution policy.* Resources for the Future.

Tietenberg, T. H. (2006) *Emissions Trading: an exercise in reforming pollution policy.* 2nd ed., Resources for the Future.

Train, R. (1975) Letter from Administrator Train to Sen. Edmund Muskie. Available from: https://bulk.resource.org/courts.gov/juris/j1441_09.sgml [accessed 30 April 2014].

3 Politics by other means

The making of the emissions trading instrument as a 'pre-history' of carbon trading

Arno Simons and Jan-Peter Voß

Introduction

The policy studies literature on carbon trading abounds. The bulk of this literature is functionalist in orientation, treating carbon trading as a policy instrument among others in the 'toolkit' of environmental governance and assessing its effectiveness. More recently, we have also seen an increase in research on the 'politics' of carbon markets focusing *inter alia* on the processes and actors involved in constructing carbon markets. A number of authors have asked, for example, why governments have decided to implement carbon markets at a given time and in a given jurisdiction (Cass 2005; Christiansen and Wettestad 2003; Damro and Mendez 2003) or which role business coalitions have played in the advocacy of carbon markets (Meckling 2011; Pinkse and Kolk 2007). As Stephan and Patterson (2012) point out, such studies on carbon markets politics are often informed by a rather conventional and restrictive definition of politics.

This chapter contributes to the project of investigating the deeper politics of carbon markets. Ironically, we do this by zooming out from the day-to-day intricacies and practices of doing carbon markets on the ground. Our perspective is on the production and circulation of those technical policy models that orient the construction of carbon markets in the first place. The politics we see from our perspective on carbon trading in the making is one related to the 'technical' means of public policy; it plays out in distributed efforts for establishing the instrumental validity of carbon trading and the latter's superiority over alternative means. Playing with words a little, we could say that we are interested in a politics of (other) means, rather than in a politics of ends.[1]

Drawing from the Science and Technology (STS) literature on innovation processes, we suggest looking at the making of emissions trading as an 'instrument' of public policy: a process in which the articulation of policy models interacts with ongoing attempts of installing practical market arrangements. We show that such interactions gave rise to an 'instrument constituency,' a configuration of people, organizations, their tools and stories, which became assembled in continuous and distributed attempts of developing, circulating, and implementing emissions trading as a particular policy design (cf. Voß 2007a, 177–178; Voß and Simons 2014). We will reconstruct the development and spread of emissions trading as linked to the emergence and dynamics of the emissions trading constituency as it enrolled wider support from communities of science, politics, and business to create demand for the 'solution' it supplies.

There are links between our research on the making of instruments and studies on the 'performativity of economics' in the creation of markets (Callon 1998; MacKenzie *et al.* 2007). Callon (2009), for instance, views the making of carbon markets as a form of real-world experimentation in a similar way to how we look at the negotiation of practical market arrangements in particular sites. MacKenzie (2009a, 2009b) looks at carbon markets as an example of economists being involved not only in analyzing markets but also in creating them from ground up. This comes close to our perspective on the force of instrument constituencies, but we focus on a specific market design as a model that travels beyond particular sites of implementation. In contrast to the economic performativity literature on carbon markets, our focus is on the politics of carbon trading as a *policy instrument* – a design and its claimed functionality that is constituted in a network of differently located experimental activities. We are less interested in how the economic theory of carbon trading becomes enacted into specific carbon market arrangements than in how the theory itself becomes forceful. This entails a view on how the theory becomes articulated and gains validity, and also how it shapes up as a program of collective action, a design *for policy* that enters the 'toolbox' of environmental governance to circumscribe policymaking processes around the world. In a similar vein we link up with and go beyond recent approaches to a sociological study of public policy instrumentation, where a perspective on policy instruments is developed that is interested in the enactment of functional models as they are taken up for remaking institutional arrangements on site, but which takes the models themselves to be given (Lascoumes and Le Gales 2007).

In the next two sections we will lay out our conceptual approach to the study of policy instruments in the making. Here we put particular emphasis on the dynamics of policy instruments and the development and force of policy instrument constituencies. This will be followed by a case study of the making of emissions trading as an instrument, even before it became linked up with the issue of climate change.

Studying policy instruments in the making

Whereas the study of policy instruments is not a new field of research for political scientists, the study of instruments in the making is. The bulk of the policy instrument research is marked by a functionalist approach (Lascoumes and Le Gales 2007). A key concern is to understand what instrument works best under what conditions, how to choose and combine instruments, or how to achieve 'optimal fit' between instruments and particular problem contexts, a trend that is also visible in the mainstream policy studies literature on carbon trading (see the introductory chapter of this volume). Often, this perspective also implies a questionable distinction between the *normative political* task of negotiating goals and the supposedly *neutral* and purely *technical* task of determining the optimal means to achieve them.

Besides the mainstream functionalist school, there are, however, a number of other, more constructivist approaches to the study of policy instruments (for an overview see Hood 2007; Linder and Peters 1998). Such non-functionalist approaches provide valuable insights into the study of policy instruments in relation to the politics involved in governing through instruments. Most importantly, they teach us that policy instruments, instead of being neutral devices, have their own effects and can carry different meanings for different actors, which go beyond their officially stated purpose of solving defined public problems. However, these studies typically lack a genealogical perspective on how certain functional models emerge, take shape, and stabilize as instruments in the first place.

One might hope to find a historical perspective on policy instruments in the broader field of policy change studies. It turns out, however, that this literature is characterized by an opposite shortcoming for it provides a historical perspective on policymaking without a particular focus on instruments. The policy change literature offers sophisticated theories of how public policy changes over long periods of time. But normally it does not include a specific focus on policy instruments as dynamic elements in themselves other than conceptualizing policy change *as* instrument change while basically assuming that instrument choice *follows* the beliefs of policy makers (Braun and Capano 2010).[2]

The innovation dynamics of policy instruments

To broaden our view of policy instrumentation and its relation to policy change we propose taking a constructionist perspective on the making of policy instruments. We assume that instrument development is a *historically contingent* and *path dependent* process in which specific forms of governing practice are shaping up and meanings given to it. This suggests that the study of policy instruments in the making can be informed by the literature on innovation processes, particularly in the field of STS, where the contingency of innovation processes has been acknowledged for a long time (Bijker *et al.* 1987; Latour 1987, 1999; Rip 1992; Van de Ven *et al.* 1999).

In order to study policy instruments in the making we must, first of all, acknowledge their 'double life' (Voß 2007a). By this we mean that policy instruments take two different appearances, whose interaction we need to study. On the one hand, policy instruments can appear as de-contextualized models, abstract mechanisms, or generic blueprints for policy design – the principle of 'Cap and Trade' in our case. On the other hand, they can appear in the form of regulatory constellations enacted in given jurisdictions as an 'implementation' of the abstract model. Any existing (or planned) carbon market is an example of this second type of appearance.

The interaction between models and their implementations can be understood as a process of material-semiotic structuration, in which practices of making generic policy designs start to orient policymaking practices, and vice versa. Drawing from Latour (1987), we call the places that string together and orchestrate the production and circulation of models in relation to their implementation 'centers of policy calculation.' Such centers allow action at a distance on various policymaking sites by providing generic and epistemically authoritative prescriptions for policy action. At the same time, these prescriptions are '(re)calculated' in the center on the basis of 'sound' expert analysis of accumulated implementation experiences, which is nonetheless driven by certain interests (Simon *et al.* 2014). From the center, the development of policy instruments appears as an ongoing process of policy experimentation. Whereas 'in vitro' policy experiments, such as simulations, can be conducted directly in the center, implementations of policy instruments serve as 'in vivo' experiments (Callon 2009). The latter are monitored from the center as potential sites for the generation of new design knowledge, of 'lessons' to be learned for increasing the model's epistemic authority.

The centers are also the place where powerful technological visions are created that help to enroll relevant actors into the project of expanding the policy instrument in question. Van Lente and Rip have analyzed how stories and promises in innovation processes can work as 'prospective structures,' so that 'what starts as an "option" can be labeled a technical "promise," and may subsequently function as a "requirement" to be achieved, and a "necessity" for technologists to work on, and for others to support' (Van Lente and Rip 1998: 216). Actors turn promises into requirements and necessities by acting on expectations and by positioning themselves and others in a 'story,' which begins to unfold and may stabilize over time, ultimately leading to its own enactment.

In the case of carbon trading such dynamics are clearly visible. Carbon market activities around the world are overseen and partly directed by a strong transnational center of policy calculation, enacted by what we call the emissions trading constituency (see the next subsection). A strong narrative has been created, in which carbon trading appears as a policy instrument, a 'technology' that is readily available for governments around the world to acquit themselves of the right to act on behalf of the public, and in the most effective way available. Emanating from the center, the carbon trading model and the technological narrative in which this model is embedded prescribe regional, national, and international policymaking processes (cf. Simons *et al.* 2014). This is where the politics of carbon markets begin: in the production of collectively supported views of the conditions for effective governance, and the furbishing of particular designs which are taken for granted with regard to the functionality they claim.

A final point to make here is that the double life of policy instrument in the making must be seen as a process taking place in the context of wider social, political, or economic developments. This is especially so because policy instruments develop over longer periods of time (i.e. decades) and in and across different jurisdictions. In our case study, for example, the broader trend toward economics-based policymaking in the US and the OECD (Organization for Economic Co-operation) countries had a major influence on the 'success' of emissions trading.

Policy instrument constituencies

The argumentation above suggests that the making of policy instruments not only results in a 'product' (the ready-made policy instrument) but also in the emergence of a socio-technical (infra)structure – the center and its links to the outside world – in which the policy instrument emerges, circulates, and gains epistemic authority *as* a policy instrument. But what is this infrastructure made of? Where do we find centers of policy calculation and who acts from there?

A key assertion of this chapter is that the emergence, spread, and expansion of policy instruments has to be understood and analyzed through the formation of policy instrument constituencies. (Policy) instrument constituencies are specific configurations of people, organizations, their tools and stories, which become assembled in continuous and distributed attempts of developing, circulating, and implementing a particular policy design. Conceptually speaking, we can analyze policy instrument constituencies as 'agencements' (in the making): combinations of heterogeneous elements that have been, and constantly are, adjusted to one another in a way that produces distinct effects. Whereas individual carbon markets have already been referred to as agencements (Callon 2009; MacKenzie 2009b) we suggest applying this concept also to the socio-technical infrastructure that connects these markets under the heading of a universally applicable policy design. In other words, we suggest analyzing the work done in and from centers of policy calculation as the effect of a type of agencement that we call instrument constituency (Voß and Simons 2014).

The French term 'agencement' implies two notions: assemblage (*agencer, agencement*) and agency (*agence*). Agencements are thus 'arrangements endowed with the capacity of acting in different ways depending on their configuration' (Callon 2007: 320). Agency is created within agencements and distributed among its constituent parts. This does not mean that agency is distributed equally among these parts, nor even that it becomes attributed in a 'fair' way. MacKenzie sees one of the virtues of the agencement concept in directing our attention to the 'conditions of possibility' of action and not only on what he calls 'action's glamorous agential peaks' (MacKenzie 2009b: 57). The peaks are what everyone sees: President Bill Clinton brought carbon trading into the Kyoto Protocol. The necessary infrastructure for him to perform this act becomes backgrounded in this sort of explanation. From our perspective, things look more complicated: it was not the President *alone* who got carbon trading into the protocol – what helped him was a whole existing infrastructure consisting of people, organizations, economic 'facts,' theoretical models, policy reports, and a number of 'successful' market experiments. This infrastructure is what we call the carbon trading constituency.

Such constituencies market their solutions to problems they help to define and through political support they actively build. As such they should be recognized as an important driving force in transnational policymaking processes, up to now neglected by most of the policy studies literature. The study of instrument constituencies becomes particularly important in connection to the observed trends of transnational governance (Djelic and Sahlin-Andersson 2006) or policy mobility (Peck and Theodore 2010).

Constituency dynamics can further lead to self-reflexivity. New identities that are directly tied to the success of an instrument may form. This can happen, for example, through the creation of specialized organizations that provide coordination services for constituency development. It might, further, involve the development of deliberate strategies of how to pursue the politics by other means: how to recruit strategically important supporters or how to engage in competition with other instruments in order to render one's own instrument as superior.

An important implication, especially for research on the life of policy instruments and the politics linked to them, is that constituency dynamics can lead to a very pragmatic redefinition of success of an instrument, the latter being defined in terms of spreading of the instrument more than in terms of actual performance. This might lead to a shift of interest: the development of the instrument could increasingly become seen as an end in itself and thus more than just a means to achieve certain policy goals. In other words, the constituency, in an attempt to secure its own existence, starts looking for problems to the solution it supplies.

The making of emissions trading as the 'pre-history' of carbon trading

The history of carbon trading begins long before the first carbon markets were developed or even planned. It is no secret that the policy design on which carbon markets are based, emerged in economic theory during the 1960s. What is more, this design had already been 'put to practice' in a number of differently sized market experiments. Thus, when the idea of carbon trading began to take off in the 1990s, there was already a significant infrastructure in place in support of these developments. In the following sections we will (re)tell the history of this infrastructure by focusing on the growth of the emissions trading constituency and the latter's epistemic influence on national and transnational policymaking processes. Our case covers a period of roughly three and a half decades, starting with the emergence of the emissions trading theory in environmental economics in the late 1960s and ending with the establishment of (international) carbon trading as a key option for fighting climate change during the 1990s. Our study is based on an in-depth analysis of primary and secondary documents. Background case knowledge has been attained through expert interviews.

Sketching a first policy design

When economists Crocker and Dales suggested markets for emission (Crocker 1966) or pollution rights (Dales 1968) they contributed to an ongoing debate among resource economists that was strongly influenced through the writings of Pigou (1920), Gordon (1954), and Coase (1960). Yet, the theoretical mainstream at the time favored Pigouvian charges to the use of environmental resources and viewed markets for permits 'as an intriguing, but somewhat eccentric and uninspiring alternative' (Meidinger 1985: 455). Only gradually and particularly in the context of its mathematical formalization (Baumol and

Oates 1971; Montgomery 1972) the emerging theory gained more legitimacy and momentum among environmental economists (Tietenberg 2006).

Two developments helped this theory to proliferate and form the heart of a potential center of policy calculation. The first was the stabilization of environmental economics as an institutionalized subfield of economics. The *Journal of Environmental Economics and Management* was founded in 1974 and the Association for Environmental and Resource Economics in 1979. The second development was the general expansion of economics as a key competency, increasingly valued in public administration. Fields of specialization in environmental and natural resource economics appeared in US graduate programs over the 1970s and 1980s and economics became an important element in the curricula of US public policy and law schools (Oates 2000). This meant that economic thinking gained epistemic authority also with regards to its application to policy analysis.

Through these channels the theory of emissions trading spread to ever-wider audiences, including future policymakers. Emissions trading was more than an economic theory, however. It was a sketch for how to *design* policy. And a key goal of the economists in favor of this design was to *change* US environmental policymaking. In other words, this was the birth of a new center of policy calculation and the instrument constituency that enacted it. At this point in time the latter consisted of the generic emissions trading model and the (growing) number of economists and academic publications that surrounded this model. In order to really act at a distance on policymaking sites with their model this early instrument constituency needed to convince others of the instrumentality of emissions trading.

Ironically, the existing US environmental policy regime at the time was not exactly pro-market. A response to the environmental movement of the 1960s, US environmental policy was at least partly based on the belief that strict (non-market) regulation had to correct for market failures as the source of environmental problems. In terms of air pollution regulation this had meant fixed emission limits for individual emission points and instructions on the type of control technologies to employ. In order to overcome 'the inertia of the status quo' advocates of emission trading thus had to 'promise benefits substantial enough to outweigh any frictional costs of moving away from a more traditional approach' (Tietenberg 2006: 48).

Practically this meant to invest in a series of 'translations' (Lane 2012). A first such translation was to (re)define the playing field; that is, the conception of challenges to which policy has to respond. Environmental economists were eager to construct environmental policymaking as a sphere of cost-benefit analysis, arguing that '(t)he best policy is the one which can achieve the objective at minimum cost' (Tietenberg 1973: 194). The next step was then to demonstrate that emission trading schemes would come at much lower costs than traditional approaches of environmental regulation by conducting 'in vitro' experiments. Since no emission markets existed at the time, a number of *ex ante* simulation studies were conducted to assess the potential cost savings of market approaches compared to the status quo. Quantifications of 70 percent cost saving and more were the rule rather than the exception (for a list of studies see Tietenberg 2006: 58). More than anything else, such studies worked to raise functional promises. According to Oates, 'these studies suggested that the potential savings were enormous' and 'provided an important empirical base to support regulatory reform in environmental policy' (Oates 2000: 146). Strikingly, he adds that 'the methodology used in these studies likely resulted in some exaggeration of the extent of savings' (Oates 2000: 146). Thus it remains an open question whether these studies had an impact despite or because of this latter fact. In any case, these stories enriched the arsenal of a young emissions trading constituency (cf. Lane 2012).

The first 'in vivo' experiment

The regulatory reform envisioned by environmental economics first took place in US air pollution policy during the 1970s and 1980s. It marked a departure from the regulatory system that had been established as a response to the public pressure for strict environmental regulations from social movements of the 1960s and that had been characterized as a 'narrow discretion–uniform-rule system' (Meidinger 1985: 455). Under the US Clean Air Act (CAA) of 1970, a newly created agency, the Environmental Protection Agency (EPA) was directed to determine and enforce nationwide ambient air standards. When, in 1975, states had missed their deadlines for bringing major locales into 'attainment' with the standards, EPA's policy became perceived as non-working. As a consequence, EPA implemented even stricter rules, which implied a restriction of additional industrial activity for 'non-attainment areas.' This was quickly contested as a 'growth ban' (see also Lane, this volume) and urged EPA to develop a number of new policies or mechanisms to achieve a flexibilization of the CAA statutes: 'offsets,' 'bubbles,' 'banking,' and 'netting.' In order to integrate these measures into a unified framework EPA, in 1982, introduced the 'Emission Reduction Credit' as a new legal entity, a common currency as it were, applicable to all four mechanisms and allowing for the trading of emission rights under what was now called the US Emission Trading Program (US-ETP).

It is difficult to find historical accounts of emissions trading that would not mention the US-ETP as the first 'in vivo' experiment of emissions trading. Our interpretation is a little more complex. On the one hand, we acknowledge that the making of the US-ETP has to be explained in part through economic performativity, i.e. the direct influence of economists on the establishment of 'markets.' According to Meidinger (1985: 447), the 'most important explanatory proposition [for the US-ETP] is that market mechanism regulation may reflect the formation and rise of a new "regulatory culture" likely to affect the form and substance of regulation more generally.' This culture was carried into EPA by graduates from economic, public policy, and law programs who after having 'received direct exposure to the economic approach to environmental decision making' pushed into the booming field of environmental regulation, casting about 'for new initiatives which they could hook their stars to and use to separate themselves from the crowd' (Meidinger 1985: 146, 463). Finding an 'organizational home' in the EPA (Cook 1988: 10), these young EPA staffers began experimenting with more flexible approaches, especially since the CAA policy failure had opened a window of opportunity for them.

On the other hand, we agree with Lane (2012) that the US-ETP was constructed as an implementation of the theory of tradable permits only in hindsight. We identify this as an active translation of an early emissions trading constituency. Neither was the US-ETP ever planned *as* an implementation of the prescriptions Crocker, Dales, and others had developed in the years before, nor did the program look much *like* these prescriptions. But it was translated into an 'ad hoc' implementation of these prescriptions in the evaluation studies, which soon emanated from a center of policy calculation that was about to gain strength by a first grip on the real world.

This translation further meant turning the US-ETP into a first 'proof-of-principle' (Voß 2007b). That the actual performance of the US-ETP lagged behind expectations might have been seen as a falsification of the model. But it was framed as a result of the fact that the US-ETP was still very much grounded in the old (non-working) regulatory regime. As Lane (2012: 598) notes, 'environmental economists were able to have their policy and critique it.'

Expanding the center, looking for new problems

Having had a proof-of-principle at hand, the emissions trading constituency could quickly gain additional support, and the idea of emissions trading started to diffuse more widely. Already in parallel to the implementation of the US-ETP a number of developments led to the strengthening of the constituency. First of all, the construction of the US-ETP produced a new actor group with an inherent interest in emissions trading. This group consisted of market intermediaries catering to the actual operation of the program and developing a reflexive interest in widening their business opportunities. AER*X, for example, was founded in 1984 as the first emissions trading company and served as a broker in the Los Angeles offset market. The company's founder, John Palmisano, became an influential spokesperson for emission trading and helped to push for its wider application (Dudek and Palmisano 1987). Later, in 1996, such service providers founded the Emissions Trading Association (EMA, today Environmental Markets Association).

By the mid-1980s also a number of environmental think tanks had become interested in emissions trading. Among them was the Environmental Defense Fund (EDF), which had turned toward environmental economics in order to become distinguishable from other groups (Pooley 2010). By hiring a number of economists and through cooperation with high-ranking politicians, EDF emerged as a key constituency member and driver of emissions trading and other types of environmental markets.

A task for the growing emissions trading constituency now was to expand the scope of application for their emerging instrument. The US-ETP was a good start but surely many more opportunities could soon be created. And in order to create these opportunities it would always be helpful to point to practical experience with emissions trading. Already in the preface of his influential evaluation study of the US-ETP, economist Tom Tietenberg (1985: xii) described the US-ETP 'not only as a rationalization of existing policy, but also as a harbinger of things to come in other areas of environmental policy and in other parts of the world.' EDF staffer Daniel Dudek and his co-author Palmisano, both with a major interest in expanding the scope of emissions trading, published their evaluation of the US-ETP in a highly cited 1987 *Columbia Journal of Environmental Law* article. The authors concluded not only 'that the successes of emissions trading cannot be denied,' but also that '[t]he transboundary problems of acid rain, stratospheric ozone depletion and climate change are major contemporary environmental problems to which emissions trading could be effectively extended' (Dudek and Palmisano 1987: 219).

A key event in the process of finding new applications for emissions trading was the bipartisan Project 88. In May 1988 two Congressmen, in close interaction with people from and around EDF, initiated Project 88 as an effort to enroll key individuals from academia, private industry, environmental organizations, and government in order to have them speak with a united voice for the use of economic incentives in environmental regulation (with a clear emphasis on tradable permits) in the run up to the US presidential elections. The final 96-page report, which was circulated among policymakers and the press, described how market-based policies could be applied to several major environmental problem areas – among them, again, acid rain, stratospheric ozone depletion, and climate change. Dudek wrote the section on 'global air pollution problems,' in which he *inter alia* proposed a US carbon offset market, debt-for-forest swaps between the US and less developed countries, and, as the ultimate goal, a system for international carbon trading.

To our knowledge, Dudek's 1987/1988 proposals for carbon trading were the first of their kind in the emissions trading literature. They were revolutionary for at least two reasons. First of all, at that time, the issue of climate change was just beginning to receive wide public attention. Though climate change had been discussed for years in academic circles and among dedicated policymakers, the breakthrough occurred in the summer of 1988. On June 23, NASA physicist James Hansen testified during a Senate hearing first that global warming was man-made, and second that the problem was directly linked to the devastating droughts the US Midwest had been experiencing at the time, triggering a wave of international media attention (Weart 2003). Only a few days later the press covered the proceedings of the 'World Conference on the Changing Atmosphere: Implications for Global Security' in Toronto, 'the first international meeting of governments on climate change to receive widespread media attention' (Corfee-Morlot *et al.* 2007: 2762).

Another reason why proposals for carbon trading were revolutionary in 1987/1988 was the general lack of practical experience with emissions trading. In the Project 88 report, Dudek deemed it 'unlikely that governments, including the United States, would be willing to adopt [carbon] trading schemes without the benefit of practical experience in implementation' (Stavins 1988: 20). As Pooley (2010: 73) notes, 'they [Dudek and his fellow economists in and around EDF] thought they knew how to get the carbon dioxide out of the smokestacks. They just needed a test case to prove it.'

During and after Project 88 was on its way the emissions trading constituency expanded rapidly in several directions at once. In hindsight, three processes were especially relevant for the future path of carbon trading. Two of them did not concern carbon trading directly but resulted in a strengthening of emissions trading – and thus the center of policy calculation that also supported the rise of carbon trading. The first such process included the Montreal Protocol in 1987 – setting phase-down targets for chlorofluorocarbons (CFCs) – and EPA's subsequent installation of a CFC emissions trading system to comply with the protocol. Constituency members were not only directly involved in this process but also helped to frame it as a precedent for international carbon trading. Dudek, for example, argued in his Project 88 chapter that EPA's CFC trading system serves as 'a guide to how such policies can operate in diverse political and economic settings. The same approach can then be extended to the climate change problem through offsets for new sources of greenhouse gas emissions' (Stavins 1988: 26).

The second important process was the making of the US acid rain program (US-ARP), an indirect result of Project 88's impact. Arguably, the most advanced policy proposal made in the Project 88 report was that to install 'a market-based approach to acid rain reduction,' and this proposal was connected to the promise that such an approach 'could save [the US] $3 billion per year, compared with the costs of a dictated technological solution' (Stavins 1988: 5). The timing of this proposal was crucial. In his ongoing election campaign, George H. W. Bush had promised significant cuts in sulfur emissions since he was looking for a way of 'using the environment to signal that he was Reagan without the sharp edges' (Pooley 2010: 75). After its victory, the Bush administration, which had become interested in the proposal of Project 88, invited EDF staffers to draft the US-ARP, which eventually became implemented under the 1990 CAA Amendments.

Once in place, the US-ARP quickly became constructed and perceived as a major success, primarily because emission reductions were achieved at very low costs (Ellerman *et al.* 2000; Kerr 1998; Stavins 1998) – and regardless of the fact that a bulk of these savings, at least in the first trading phase, may actually have to be attributed to independent economic factors, such as the lower-than-predicted costs of using control technologies or the deregulation of rail rates (Burtraw 1996; EPA 2001). In any case, the emissions trading constituency now had its first 'working exemplar' (Voß 2007b).

The route toward carbon trading

Both the CFC trading scheme and the US-ARP supported the third, most important, process leading toward carbon trading: the creation and implementation of the United Nations Framework Convention on Climate Change (UNFCCC). The suggestion for adopting such a framework was one of the outcomes of the 1988 Toronto conference, and it was made real during the 1992 Earth Summit in Rio de Janeiro. From early on, the emissions trading constituency used this process to advocate the use of carbon trading. At the turn of the decade more and more authors began to make proposals for some form of carbon trading (Dudek and LeBlanc 1990; Grubb 1989; Swisher and Masters 1989, 1991; Trexler *et al.* 1989). In 1989 the Royal Institute of International Affairs issued 'The greenhouse effect: Negotiating targets,' a major report that received international attention because it discussed a number of political strategies to deal with climate change. Michael Grubb, author of this report, concluded that '[a] system of marketable carbon emission permits offers by far the most promising approach' (Grubb 1989: vii).

A crucial step was when the growing carbon trading constituency started piggybacking directly on the institutional infrastructure of two international organizations: the OECD and the United Nations Conference on Trade and Development (UNCTAD). The OECD had picked up on emissions trading already since the mid-1980s in the context of proposing and evaluating the use of economic policy instruments (Opschoor and Vos 1989). In 1991 the OECD convened two expert workshops 'to analyse the practical dimensions of the two market instruments most likely to be used in abating greenhouse gases – tradeable permits and emission taxes' (OECD 1992: 18). The list of participants of the emissions trading workshop included Dudek, Tietenberg, and Grubb. One year later, an edited volume was produced from the workshop papers, which reviewed the US emissions trading experience and discussed details of a potential international carbon trading scheme (OECD 1992).

UN bodies had been concerned with climate change since the 1980s, when the UN Environment Program and the World Meteorology Organization convened a series of international workshops, which also led to the creation of the Intergovernmental Panel on Climate Change in 1988. But UNCTAD was the first UN body to advocate emissions trading. The proposal for international carbon trading was first made during a pre-Rio ministerial conference on sustainable development, held in Bergen, and subsequently published in a number of edited volumes, gathering some of the same authors who also drafted the OECD reports (UNCTAD 1992, 1994, 1995).

Writing under the names of the OECD and UNCTAD yielded enormous legitimacy gains for the constituency. It is a perfect example of what we mean by politics by other means. Appropriating the authoritative status of OECD and UNCTAD helped to render carbon trading as a purely technical means, largely independent from policy goals and values, and whose superiority to other means can be determined solely on a scientific (economic) basis. Only carbon trading, it was argued, would create enough 'flexibility' needed to achieve a limitation of carbon emissions at the lowest possible economic costs. In the overview of UNCTAD's first big report on carbon trading, for example, UNCTAD Secretary General Kenneth Dadzie argued that

[a]ny search for a control mechanism that could be used in a broad-based international agreement to limit greenhouse gas emissions must give high priority to the need to achieve such a limitation at the lowest possible economic cost. . . . A successful control mechanism must . . . hold out the prospect that least-cost solutions to abatement objectives will be found, even when information on costs available to regulators is incomplete, and must also provide governments with flexibility in meeting their objectives. . . . Transferable entitlement systems meet these criteria.

(UNCTAD 1992: IVff)

What is more, the piggybacking strategy also helped to plant the idea into people's heads that international carbon trading might actually become real one day. This was important because in 1992 carbon trading was still seen as quite a revolutionary, and potentially unrealistic, strategy. Consider the following quote from the introduction to the 1992 OECD report:

There is no strong indication yet that market approaches will form the basis of very many national greenhouse gas abatement policies. Clearly, there is some danger in working out the details of how to do something, before one has decided that it will be done at all. On the other hand, discussed above, it is also true that the decision to implement a particular market instrument should hinge, to some extent, on the practical implementability of that instrument. The two problems, therefore, are not mutually exclusive.

(OECD 1992: 18)

Discussing the 'technical' implementability of an instrument before a political consensus has formed is again a form of politics by other means, not only because it anticipates a potential decision but also because it frames it as a good one.

As Braun (Braun 2009: 472) notes, the work of OECD and UNCTAD 'have set the agenda for greenhouse gas emissions trading' in the early 1990s and convinced the US government to push for the integration of carbon trading in the 1997 Kyoto Protocol. Arguably, carbon trading would have never made it into the Kyoto Protocol if it had not been for the negotiation skills of the Clinton administration. But neither would carbon trading have made it if it had not been for the material-semiotic infrastructure that rendered carbon trading a promising policy instrument. By 1997 the (carbon) emissions trading constituency consisted of individuals and organizations from various social sectors as well as their (success) stories and narratives, published and circulated in thousands of documents and enrolling actors in meetings all around the world. This was an agencement capable of orienting transnational policymaking by way of establishing a strong center of policy calculation.

Though 'real' carbon trading only began in the 2000s our little historical account ends here. The purpose of this chapter was mainly to show that the history of carbon trading had already begun in the 1960s and must be analyzed as a process of constituency formation involving politics by other means.

Conclusion

We have studied the politics of carbon markets by following the making of emissions trading as an instrument of public policy: how it became furbished as a model that is taken as a ready-made option with a given functionality for public policy. We have shown how the expansion of the idea and practice of carbon trading has been connected with the growth and influence of an emissions trading constituency, which formed well before carbon markets became an option in climate change policy. This constituency formed in heterogeneous interactions, which from the late 1960s to the late 1980s led to the establishment of air emissions trading as a 'working' policy instrument in theory and practice. By the late 1980s the constituency already included actors from the worlds of academia, policy, and business, who for different reasons came to push emissions trading. Thus, when in 1987 climate change surfaced as a global problem in the public eye, the solution was already there – advocated and supplied in conceptual form by the existing emissions trading constituency, which was seeking to expand the application of 'its' instrument (cf. Voß and Simons 2014; Voß 2007b). Our analysis thus shows that the making of carbon markets did not start with, and was never confined to, any particular situation and locale of 'implementation' (like the later cases of corporate, national, or international carbon market systems). Rather, it comprises a whole network of interlinked experiments, both 'in vivo' and 'in vitro,' linked up with each other through circulating evidence (Latour 1999).

Theoretically, we argued that this process of establishing *epistemic* authority over the definition of the conditions of effective policymaking exhibits a *politics by other means*. The means of establishing epistemic authority are different from the ones of establishing political authority. Whereas the latter involve the definition of particular goals that afford action in the name of the public, the former involve the construction of a certain representation of political reality and the conditions for effective governing action that must be met, if policy wants to claim to be rational (cf. Ezrahi 1990).[3]

Epistemic authority plays together with political authority in that it prescribes certain corridors of feasibility for political action and thus preconfigures the range of options available for rational decisions of the political collective (cf. Simons *et al.* 2014; Voß 2007a). The politics of making collectively binding decisions on how to view the given reality of governing and what to accept as conditions and functional options, however, are not visible from a narrow perspective on choice of policies from a given set of taken-for-granted options. We see the broader politics of policy instruments in the establishment of a view of public policymaking as a process of policy choice, which implies a detachment of ends and means of public policy, and with the displacement of negotiations over the design of collective action programs from open public arenas to the confined worlds of experts.

Acknowledgements

We acknowledge funding by the German Federal Ministry of Education and Research (BMBF) through grant no. 01UU0906 (Innovation in Governance research group). Thanks go to Heather Lovell and Johannes Stripple for comments on earlier versions of this paper, to Nina Amelung, Carsten Mann and Sebastian Ureta for inspiring discussion on the more general issue of innovation in governance, and to Jan Hussels and Thomas Crowe for practical support.

Notes

1 This perspective is compatible with newer trends in the critical policy studies literature, e.g. with the governmentality studies literature (Bruno 2009; Dean 1999; Rose and Miller 1992), to broader accounts of technological governing (Barry 2001, 2002) as well as to an emerging sociology of policy instruments (Bruno *et al.* 2006; Lascoumes and Le Gales 2007).
2 A remarkable sensitivity for the life of policy options and 'solutions' independent of ongoing processes of problem articulation and struggle for political power is reflected in Kingdon's (2003) famous multiple stream model. As far as we can see, Kingdon's model even includes a rudimentary perspective on the local subsidiaries of what we call (transnational) instrument constituencies. But Kingdon's 'policy entrepreneurs' act individually and pick from ideas that float freely through the 'policy primeval soup.' In contrast, we try to show that the emissions trading constituency developed a transnational agency, which built up along the innovation journey of the policy instrument.
3 Methmann and Stephan (in this volume) collapse this distinction to some extent and call both aspects 'political.'

References

Barry, A. (2001) *Political machines: Governing a technological society.* London: Athlone Press.

Barry, A. (2002) The anti-political economy. *Economy and Society*, 31 (2), 268–84.

Baumol, W.J. and Oates, W.E. (1971) The use of standards and prices for protection of the environment. *The Swedish Journal of Economics*, 73 (1), 42–54.

Bijker, W.E., Hughes, T.P. and Pinch, T., eds. (1987) *The social construction of technological systems. New directions in the sociology and history of technology.* Cambridge: MIT Press.

Braun, D. and Capano, G. (2010) The missing link: Policy ideas and policy instruments. Introductory paper to the ECPR Joint Session Workshop on *Policy Ideas and Policy Instruments*, March 23–27, Münster.

Braun, M. (2009) The evolution of emissions trading in the European Union: The role of policy networks, knowledge and policy entrepreneurs. *Accounting, Organizations and Society*, 34 (3–4), 469–487.

Bruno, I. (2009) The 'indefinite discipline' of competitiveness benchmarking as a neoliberal technology of government. *Minerva*, 47 (3), 261–280.

Bruno, I., Jacquot, S. and Mandin, L. (2006) Europeanization through its instrumentation: benchmarking, mainstreaming and the open method of co-ordination . . . toolbox or Pandora's box? *Journal of European Public Policy*, 13 (4), 519–536.

Burtraw, D. (1996) The SO_2 emissions trading program: Cost savings without allowance trades. *Contemporary Economic Policy*, 14 (2), 79–94.

Callon, M., ed. (1998) *The laws of the markets.* Oxford: Blackwell.

Callon, M. (2007) What does it mean to say that economics is performative? *In:* D. MacKenzie, F. Muniesa and L. Siu, eds., *Do economists make markets?* Princeton: Princeton University Press, 311–357.

Callon, M. (2009) Civilizing markets: Carbon trading between in vitro and in vivo experiments. *Accounting, Organizations and Society*, 34 (3–4), 535–548.

Cass, L. (2005) Norm entrapment and preference change: The evolution of the European Union position on international emissions trading. *Global Environmental Politics*, 5 (2), 38–60.

Christiansen, A.C. and Wettestad, J. (2003) The EU as a frontrunner on greenhouse gas emissions trading: How did it happen and will the EU succeed? *Climate Policy*, 3 (1), 3–18.

Coase, R.H. (1960) The problem of social cost. *The Journal of Law and Economics*, 3 (1), 1–44.

Cook, B.J. (1988) *Bureaucratic politics and regulatory reform: The EPA and emissions trading*. New York: Greenwood.

Corfee-Morlot, J., Maslin, M. and Burgess, J. (2007) Global warming in the public sphere. *Philosophical Transactions of the Royal Society A*, 365 (1860), 2741–2776.

Crocker, T.D. (1966) The structuring of atmospheric pollution control systems. *In:* H. Wolozin, ed., *The economics of air pollution*, New York: Norton, 61–86.

Dales, J.H. (1968) *Pollution, property and prices*. Toronto: University of Toronto Press.

Damro, C. and Mendez, P.L. (2003) Emissions trading at Kyoto: From EU resistance to Union innovation. *Environmental Politics*, 12 (2), 71–94.

Dean, M. (1999) *Governmentality: Power and rule in modern society*. London: SAGE.

Djelic, M.-L. and Sahlin-Andersson, K. (2006) *Transnational governance: Institutional dynamics of regulation*. Cambridge: Cambridge University Press.

Dudek, D.J. and LeBlanc, A. (1990) Offsetting new CO_2 emissions: A rational first greenhouse policy step. *Contemporary Economic Policy*, 8 (3), 29–42.

Dudek, D.J. and Palmisano, J. (1987) Emissions trading: Why is this thoroughbred hobbled? *Columbia Journal of Environmental Law*, 13, 217–275.

Ellerman, A.D., Joskow, P.L., Schmalensee, R., Montero, J.-P. and Bailey, E.M. (2000) *Markets for clean air: The US Acid Rain Program*. Cambridge: Cambridge University Press.

EPA (2001) *The United States experience with economic incentives for protecting the environment*. Washington, DC: EPA.

Ezrahi, Y. (1990) *The descent of Icarus: Science and the transformation of contemporary democracy*. Cambridge, MA: MIT Press.

Gordon, H.S. (1954) The economic theory of a common-property resource: The fishery. *The Journal of Political Economy*, 62 (2), 124–142.

Grubb, M. (1989) The greenhouse effect: Negotiating targets. *International Affairs*, 66 (1), 67–89.

Hood, C. (2007) Intellectual obsolescence and intellectual makeovers: Reflections on the tools of government after two decades. *Governance*, 20 (1), 127–144.

Kerr, R.A. (1998) Acid rain control: Success on the cheap. *Science*, 282 (5391), 1024.

Kingdon, J.W. (2003) *Agendas, alternatives, and public policies*. 2nd ed. New York: Longman.

Lane, R. (2012) The promiscuous history of market efficiency: The development of early emissions trading systems. *Environmental Politics*, 21 (4), 583–603.

Lascoumes, P. and Le Gales, P. (2007) Introduction: Understanding public policy through its instruments – From the nature of instruments to the sociology of public policy instrumentation. *Governance: An International Journal of Policy, Administration, and Institutions*, 20 (1), 1–21.

Latour, B. (1987) *Science in action: How to follow scientists and engineers through society*. Cambridge: Harvard University Press.

Latour, B. (1999) *Pandora's hope: An essay on the reality of science studies*. Cambridge: Harvard University Press.

Linder, S.H. and Peters, B.G. (1998) The study of policy instruments: Four schools of thought. *In:* B.G. Peters and F.K.M. van Nispen, eds., *Public policy instruments: Evaluating the tools of public administration*. Cheltenham: Edward Elgar Publishing, 33–45.

MacKenzie, D. (2009a) Making things the same: Gases, emission rights and the politics of carbon markets. *Accounting, Organizations and Society*, 34 (3–4), 440–455.

MacKenzie, D. (2009b) *Material markets: How economic agents are constructed*. Oxford: Oxford University Press.

MacKenzie, D., Muniesa, F. and Siu, L., eds. (2007) *Do economists make markets?* Princeton: Princeton University Press.

Meckling, J. (2011) *Carbon coalitions*. Cambridge: MIT Press.

Meidinger, E. (1985) On explaining the development of emissions trading in US air pollution regulation. *Law & Policy*, 7 (4), 447.

Montgomery, W.D. (1972) Markets in licenses and efficient pollution control programs. *Journal of Economic Theory*, 5 (3), 395–418.

Oates, W.E. (2000) From research to policy: The case of environmental economics. *University of Illinois Law Review*, (1), 135–154.

OECD, ed. (1992) *Climate change: Designing a tradeable permit system*. Paris: OECD Publishing.

Opschoor, J.B. and Vos, H.B. (1989) *Economic instruments for environmental protection*. Paris: OECD Publishing.

Peck, J. and Theodore, N. (2010) Mobilizing policy: Models, methods, and mutations. *Geoforum*, 41 (2), 169–174.

Pigou, A.C. (1920) *The economics of welfare*. London: Macmillan.

Pinkse, J. and Kolk, A. (2007) Multinational corporations and emissions trading: Strategic responses to new institutional constraints. *European Management Journal*, 25 (6), 441–452.

Pooley, E. (2010) *The climate war: True believers, power brokers, and the fight to save the earth*. New York: Hyperion.

Rip, A. (1992) A quasi-evolutionary model of technological development and a cognitive approach to technology policy. *RISESST. Rivista di studi epistemologici e sociale sulla scienza e la technologia*, 69–102.

Rose, N. and Miller, P. (1992) Political power beyond the state: Problematics of government. *The British Journal of Sociology*, 43 (2), 173–205.

Simons, A., Lis, A. and Lippert, I. (2014) The political duality of scale-making in environmental markets. *Environmental Politics*, 23 (4): 632–649.

Stavins, R.N. (1988) *Project 88. Harnessing market forces to protect our environment: Initiatives for the New President*. Washington, DC: Environmental Policy Institute.

Stavins, R.N. (1998) What can we learn from the grand policy experiment? Lessons from SO_2 allowance trading. *The Journal of Economic Perspectives*, 12 (3), 69–88.

Stephan, B. and Paterson, M. (2012) The politics of carbon markets: An introduction. *Environmental Politics*, 21 (4), 545–562.

Swisher, J.N. and Masters, G.M. (1991) Buying environmental insurance: Prospects for trading of global climate-protection services. *Climatic Change*, 19 (1), 233–240.

Swisher, J.N. and Masters, G.M. (1989) International carbon emission offsets: A tradeable currency for climate protection services. *In: Proceedings of the Conference on Technology-Based Confidence Building: Energy and Environment*, 9–14 July 1989 St. Johns College Santa Fe, New Mexico, p. 464ff.

Tietenberg, T.H. (1973) Controlling pollution by price and standard systems: A general equilibrium analysis. *The Swedish Journal of Economics*, 75 (2), 193–203.

Tietenberg, T.H. (1985) *Emissions trading: An exercise in reforming pollution policy*. Washington, DC: Resources for the Future.

Tietenberg, T.H. (2006) *Emissions trading: Principles and practice*. 2nd rev. ed. Washington, DC: Resources for the Future.

Trexler, M., Faeth, P.E. and Kramer, J.M. (1989) *Forestry as a response to global warming: An analysis of the Guatemala Agroforestry and Carbon Sequestration Project.* Washington DC: World Resources Institute.

UNCTAD (1992) *Combating global warming: Study on a global system of tradeable carbon emission entitlements.* New York: UNCTAD.

UNCTAD (1994) *Combating global warming: Possible rules, regulations and administrative arrangements for a global market in CO_2 emission entitlements.* Geneva: United Nations.

UNCTAD (1995) *Controlling carbon dioxide emissions: The tradable permit system.* Geneva: United Nations.

Van de Ven, A., Polley, D., Garud, R. and Venkataraman, S. (1999) *The innovation journey.* New York: Oxford University Press.

Van Lente, H. and Rip, A. (1998) Expectations in technological developments: An example of prospective structures to be filled in by agency. *In:* C. Disco and B. van der Meulen, eds., *Getting new technologies together: Studies in making sociotechnical order.* New York: Walter de Gruyter, 203–231.

Voß, J.-P. (2007a) *Designs on governance: Development of policy instruments and dynamics in governance.* Dissertation. Twente: University of Twente. Available online: http://doc.utwente.nl/58085/1/thesis_Voss.pdf.

Voß, J.-P. (2007b) Innovation processes in governance: The development of 'emissions trading' as a new policy instrument. *Science and Public Policy*, 34 (5), 329–343.

Voß, J.-P. and Simons, A. (2014) Instrument constituencies and the supply-side of policy innovation: The social life of emissions trading. *Environmental Politics*, 23 (5). http://dx.doi.org/10.1080/09644016.2014.923625.

Weart, S.R. (2003) *The discovery of global warming.* Cambridge: Harvard University Press.

4 Allometric equations and timber markets

An important forerunner of REDD+?

Heather Lovell and Donald MacKenzie

Introduction

Allometry describes the practice of measuring one part of something in order to estimate the size of another (larger, harder to measure) part. For forests, allometry typically takes the form of a measurement of a tree's diameter (at 'breast height', shortened to 'DBH') – a task easily performed in the field with a tape measure. Forest allometric equations have been used in various forms since the start of the twentieth century to provide an estimate of timber volume, biomass, and most recently carbon. In this chapter we explore how and why allometric equations have been adapted for use in forest carbon markets. To do so we draw on two interrelated strands of scholarship. First, science and technology studies ideas about 'action-at-a-distance' and the role of technologies and techniques in maintaining stability in a network, principally the idea of 'immutable mobiles'; objects that move through space without changing (Latour 1987; Law 1987). Second, we consider conceptualisations of how markets are made, the practices of markets, and their technical, material components, drawing on the notion of market devices (Callon *et al.* 2007), defined as 'the material and discursive assemblages that intervene in the construction of markets' (Muniesa *et al.* 2007: 2). Our aim is to investigate the multiple, and sometimes unexpected, practices, rationalities, techniques and technologies that have turned greenhouse gas emissions into tradeable permits, using the case study of one particular carbon market, that of forest carbon, or REDD+. Further, we wish to understand how existing market practices that predate the creation of forest carbon as a commodity have shaped this market.

REDD+ (in full 'Reducing Emissions from Deforestation and forest Degradation and the conservation, sustainable management of forests and enhancement of forest carbon stocks') is a market-based policy initiative of the UN Framework Convention on Climate Change (UNFCCC). The core objective of REDD+ is to financially compensate developing countries (termed 'non-Annex One' countries, under the UNFCCC) that are able to prevent deforestation and degradation of their carbon-rich tropical forests. It was the 2007 Bali Action Plan at the UNFCCC Conference of the Parties (COP) 13 that marked the formal agreement to reinvigorate the role of forests within the UNFCCC, stating that a comprehensive approach to mitigating climate change should include: 'Policy approaches and positive incentives on issues relating to reducing emissions from deforestation and forest degradation in developing countries; and the role of conservation, sustainable management of forests and enhancement of forest carbon stocks in developing countries' (UNFCCC 2007, decision 1/CP.13, para. 1b iii). While much has been researched about REDD+, including in-depth analysis of its politics (Boyd *et al.* 2007), our focus here is limited to the practices of 'Measurement, Reporting and Verification' (known within REDD+ and wider carbon market networks as 'MRV'). MRV is an interesting focus of social science enquiry because it is viewed as the cornerstone of carbon markets: accurate measurement and tracking of carbon are seen as central to making carbon markets work, and this is equally true for the subset of forest carbon markets (Grainger 2009; Lovbrand and Stripple 2010). For example, one recent UN REDD Programme publication notes that:

> the potential benefits [of REDD+] for non-Annex One parties will be based on results that must be measured, reported and verified. *The precision of these results therefore has a major impact on potential financial compensation,* and the capacity to measure forest carbon stocks is thus of increasing importance for countries who plan to contribute to mitigating climate change through their forest activities.
>
> (Picard *et al.* 2012: 17, emphasis added)

Although REDD+ MRV is perhaps not an obvious site for analysis of the politics of carbon markets, we aim to demonstrate here the intermingling of the political with the scientific within forest carbon MRV. Examining the politics before carbon, our chapter tackles (to our knowledge) a hitherto overlooked question for forest carbon market analysis, namely, 'How have forest carbon markets been prefigured or structured through the earlier market for commercial timber?' Knowing precisely how much you have of any commodity – be it wheat, rice, timber or carbon – is a central and well-researched element of market function (Cronon 1991). The allometric equation functions in REDD+ in this regard by allowing the calculation of the carbon content of forests through relatively straightforward measurement of certain tree characteristics (diameter, height) in the field. Using these measurements a series of calculations and conversions are then performed (about which, more below) in order to derive a figure for carbon content for the sample of trees measured, and hence an estimate for the whole forest. Allometric equations are verified for use in REDD+ MRV: they are the central component of field-based local and regional REDD+ MRV, plus are the main method for 'ground-truthing' remote sensing data (Chave *et al.* 2005; Gibbs *et al.* 2007; Picard *et al.* 2012).

While, in many respects, the 'travelling' of allometric equations from timber to carbon markets is straightforward – to translate dry tree biomass to mass of carbon one simply divides by two (indeed the simplicity of this conversion is crucial, about which, more below), in other ways the journey has some hidden shortcuts. Issues about the historical legacy of where allometric equations were developed (based on managed forests in particular countries, with single-species uniform stands of trees) are important. These origins, and the particular fieldwork practices that developed around them, place technical limitations on the applicability of allometric equations to REDD+, which, crucially, have not been widely taken up within REDD+ MRV debates and documentation. This apparent reluctance to acknowledge and resolve the technical limitations of allometric equations for REDD+ is argued here to stem from: first, a desire among the forestry and forest ecology expert community to reposition their expertise for carbon markets; and, second, a drive from policy makers to have workable MRV solutions that allow REDD+ to be operational.

The chapter is based on a number of different strands of empirical research undertaken as part of a three-year Nuffield Foundation Fellowship, held jointly by Lovell and MacKenzie, including: an online survey of authors (#44) of a lead handbook for forest carbon measurement (the 'Sourcebook', published by the international organisation Global Observation of Forest and Land Cover Dynamics (GOFC-GOLD)) in late 2011; a follow-up online survey of users of the GOFC-GOLD Sourcebook (#116) in August 2012; participant observation and work shadowing of a remote sensing scientist at the University of Edinburgh for two months (summer 2010); semi-structured interviews (#21) (2010–2012) with leading international forest scientists in academia, NGOs, the UN and the private sector; and attendance at a number of REDD+ and forestry seminars and conferences.[1] This primary empirical research has been complemented by a policy 'grey' literature review of REDD+ documents related to MRV, and an academic science literature review of the use and application of allometric equations (with respect to both timber and carbon markets).

The chapter is structured as follows. First, two strands of what is judged to be the most relevant scholarship are examined, namely immutable mobiles from science and technology studies, and market devices, from economic sociology. Second, in the main empirical section of the chapter an overview of the international timber industry is given, followed by further discussion of forest carbon markets and evaluation of the assumptions and expert knowledge embodied in allometric equations. The ways in which allometric equations have struggled in their new role within carbon markets are explored. In conclusion, a number of theoretical insights are proposed regarding the migration of market devices and the politics of expertise.

Immutable mobiles and market devices

Immutable mobiles and action-at-a-distance

An area of scholarship with relevance for our case study is that of calculation and the construction of international science 'actor-networks'. This literature is, in the main, based on historical case studies (often in imperial/colonial times and therefore with a close fit to the colonial-era commercial forestry market introduced in the next section of the chapter), and sits broadly within the field of science and technology studies. A notable text is Law's analysis of Portuguese imperial expansion in the fifteenth and sixteenth centuries, which investigates how 'long-distance social control' came about, essential to the imperial aims and capabilities of Portugal. The main theoretical claim is 'that the undistorted communication necessary for long-distance control depends upon the generation of a structure of heterogeneous elements containing envoys which are mobile, durable, forceful and able to return' (Law 1986: 257). The object of analysis is the 'envoys' – the role played by technologies, documents and techniques in creating new international markets and ways of operating. Law's ideas about long-distance control and devices were developed further by Latour (1987) into the notion of an *immutable mobile*, namely 'that which moves through . . . space while holding its shape' (Law and Mol 2001: 619). Latour also further developed the idea of scientific centres of calculation; institutions vital to initiating, enabling and maintaining immutable mobiles (Latour 1987). In *Science in Action* Latour uses several different examples (including Law's case study of the Portuguese imperial expansion) to demonstrate how new types of space-time are constructed by centres of calculation and immutable mobiles.

Elements of these ideas have flowed into more recent economic sociology scholarship on the making of markets (Caliskan and Callon 2010; MacKenzie 2008), where there is a similar emphasis on the heterogeneous composition of markets, and how processes of calculation are performed and controlled. A central mechanism for 'action-at-a-distance' common to both areas of scholarship are the tools and types of measurement used to make new networks stable:

> To be sure, expeditions, collections, probes, observatories and enquiries are only some of the many ways that allow a centre to act at a distance. Myriads of others appear as soon as we follow scientists in action, but they all obey the same selective pressure. *Everything that might enhance either the mobility, or the stability, or the combinability of the elements will be welcomed . . . .*
> (Latour 1987: 227, emphasis added)

These ideas speak well to the case under consideration, for there are clear parallels in the spread of (western) forest techniques such as the allometric equation through centres of calculation (e.g. the UN Food and Agriculture Organisation (FAO) and national forestry research centres). Notably, there is a strong practical element to the conceptualisation of immutable mobiles: technologies, objects and devices only become part of the network if they reliably perform the task required – in short, if they 'work'. In Law's Portugal case study, for instance, new techniques of navigation were essential as ships were going further from shore than before and so were no longer able to navigate via tides and sea depth. New methods of navigating via the stars and Sun were badly needed, and a workable method emerged, embodied in one important table: the table of solar declination, as Law explains: 'The table of solar declination . . . represented the distillation of many years of astronomical expertise, of thousands and thousands of calculations, of correspondence, of argument and innovation' (Law 1986: 252). As is explored in the core empirical section below, there are again parallels here with allometric equation data tables comprising the equation parameters which embody decades of careful measurement, data gathering and preparation, distilled into one set of figures, essential for accurately measuring volume or biomass. Crucially, as Law notes, for the table of solar declination to work and have practical value it had to be in a transferable form, able to be placed on every ship. This is also true for tables of tree volume data, derived from allometric equations, which could easily be taken into the field (the forest), and distributed among national forestry agencies and research centres.

But while the literature on immutable mobiles has undoubtedly enriched the field of science and technology studies, it does have weaknesses, especially in its neglect of the politics of technology development, innovation and diffusion. Surprisingly, for a literature with a strong focus on power and control – illustrated, for example, by Latour's reflection that there is a need to 'understand how different spaces and different times may be produced *inside the networks built to mobilise, cumulate and recombine the world*' (Latour 1987: 228, emphasis added) – the broader more conventional politics of colonialism, international relations, and corporate politics is largely absent. Indeed, this is the objective of such 'bottom-up' heterogeneous approaches to the study of technology – that the operation of politics is laid bare through revealing the day-to-day hybrid workings of technology change and innovation. While this grounded approach is undoubtedly important (and indeed a central aim of our analysis here), what it misses is more structured, institutionalised forms of power and politics.

Market devices

A second set of (interrelated) ideas which is relevant for carbon markets, particularly when considering the origins of these markets, is a body of research from economic sociology that thinks critically from first principles about how things get measured, classified, and accounted for within markets (Barry 2005; Callon 1998; Fligstein 1996; Hardie and MacKenzie 2007; MacKenzie 2008; Pryke 2007; White 1981). Its principal applications so far have been to financial markets (e.g. the markets in equities, bonds and derivatives), but there are a number of examples of carbon market case studies (Lovell and MacKenzie 2011; MacKenzie 2009). This interdisciplinary approach to the study of markets asserts that markets comprise a mix of people, technology, objects and things. Further, markets should be viewed and analysed as integral to society, rather than a separate sphere of activity (Caliskan and Callon 2010; Callon 1998; Foucault 2007; MacKenzie 2008). In this context, examining markets for carbon focuses attention on the intricate networks of people and 'things' (including allometric equations) that constitute carbon markets, thereby explaining why abstract models rarely fit the specifics of particular times and places, and, in our case study therefore helps us understand why the development of the allometric equation for one market – timber in the 1940s – might struggle when transposed to an entirely different time and place, REDD+ in the 2010s.

One idea of this field of scholarship on making markets with particular resonance for our empirical case, and that we focus on here, is that of market devices. Market devices are:

> the material and discursive assemblages that intervene in the construction of markets . . . From analytical techniques to pricing models, from purchase settings to merchandising tools, from trading protocols to aggregate indicators, the topic of market devices includes a wide array of objects.
>
> (Muniesa *et al.* 2007: 2)

It is the 'object' of the allometric equation that we view as a type of market device. While equations in this field of scholarship are typically financial market equations (see, for example, MacKenzie's (2005) account of the Black-Scholes equation), the allometric equation is distinctive because of its natural science origin. Although science, research and experimentation are highlighted as critical to market devices, it is not typically the natural sciences that are referred to but rather 'financial engineering' and 'marketing research' knowledge and expertise which are denoted 'the sciences of the market' (Muniesa *et al.* 2007: 5). Here we extend analysis of market devices to encompass the natural sciences.

There is clear evidence (detailed below) showing how the allometric equation, whether applied to measure timber or carbon, 'renders things, behaviours and processes economic' (Muniesa *et al.* 2007: 3); the key signifier of a market device. Market devices enable calculation through 'configuring economic calculative capacities and in qualifying market objects', and thus '[t]he ways in which market devices are tinkered with, adjusted and calibrated affect the ways in which persons and things are translated into calculative and calculable beings' (Muniesa *et al.* 2007: 5). For forest markets, the allometric equation is a market device that allows (along with other measurement techniques) the forests to become a market, through estimation of the resource in question (timber or carbon). It abstracts from the living, complex forest ecosystem an estimation of the quantity of timber yield or carbon stored, and thereby simplifies and renders the ecosystem calculable.

A number of market device empirical investigations have shown that new markets typically (and unsurprisingly) comprise elements of what has gone before (see, for example, Hardie and MacKenzie 2007; Pryke 2007). However, for the most part the migration of market devices between markets – as with our case – is an issue that has been rather overlooked. The concept of immutable mobiles is, therefore, a useful complement to market devices, because of its attention to how things (such as equations) travel, whether they alter in the process, and how new sociotechnical relationships form once relocated. Furthermore, immutable mobiles and market devices share a science and technology studies heritage, with a common conceptualisation of heterogeneous networks of people, objects and technologies including centres of calculation (see, for example, Callon 2007; Law 1986). We now turn to the empirical case and discuss how well these theories help us understand the operationalisation of REDD+ forest carbon markets, starting with an overview of the international timber industry.

International timber markets: the origins of the allometric equation

In considering how carbon markets were prefigured or structured through earlier market forms, here we briefly summarise the rich history and politics of colonial forestry in order to contextualise the allometric equation. We introduce the practices and institutions of forest measurement developed by and used within the timber industry since the late nineteenth century. Colonial forestry, in the period from the nineteenth to early twentieth century, has been characterised as the spread of a common set of forestry practices from Europe at the centre, out to the peripheral colonies (Scott 1998). Indeed, analysis has in some cases drawn on Latourian ideas to characterise 'Europe as a centre of calculation and the producer of hegemonic classificatory schemes' (Vandergeest and Peluso 2006b: 360). In these accounts there is an idea of 'a basic model for professional forestry (Germany/France) . . . that central powers in colonial empires sent out directions and models to the periphery' (Vandergeest and Peluso 2006a: 32). While Vandergeest and Peluso, among others, are critical of this uncomplicated transfer of forestry expertise and methods from Europe to its colonies, it is, broadly, a reasonably accurate characterisation of how the international timber industry developed (Scott 1998). Germany is seen as the origin of forestry science, for it is in Germany from the mid-sixteenth century onwards that the discipline of forest science first emerged (Grainger 2009), with its strict attention to geometrical and statistical methods to govern the harvesting of timber from forests (Hölzl 2010). Hölzl, for example, describes the development of forest management planning and experimental forestry in Germany in the early nineteenth century, noting the focus of the German state on orderly, managed forests, designed to maximise timber harvest: 'Producing high-quality timber for the hungry . . . markets became the prime goal of state forestry' (ibid.: 444).

Forests with a low number of species were seen as most desirable, as this made timber yield estimates and harvesting much more straightforward than for mixed-species forest. As Hölzl (2010: 454) describes: '"Forest weeds" were eliminated, pastures forested, coppices changed into high timber forest, and mixed forests turned into [single] age-group spruce or oak plantations.' Crucially, these ideas about the desirability of heavily managed 'disciplined', single-species forests travelled beyond Germany and out to the colonies and 'became influential on a global scale' (ibid.: 454). There are clear parallels here with the concept of immutable mobiles, where ideas, technologies and practices diffuse from a (typically European) centre of calculation, namely centres of forestry training and expertise across Europe, notably in the UK, France and Germany. Cleary (2005), in his analysis of forests in the development of France's colonial territories in Indo-China, 1900–1940, points to the ways in which the scientific practices of Western forestry were implemented in French Indo-China, noting that '[v]irtually all senior French foresters working in Indochina were products of the powerful School of Forestry at Nancy in France' (Cleary 2005: 270) and 'French forest officials . . . were quick to apply concepts of "rational" and "modern" exploitation to the tropical forest environment. This epistemology was rooted in the developing science of ecology' (ibid.: 280). It was at these forestry schools that norms about 'ideal' forests as well as practical techniques for estimating timber volume – including an early version of the allometric equation – were developed and then applied to colonial tropical forest management.

As the colonial era came to an end, increasing attention was given to the collection of global scale forest statistics by international institutions (Grainger 2009; Mather 2005). In the post-Second World War period the United Nations FAO was established, coming into being at a Conference on Food and Agriculture in Virginia, US in 1943. At the time there were significant concerns about global timber resource shortages, and this was a central factor in establishing the FAO. In its formative decades (the 1950s and 1960s) the FAO had a clear objective of maximising timber production. Thus in the FAO's first global forest inventory, published in 1948 as *Forest Resources of the World*, forest data were gathered and presented in tables showing, for example, area of productive forests, accessible and inaccessible forests, and volumes of standing timber.

A key element of the international timber industry's toolkit for measuring timber resources in the field is the allometric equation, as follows:

$$\text{Stem volume} = a \times D^b$$

where D is tree diameter at breast height (either 1.3 metres above ground or 1.4 metres, depending on the particular forest school or 'centre of calculation' that is being followed (Chave *et al.* 2005)); and a and b are species-specific parameters, derived from field trials. So, for this particular allometric equation, developed for the *Abies sibirica* (fir) tree in Germany, a is 0.0001316 and b is 2.52 (see Zianis and Seura 2005).

In its earliest form, in the first half of the twentieth century, the allometric equation was employed to give an easily obtained estimate of timber volume (often termed 'merchantable timber') for a given area of forest. The field measurement taken is either of the tree diameter at 'breast height', or the tree height, or in some cases both. Of interest to the timber merchants and foresters is the straight part of wood – the trunk of the tree (known as 'the bole') – so the equation rests on the fact that there is a relationship between timber volume and something much easier to measure in the field (i.e. tree diameter and/or height). However, allometric equations do vary from species to species and geographically (because growth rates of trees alter with climate). So, although there is a generic 'template' equation (as shown above, Vol = $a \times D^b$), in practice species- and country-specific allometric equations are used – with slight modification to the form of equation, and with their own specific parameters, a, b and sometimes c, calculated from field trials – in order to maximise the accuracy of results (Brown 2001; Chave *et al.* 2005; Zianis and Mencuccini 2004).

Allometric equations form the basis of forest inventories, conducted regularly and in large numbers by the FAO, timber merchants and national forestry agencies to measure and document the forests they manage (Picard *et al.* 2012). The allometric equation thus 'renders things, behaviours and processes economic' (Muniesa *et al.* 2007: 3), in this case turning living forests into the commodity of timber. It is a vital market device because in its absence volumes of 'standing' or 'merchantable' timber could be only very poorly estimated, leading to market fluctuations and instability because of inaccurate data.

Investigating the migration of allometric equations from timber to carbon markets

Background on forest carbon markets and REDD+ MRV

There is widespread agreement about the value of including forests in a post-Kyoto United Nations (UN) climate treaty, as tropical deforestation and other land use change constitutes an estimated 8 per cent of global greenhouse gas emissions (Global Carbon Project 2013). However, the translation of this overarching policy objective into a workable set of policy guidelines, standards and practices has proved much more difficult, and work is still ongoing to decide upon key issues. As noted, accurate measurement of forest carbon, or MRV, has become a key element of attempts to integrate forests more fully into international climate policy. The UNFCCC has agreed that the 2003 IPCC Good Practice Guidance (GPG) should form the basis for measuring forest carbon under REDD+ (see UNFCCC 2007, Decision 2/CP.13(6)), but there remains a great deal of uncertainty, and hence a lot of research and policy activity, about how to measure forest carbon to a reasonable degree of accuracy at a global scale. This uncertainty in REDD+ MRV has fostered a politics of expertise between different types of forest carbon measurement, most notably between remote sensing and 'on the ground' methods used by foresters and forest ecologists, including the allometric equation (Lovell 2013).

Much could be said about the history of forests within the UNFCCC (see, for example, Boyd 2010; Fogel 2005). In short, forests have had a rather controversial role, largely stemming from unease about the implications of including tropical forests in an international agreement, because of the potential for it to absolve certain countries from undertaking climate mitigation in other sectors. Indeed, to date, this arena of international relations is where most of the work on the politics of forest carbon markets can be found (see

Bäckstrand and Lövbrand 2006; Baldwin 2003). Under the Kyoto Protocol's Clean Development Mechanism some forestry projects were allowed, but in practice only a handful of projects came to fruition (because these projects were quickly discovered to be less profitable, more complicated and risky than other types of Clean Development Mechanism project; see Fogel 2005). It was hoped, therefore, that the initial REDD agreement at the UNFCCC COP-13 meeting in Bali signalled a new era of forest protection under the UNFCCC. Since the 2007 Bali COP more detail has followed at each Conference of Parties and interim UNFCCC meetings (notably meetings of the Subsidiary Body on Scientific and Technical Advice (SBSTA)), including about how to measure carbon in forests. It is at the national level that forest MRV/REDD+ is being implemented, in keeping with the overall thrust of the UNFCCC, that responsibility for meeting climate change mitigation targets, and monitoring progress against them, is the remit of nation-states.

However, as noted, progress in getting REDD+ actually operational has been slow – held up mostly by problems with the overall climate negotiations – leading to the rather curious current situation characterised by extensive, detailed debate and activity about how to do REDD+ (i.e. how to operationalise it), plus the implementation of numerous local projects, yet without a final REDD+ policy agreement being in place (Pistorius 2012). In this complex and fast-changing field of policy development and technology innovation allometric equations represent one strand of REDD+ MRV. Several different methods are being advocated for measuring forest carbon, stemming from different areas of scientific expertise – remote sensing, forest ecology and forestry – and allometric equations are therefore jostling for attention within this crowded policy arena (Lovell 2013).

Altering the allometric equation for REDD+ MRV

We now turn to investigate in more detail the assumptions and knowledges embodied in the allometric equation, thereby revealing how and why its application to REDD+ needs careful scrutiny, as there are a number of technical limitations. One vital 'fact' that underpins the transfer of the allometric equation from timber to carbon markets is the simple, unvarying relationship between tree biomass and carbon mass: for nearly all tree species carbon mass is equal to 50 per cent of dry weight biomass (Brown and Lugo 1992; Penman *et al.* 2003) (although see Elias and Potvin (2003) for a critique, there is a small degree of species-to-species variation). That this 50 per cent biomass-carbon conversion rule holds constant across tree species globally has made it possible to convert historical inventories of forest timber volume to estimates of carbon. Crucially, these historical studies – the many, extensive forest inventories conducted to estimate timber yields – are the most extensive, reliable set of historical data we have about global forests. The allometric equation, therefore, potentially lies at the heart of REDD+ MRV, because of its ability to open up historical forest inventories in order to provide carbon data (Gibbs *et al.* 2007).

However, there are a number of steps that must be taken in order to obtain a figure for the dry weight biomass of a tree. These are additional steps to do with the allometric equation that are not required for timber markets, but which are crucial for REDD+. As previously explained, allometric equations are used by the timber industry to calculate the volume of the tree stem (the 'bole'), from which timber planks – the commodity in question for timber markets – are derived. So the original (timber) allometric equation effectively ignored other parts of the tree, i.e. branches, leaves and roots. But these tree parts, of course, all contain carbon, so an estimate of the total biomass of the tree is required for the allometric equation to work as a successful market device for REDD+. Thus a so-called 'expansion factor' was added to the allometric equation, to calculate total tree biomass, and not just the volume of the bole. The expansion factor was first developed by forest ecology research scientists, who have always been interested in the more comprehensive biomass measurement of the tree (Brown 2001).

However, in order to obtain the data to calculate an expanded (biomass) allometric equation, i.e. to establish the value of the equation parameters, a time-consuming and expensive process known as 'destructive harvest measurement' (Gibbs *et al.* 2007: 4) must take place. Destructive harvest measurement involves identifying sample trees which are measured then harvested, and a subsample dried and weighed (including roots, stems, branches and leaves if calculating total biomass) to derive the relationship between tree diameter and/or height and biomass for that particular tree species, or groups of species, growing in that location. Because harvesting and measuring whole trees in this way is difficult and expensive (see Figure 4.1) it is not done very often. For example in Zianis and Seura's (2005) review of European allometric equations, only four out of the 607 biomass allometric equation field studies evaluated comprised measurement of the full tree roots. This explains the value in having allometric equation tables, where the parameters (derived from previous destructive harvest measurement field studies) can simply be looked up, based on the field tree diameter or height measurements, and a figure for dry biomass estimated. So, given the difficulty and expense of destructive tree harvesting, the value of allometric equations within REDD+ markets is arguably higher than for timber markets. This is because in timber markets allometric equations for the bole measurements can be relatively easily and routinely tested, because timber is being harvested anyhow. In contrast, for forest carbon, destruction and harvesting is not a routine element of this market (indeed, the objective is forest conservation, so quite the opposite).Thus, as Zianis and Seura (2005) report, bole or 'stem' volume allometric equations tend to be based on much larger more comprehensive field studies: most of the 230 European stem volume equations they studied were based on a sample size of several hundred felled trees, and eight equations were based on as many as five thousand trees. While these stem volume allometric equations could potentially be converted to give an estimate of biomass (using the expansion factor), the limitation is that the field data are based on commercial timber species, not usually found in the forests measured for REDD+, a point returned to below.

Figure 4.1 Backhoe in the field

Source: Photograph courtesy of Dr Casey Ryan

A second additional modification of the allometric equation essential for it to act as a market device within forest carbon markets is to calculate the tree *dry weight* biomass, in order to adjust for the different densities of wood across different species. Some species of tree have very dense wood, with little moisture content ('hardwood' species), whereas other species are much more porous and comprise lots of water ('softwood' species). By removing the water from tree samples – using a form of oven, either in the field or the laboratory – what remains are the core chemical building blocks of wood, which hold true regardless of species, namely cellulose, hemicellulose and lignin. The chemical mass of carbon within these three substances that comprise wood is 50 per cent of the total dry weight. As noted, this simple end calculation to convert biomass to carbon mass – valid across all species – has facilitated the travelling of allometric equations from timber to carbon markets.

These two additional steps required to make allometric equations work for forest carbon markets suggest a degree of mutability in the equation, thereby somewhat challenging the 'immutable mobile' conceptualisation posited above. However, the core stability and durability of immutable mobiles has evolved as a concept over time, called into question by some of its founding scholars, as Law and Mol argue:

The focus [of the early scholarship] was on control: on the work needed to hold a configuration stable; on the effort required to create a wider network fit for the transmission of immutable mobiles. But there is another problem; it is, quite simply, that often enough ideas, facts, information, even technologies, turn out to spread in a manner that is much more *fluid*. It is precisely a lack of rigidity that most helps movement.

(Law and Mol 2001: 619, emphasis in original)

The case of the allometric equation appears to back up this conceptual relaxation of immutability. Allometric equations have changed as they have spread from timber markets to forest carbon markets. Indeed, their 'lack of rigidity' – the ability for extra steps and processes to be added to the original core allometric equation – has been essential to them being able to function as a market device in REDD+.

Analysing the allometric equation network

There are further important differences in applying the allometric equation to forest carbon, instead of timber, which go beyond the allometric equation itself to the wider network (or sociotechnical context) in which it was developed and deployed. Politics comes to the fore here in terms of how the technical limitations of the allometric equation as applied to REDD+ have been somewhat underplayed by forestry and forest ecology experts working on REDD+ MRV, as well as neglected by policy makers. We consider these technical limitations (termed 'relational difficulties') in turn, including: mixed-species forest, diverse age profiles, sampling procedures, and incomplete/absent data sets for tree species. The empirical findings largely support the notion of heterogeneous networks – a central plank of theories of both immutable mobiles and market devices – for, as our analysis reveals, the work done by allometric equations cannot be explained solely by evaluation of the equation itself, but rather through the relations it has with other things and people, and to what extent these relations have also migrated to REDD+, and, if so, whether they have altered in the process.

The first of these relational difficulties is that REDD+ forests tend to be mixed-species, while timber market forests tend to comprise homogeneous stands, as Mohren explains:

Forest inventory essentially started with mapping of forest stands as homogeneous units with regard to species composition, density and size, and strongly developed during the 20th Century with the establishment of a wide array of sampling techniques; focusing on timber volume and forest stand structure.

(Mohren *et al.* 2012: 686)

Allometric equations historically, therefore, were based on a small set of commercially valuable tree species, and thus struggle to calculate the mixed-species 'natural' forests that populate carbon markets, as one interviewee explains:

For some species in the tropical rainforest you can only take a general [allometric equation], because there are so many species, and sometimes it is very difficult to identify each species in the field . . . in the tropical rainforest we have more than two or three hundred different species – when you are making the measurement it can go up to five hundred [species], so it is very complicated!

(Interview, Forest carbon/REDD+ researcher, November 2012)

Market devices are conceptualised as particular to the specifics of times and places, and can therefore only be understood in context (Callon *et al.* 2007). From a market device perspective, therefore, it is perfectly understandable that these difficulties arise because the allometric equation has been transposed to a different time and place (mixed-species natural forests, rather than plantations) and, further, is operating within a market that has a different objective (preservation rather than harvesting of forests).

A second relational difficulty that arises when dealing with tropical natural forests is the diverse age profile of the trees. Timber forests are often managed to generate a relatively even-aged stand of trees, but in natural unmanaged forests there will, of course, be a far greater age range of trees. It is for the large, older trees that allometric equations have recognised weaknesses in accurately measuring tree biomass (Chave *et al.* 2005). This is simply because as a tree ages it is more likely to become misshapen (e.g. lose branches), be rotten in parts, and/or be hollow (through the action of termites, and other forest species making their homes in the trees). While this weakness of allometric equations is not so much of an issue for timber market forests (because there are usually no very old trees), it does pose problems for accurately measuring carbon in REDD+ forests. Thus, for instance, detailed investigation of tropical African Miombo woodlands has discovered that the majority of large trees are hollow (see Figure 4.2), and this could have an important bearing on the overall carbon content of the forests (Ryan 2009). A further characteristic of mixed-aged unmanaged forests is non-circular trees which causes problems in accurately gaining a 'diameter at breast height measurement'; again, something which does not arise in managed, more standardised timber forests.

A third relational difficulty in applying the allometric equation to carbon, instead of timber, is in norms around sampling procedures, i.e. the number and type of plots used to generate biomass data for a whole forest. If one has an even-aged, single-species forest managed for timber then a sensible sampling procedure (and the one that has dominated forestry training and centres) is a large number of small plots (because in every plot there will be a large tree; there is little variation between plots). However, in an unmanaged, more diverse REDD+ forest (e.g. the patchy, sparse Miombo woodland in Africa) this type of sampling strategy can easily lead to the big trees (which are the large carbon stores) being missed (Ryan *et al.* 2011). This sampling mismatch also extends further to larger geographical scales: allometric equations were only ever intended to provide data on local forests, for local or regional timber

Figure 4.2 Hollow tree

Source: Photograph courtesy of Dr Casey Ryan

companies or government forestry agencies. The challenge for REDD+ is in gathering data at a global scale, and there are some tensions here in scaling up, as Gibbs *et al.* explain:

> Forest inventory data can provide high-quality information for a particular region, but existing inventories were generally not collected using sampling schemes appropriate for the biome scale . . . the compilations of studies used to develop the biome averages generally focused on mature stands and were based on a few plots that may or may not adequately represent the biome or region.
>
> (Gibbs *et al.* 2007: 4)

Fourth, and finally, many parts of the world are not covered by existing allometric equations. The time-consuming and expensive destructive tree harvesting required to cross-check allometric equations, and generate new equations for new species (as discussed earlier in the chapter), acts as a disincentive to improving the global coverage of the allometric equation. Africa in particular, with its lack of well-developed colonial forestry industry, is a continent where mostly only generalised allometric equations are applied, i.e. ones which have travelled in from elsewhere (Gibbs *et al.* 2007). As Chave *et al.* (2005) observe: 'The use of allometric regression models is a crucial step in estimating AGB [above ground biomass], yet it is seldom directly tested. Direct tree harvest data are difficult to acquire in the field, and few published studies are available' (Chave *et al.* 2005: 88). In this way the labour, cost and effort of destructive tree harvesting has, in effect, greatly facilitated the travel of the allometric equation: if it were easy to establish the allometric equation parameters for new species, in new locations, the allometric equation tables would not be valued so highly, and would not migrate. As one interviewee explains:

> We are using a general allometric equation instead of developing one for the site [in Mexico], and *then we have a lot of inaccurate results of forest carbon.* We cannot be sure how close or far we are from the real value unless we cut down the trees. And in the end that is very tricky because we are saying that we want to protect the trees, but to get the accurate measurement we would have to cut down the trees!
>
> (Interview, Forest carbon/REDD+ researcher, November 2012)

The interviewee here hints at a fifth and final relational difference in the purpose of allometric equations as applied to carbon markets compared with timber markets, again, an issue touched on earlier in the chapter; namely, that for carbon markets the aim is to preserve the trees, whereas for timber markets of course the tree will eventually be cut down and destroyed. The central role of the tree within the network is different. The conservation component of carbon markets acts as a further disincentive to carrying out the destructive harvest measurement necessary for improving the accuracy of allometric equations, in situations where local, species-specific allometric equations are not available.

What these relational difficulties demonstrate is the path dependency of the allometric equation: it carries with it aspects of its relations developed within the international timber industry and colonial forestry markets. It therefore struggles to compete in the complex, technical politics of REDD+ MRV. Its very longevity disadvantages it vis-à-vis other newer measurement techniques, such as from remote sensing science. Further, the devices, practices and tools of international timber markets have carried with them aspects of their political and institutional origins in colonial forestry, and this has shaped their acceptability and perceived role with contemporary REDD+ markets. This politics of expertise has manifested in different ways, but a predominant discourse is about the revitalisation of forestry expertise because of REDD+, albeit with tensions, as an interviewee explains:

> I think it [REDD+] is seen as an opportunity among some [forestry] institutions, to find a use for their work, that's a sceptical view of it. Obviously, it is important work. But I think if you look at forestry commissions and forestry agencies in different countries, they, you know, *they now see a new role for their knowledge and their information.* So it is, it is quite an interesting new opportunity for some of these forest agencies in developing countries. And it's, I mean, it's a great opportunity, *but it is not necessarily meant that they have tried to adopt new techniques.*
> (Interview, Head of Science at a forest NGO, August 2010)

The simplicity of measuring diameter at breast height, most often with a tape measure – is a distinct disadvantage when trying to capture the attention of policy makers, where such forestry techniques struggle to compete against the seductive, high-tech visual imagery of remote sensing. There is evidence of a somewhat sceptical view about the ability of this longstanding forestry profession to innovate, and for its techniques and practices – such as the allometric equation – to adapt and change, as an interviewee comments:

> Foresters typically, I mean they are a lot more nowadays asking other questions. But typically, they want to know how much timber was there. If you look at the old inventories in FAO, they always are often called, pre commercial or pre harvest . . . I think they use other terms. But the bottom line is, they are assessing how much timber is there to decide whether to commercialise it or not . . . and those techniques, they have just then reapplied to carbon.
> (Interview, Director of a forestry consultancy, November 2012)

In this way forestry techniques and practices, such as the allometric equation, are sometimes seen as old fashioned and out-dated (Picard *et al.* 2012). There is a politics of expertise at work here in the positioning of certain types of forest MRV as better equipped to do the work of monitoring and measuring forest carbon. It is remote sensing science that the allometric equation and associated forestry and forest ecology techniques are most often contrasted to, for example, a remote sensing scientist argues that '[p]eople running tape measures round trees, this is what we have to get away from' (Dr Asner, quoted in Tollefson 2009).

This is despite the fact that in reality allometric equations and remote sensing are complementary forms of MRV, as one interviewee explains:

> [W]hilst there does seem to be a lot of attention on remote sensing, the remote sensors are very aware that you need basic standard routine forestry practices as well, that have been done for many years. *And they are not so fashionable and popular, but it can't be done without that.*
>
> (Interview, remote sensing scientist at a
> Space Company, August 2010)

And as a recent UN REDD technical report about the use of allometric equations states by way of introduction:

> Ultimately, any forest carbon stock measurement at some point needs trees to be weighed in the field. This constitutes the keystone on which rests the entire edifice of forest carbon stock estimations, whatever the scale considered.
>
> (Picard *et al.* 2012: 17)

Summary and conclusions

In this chapter we have analysed how forest carbon measurement for REDD+ has drawn in a measurement technique called the allometric equation that previously inhabited international timber markets. The devices, tools and techniques of measuring timber have been adopted to measure carbon, instead of timber volume. The case is an example of the (partial) mutability of allometric equations. There are subtle but important technical difficulties in transferring the allometric equation from measuring timber to carbon, including how to deal with mixed-species forests and old, hollow trees. These are termed 'relational difficulties' because they are not directly related to the allometric equation itself, but rather the relations it has with people and things within the new market network of forest carbon. These relational difficulties can only be explained and understood by paying attention to the timber industry origins of the allometric equation. The substantial work involved in constructing a biomass allometric equation from scratch – the destructive harvesting technique – has encouraged the migration of already existing allometric equation tables (where the equation parameters are predefined), from country to country and species to species. There are, however, notable gaps in this travelling – e.g. the sparse Miombo forests of Africa – where the timber industry has never been much present, and hence locally relevant allometric equations are rare.

A number of new theoretical insights based on our case have been proposed. First, that market devices can and should be analysed as operating across different markets, and not just particular to one type of market. Second, that in travelling between markets, market device relations also move with them – certain ways of doing, norms and expertise – and this causes tensions because these 'relational difficulties' are not explicitly recognised, discussed and acted upon. Third, there is value in bringing together the concepts of market devices and immutable mobiles when considering cases of the migration of market devices. Fourth, that natural science equations are a type of market device, and not just the financial equations that populate current scholarship on market devices. Fifth, and relatedly, that the natural sciences are a type of market science and expertise, especially when the markets under consideration comprise natural commodities such as cotton, wheat or carbon.

The politics of expertise in REDD+ MRV is hard to discern, because it is so closely entangled with the technical capabilities of the allometric equation, and hence our observations on the politics are somewhat provisional and deserve further exploration. First, a general observation that the long history of forestry and forest ecology expertise and the rather simple field measurement techniques involved in the allometric equation have, by and large, acted to disadvantage its knowledge claims within contemporary REDD+ policy circles. There is evidence that the forestry profession is seen as slow to adapt and less able to respond to the MRV demands generated by REDD+ than other newer more high-tech MRV methods, such as remote sensing. A politics of expertise is at work in forest carbon markets, revealed through the case study of allometric equations. Second, there is evidence of a lack of disclosure and discussion around the technical limitations of allometric equations when applied to measure carbon rather than timber. This stems in part from a desire of the forestry and forest ecology professions to find new ways of using their skills and expertise in REDD+, but also relates to the strong need for those involved in REDD+ policy to have workable REDD+ MRV methods, in order to operationalise the policy. Third, the continuing strong policy focus on REDD+ MRV, with a multitude of reports published in recent years (see Pistorius 2012 for a review), could be because tackling the underlying causes of tropical deforestation – agriculture and poverty – is so difficult. REDD+ MRV is, according to this view, a distraction from the main task and hence serves a wider political function.

Acknowledgements

Heather Lovell and Donald MacKenzie would like to acknowledge the financial support of the UK Nuffield Foundation for a New Career Development Fellowship – Fungible Carbon (NCF/35037), 2008–2013. The content of the chapter is based on empirical research completed under this grant. Many thanks are due to the editors of the book and Matthew Paterson and Jan-Peter Voß for their helpful comments on earlier drafts, as well as to Dr Casey Ryan, a Lecturer in Ecosystem Services at the School of GeoSciences, University of Edinburgh, who kindly provided a technical review. Any remaining errors are the responsibility of the authors. Lastly, the generosity of all interviewees and survey participants is much appreciated, as is the valuable assistance of Dr Brice Mora and Professor Martin Herold at the Global Observation of Forest and Land Cover Dynamics (GOFC-GOLD).

Note

1 Including the Miombo Conference, Edinburgh, 2008; the Commonwealth International Forestry Conference, Edinburgh, July 2010; the UK Government Earth Observation Workshop, Reading, June 2010; Forest Day 6 at the UNFCCC annual meeting (COP/MOP), Doha, Qatar, November 2012.

References

Bäckstrand, K. and Lövbrand, E. (2006) Planting trees to mitigate climate change: contested discourses of ecological modernization, green governmentality and civic environmentalism. *Global Environmental Politics*, 6 (1), 50–75.

Baldwin, A. (2003) The nature of Boreal Forest: governmentality and forest-nature. *Space & Culture*, 6 (4), 415–428.

Barry, A. (2005) The anti-political economy. *In:* A. Barry and D. Slater, eds., *The Technological Economy*. London/New York: Routledge, 84–100.

Boyd, E., Gutierrez, M. and Chang, M. (2007) Small-scale forest carbon projects: adapting CDM to low-income communities. *Global Environmental Change*, 17, 250–269.

Boyd, W. (2010) Ways of seeing in environmental law: how deforestation became an object of climate governance. *Ecology Law Quarterly*, 37, 843–916.

Brown, S. (2001) Measuring carbon in forests: current status and future challenges. *Environmental Pollution*, 116, 363–372.

Brown, S. and Lugo, A. E. (1992) Aboveground biomass estimates for tropical moist forests of the Brazilian Amazon. *Interciencia. Caracas*, 17 (1), 8–18.

Caliskan, K. and Callon, M. (2010) Economization, part 2: a research programme for the study of markets. *Economy and Society*, 39 (1), 1–32.

Callon, M., ed. (1998) *The Laws of the Markets*. Oxford: Blackwell Publishers/The Sociological Review.

Callon, M. (2007) What does it mean to say that economics is performative? *In:* F. MacKenzie, F. Muniesa and L. Siu, eds., *Do Economists Make Markets? On the performativity of economics*. Princeton: Princeton University Press, 311–357.

Callon, M., Millo, Y. and Muniesa, F., eds. (2007) *Market Devices*. Oxford: Blackwell Publishing.

Chave, J. et al. (2005) Tree allometry and improved estimation of carbon stocks and balance in tropical forests. *Oecologia*, 145 (1), 87–99.

Cleary, M. (2005) Managing the forest in colonial Indochina c. 1900–1940. *Modern Asian Studies*, 39 (2), 257–283.

Cronon, W. (1991) *Nature's Metropolis: Chicago and the Great West*. New York: Norton and Company.

Elias, M. and Potvin, C. (2003) Assessing inter-and intra-specific variation in trunk carbon concentration for 32 neotropical tree species. *Canadian Journal of Forest Research*, 33 (6), 1039–1045.

Fligstein, N. (1996) Markets as politics: a political-cultural approach to market institutions. *American Sociological Review*, 61 (4), 656–673.

Fogel, C. (2005) Biotic carbon sequestration and the Kyoto Protocol: the construction of global knowledge by the Intergovernmental Panel on Climate Change. *International Environmental Agreements: Politics, Law and Economics*, 5 (2), 191–210.

Foucault, M. (2007) *Security, Territory, Population*. Basingstoke/Hampshire: Palgrave Macmillan.

Gibbs, H. K. et al. (2007) Monitoring and estimating tropical forest carbon stocks: making REDD a reality. *Environmental Research Letters*, 2 (4), 045023.

Global Carbon Project (2013) *Global Carbon Budget 2013 – Summary*. Available from: http://www. globalcarbonproject.org/carbonbudget/13/hl-compact.htm [accessed 11February 2014].

Grainger, A. (2009) Towards a new global forest science. *International Forestry Review*, 11 (1), 126–133.

Hardie, I. and MacKenzie, D. (2007) Assembling an economic actor: the *agencement* of a Hedge Fund. *The Sociological Review*, 55 (1), 57–80.

Hölzl, R. (2010) Historicizing sustainability: German scientific forestry in the eighteenth and nineteenth centuries. *Science as Culture*, 19 (4), 431–460.

Latour, B. (1987) *Science in Action*. Cambridge: Harvard University Press.

Law, J. (1986) On the methods of long-distance control: vessels, navigation and the Portuguese route to India. *In:* J. Law, ed., *Power, Action and Belief: a new sociology of Knowledge?* London: Routledge, 234–263.

Law, J. (1987) Technology and heterogeneous engineering: the case of Portuguese expansion. *In:* W. E. Bijker, T. P. Hughes and T. J. Pinch, eds., *The Social Construction of Technological Systems: new directions in the sociology and history of technology*. Cambridge: MIT Press, 111–134.

Law, J. and Mol, A. (2001) Situating technoscience: an inquiry into spatialities. *Environment and Planning*, 19 (5), 609–622.

Lovbrand, E. and Stripple, J. (2010) Governing the climate from space: monitoring, reporting and verification as ordering practice. *2010 International Sociological Association Annual Convention*, New Orleans.

Lovell, H. (2013) Measuring forest carbon. *In:* J. Stripple and H. Bulkeley, eds., *Governing the Climate: new approaches to rationality, power and politics*. Cambridge: Cambridge University Press, 175–196.

Lovell, H. and MacKenzie, D. (2011) Accounting for carbon: the role of accounting professional organisations in governing climate change. *Antipode*, 43 (3), 704–730.

MacKenzie, D. (2005) Opening the black boxes of global finance. *Review of International Political Economy*, 12 (4), 555–576.

MacKenzie, D. (2008) *Material Markets: how economic agents are constructed*. Oxford: Oxford University Press.

MacKenzie, D. (2009) Making things the same: gases, emission rights and the politics of carbon markets. *Accounting, Organizations and Society*, 34 (3–4), 440–455.

Mather, A. S. (2005) Assessing the world's forests. *Global Environmental Change*, 15 (3), 267–280.

Mohren, G. *et al.* (2012) Forest inventories for carbon change assessments. *Current Opinion in Environmental Sustainability*, 4 (6), 686–695.

Muniesa, F., Millo, Y. and Callon, M. (2007) An introduction to market devices. *In:* M. Callon, Y. Millo and F. Muniesa, eds., *Market Devices*. Oxford: Blackwell Publishing/ The Sociological Review, 1–12.

Penman, J. *et al.* (2003) IPCC Good Practice Guidance for land use, land-use change and forestry. *National Greenhouse Gas Inventories Programme*. Kanagawa: Institute for Global Environmental Strategies.

Picard, N., Saint-André, L. and Henry, M. (2012) *Manual for Building Tree Volume and Biomass Allometric Equations: from field measurement to prediction*. Rome: Food and Agricultural Organization of the United Nations/Montpellier: Centre de Coopération Internationale en Recherche Agronomique pour le Développement.

Pistorius, T. (2012) From RED to REDD+: the evolution of a forest-based mitigation approach for developing countries. *Current Opinion in Environmental Sustainability*, 4, 313–324.

Pryke, M. (2007) Geomoney: an option of frost, going long on clouds. *Geoforum*, 38, 576–588.

Ryan, C. M. (2009) *Carbon Cycling, Fire and Phenology in a Tropical Savanna Woodland in Nhambita, Mozambique.* PhD Thesis. Edinburgh: University of Edinburgh.

Ryan, C. M., Williams, M. and Grace, J. (2011) Above- and belowground carbon stocks in a Miombo woodland landscape of Mozambique. *Biotropica*, 43 (4), 423–432.

Scott, J. C. (1998) *Seeing Like a State: how certain schemes to improve the human condition have failed.* New Haven: Yale University Press.

Tollefson, J. (2009) Climate: counting carbon in the Amazon. *Nature*, 461, 1048–1052.

UNFCCC (2007) *Report of the Conference of the Parties on its Thirteenth Session.* Paper presented in Bali from 3 to 15 December 2007. United Nations Framework Convention on Climate Change (UNFCCC).

Vandergeest, P. and Peluso, N. L. (2006a) Empires of forestry: professional forestry and state power in southeast Asia, part 1. *Environment and History*, 31–64.

Vandergeest, P. and Peluso, N. L. (2006b) Empires of forestry: professional forestry and state power in southeast Asia, part 2. *Environment and History*, 359–393.

White, H. C. (1981) Where do markets come from? *American Journal of Sociology*, 87, 517–547.

Zianis, D. and Mencuccini, M. (2004) On simplifying allometric analyses of forest biomass. *Forest Ecology and Management*, 187 (2), 311–332.

Zianis, D. and Seura, S. M. (2005) *Biomass and Stem Volume Equations for Tree Species in Europe.* Vantaa: Finnish Society of Forest Science/Finnish Forest Research Institute.

5 Virtuous carbon

Matthew Paterson and Johannes Stripple

Introduction

Somewhere in the financial districts of London, a trader follows the fluctuations of carbon on the screen. Only a few pieces of information make it there; the bids, the dates, the volume and the price. Two straight horizontal lines (one for selling and one for buying) circumscribe a winding line and enable the trader to bet on short-term price movements. Carbon is here fully commodified, which in its financial sense means that it is a totally standardised and commensurable unit, an asset class as money, oil or gold. But how did we end up in the financial district of London where recently Barclays bought the Swedish carbon trader Tricorona for a cash offer of US$142 million? At the time of the acquisition in 2010, the global carbon emission market was worth $144 billion and Barclays attempted to capitalise on Tricorona's pre-2012 carbon offset portfolio of 43.7 million tonnes. But what are we to make of the fact that Barclays in June 2012 sold Tricorona back to the company's management and that the world's carbon market's value declined to 40 billion euros in 2013, the lowest since 2007? On the other hand, in 2013 nine new national or regional emissions trading schemes (five of them in China) started to operate and the share of global emissions covered by emissions trading is expected to increase by 70 per cent from a 2005 basis (ICAP 2014). Despite the volatility and the different trends in the carbon economy, it is appropriate to ask the fundamental question of what kind of commodity is carbon? And what enables emission reductions to travel from, say, a hydropower installation in Chile through the digital screen in the financial circuits of London to end up neutralising the emissions of the cement industry in Sweden?

As is now widely recognised, carbon markets have emerged as the major response to climate change. From early proposals for emissions trading schemes (ETS) at the international level (Barrett *et al.* 1992; Grubb 1989; see also Simons and Voß, this volume), to the Kyoto Protocol negotiations, to the EU's crucial switch of position in 1998, favouring carbon markets, to the development of the actual markets around the EU ETS and the Clean Development Mechanism (CDM), to the roll-out of cap-and-trade schemes in various other countries. In 2013, nine cap-and-trade schemes were established around the world, from California, Switzerland, Kazakhstan to Québec to five regions in China. Three additional schemes are likely to follow in 2014–2015 (see also Engels *et al.* and Lederer, this volume). Despite what might be seen as a recent series of setbacks, with legislative failures in the US and stumbling in Australia, among others, carbon markets are still widely regarded as the cornerstone of a global response to climate change. But academic analysis is still catching up. There are huge literatures on the individual markets, especially on the EU ETS (Asselt and Biermann 2007; Peeters 2006; Skjærseth and Wettestad 2008) and the CDM (Ellis *et al.* 2007; Paulsson 2009; Streck and Lin 2008), with a smaller literature on the voluntary market (Bumpus and Liverman 2008; Lovell 2010) or newer markets like the Regional Greenhouse Gas Initiative (RGGI) (Rabe 2008). Most of this literature focuses on the questions of policy design and effectiveness; relatively little on how the markets themselves are 'put together'. There is also, however, an emerging literature on the markets themselves, with some looking from political economy approaches (Brunnengräber 2007; Bumpus and Liverman 2008; Lohmann 2006; Meckling 2011; Newell and Paterson 2010; Reyes 2011), some writing on carbon markets from within the ecological modernisation tradition (Spaargaren and Mol 2013), others from poststructuralist perspectives, including Actor Network Theory approaches (Callon 2009; Lane 2012; MacKenzie 2009), governmentality approaches (Lövbrand and Stripple 2011; Oels 2005; Paterson and Stripple 2010; Stripple and Lövbrand 2010; and various chapters in Stripple and Bulkeley 2014), theories of hegemony (Stephan 2012), or a combination of these (as in Lovell and Liverman 2010 or various contributions to Methmann *et al.* 2013). These latter poststructuralist approaches have often revolved around the question of how carbon markets become established as governable domains and tend to focus on particular elements in their construction, as in MacKenzie's (2009) analysis of the commensuration process, for example.

Our aim here is to attempt to sketch out at a more general, comprehensive level, the processes involved in the construction of carbon markets. As Ruggie (1998) once asked in a different context, how does it all hang together? Our ambition is to understand how the carbon economy 'hangs together'. Our aspiration is thus synthetic, drawing on existing research by ourselves and others. To do this, the chapter draws broadly on governmentality theory to facilitate an analysis of power and politics without the state as locus, origin and outcome. More specifically, the chapter draws on Der Derian's 'virtuous war' theory to develop an argument about carbon as a 'virtuous' commodity. By this we refer to the close affinity between virtuality and virtue – the technological and the ethical – in the construction of carbon markets. Conceiving of carbon credits as virtuous commodities draws attention to both their fictitious character (as imagined units, complex abstractions that exist only by way of agreement) and their virtue (how those units are only provisionally stabilised, and where their ethical contestation is part of their construction). This produces a distinct sort of governmentality, which aims to neutralise resistance by imbuing the commodities of carbon markets with a self-evident moral quality.

We explore virtuality and virtue at *five* moments in the commodification of carbon, suggesting that together they make up the key processes of assembling the carbon economy. Such processes have a logical rather than strictly historical relation to each other, and can, for carbon, be imagined somewhat as shown in Figure 5.1. In identifying these various moments in carbon's commodification, we synthesise the conceptualisations in existing literature on the subject, which identifies one or more of these moments. Our aim, as suggested above, is to bring all of this together in a single framework. The *first* moment is the invention of the 'tCO$_2$e', the basic unit of account, which has become adopted in almost all carbon markets to date. The *second* is the proliferation of this unit into several 'asset classes'. The *third* is about assuring that a tonne is a tonne is a tonne, i.e. the measurement devices and the methodologies of calculation that specify how any specific tonne of carbon should be created. The *fourth* and *fifth* are the differentiation of carbon commodities into what are sometimes called 'Boutique' and 'Walmart' carbon, examples of two divergent forms of carbon commodification. It should be emphasised that these are not historical moments; we do not imply a strict sequential logic between them, and

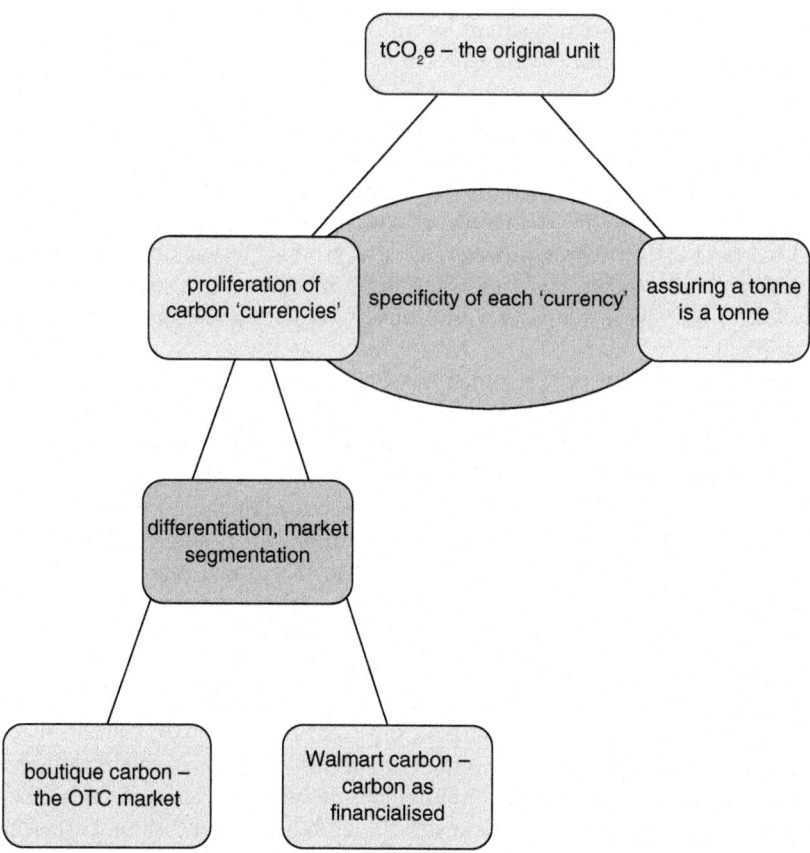

Figure 5.1 Moments in carbon's commodification

the historical elements presented in what follows is more messy than such a sequential arrangement might imply.

Carbon governmentalities and the virtue/virtuality relation

The creation of carbon markets entails a specific sort of technology of rule, a governmentality of carbon that defines the objects of government in a particular way and governs through certain techniques and devices. The government of carbon markets reflects many of the features governmentality scholars (Barry *et al.* 1996; Dean 2004; Miller and Rose 2008) already identify as key to 'advanced liberal government'. Liberal political reason underpins carbon governmentality. It is not primarily about governing in a totalising fashion, but about self-regulation. As social processes are decentralised to individuals in advanced liberalism, so achieving the goals of government entails elaborate biopolitical techniques of management at the macro-level (extensive data collection, planning processes, etc.) and at the micro-level, the engendering of ever more precise self-management by individuals. As markets, including carbon markets, decentralise decisions to individual subjects, shaping those subjects becomes ever more important for the smooth ordering of power. Carbon markets have, illustrating this dynamic well, entailed the elaboration of a dense set of power-knowledge processes to do with monitoring of and accounting for emissions, creating means of calculating emissions reductions from projects, enforcing rules (or better, designing self-enforcing rules), and so on (Stripple and Lövbrand 2010). These processes do not produce a neat and tidy design, but rather entail a heterogeneous assemblage of mechanisms, techniques and knowledges by which the natural and social world is represented, categorised and ordered.

For example, Lovell and Liverman (2010: 268) draw attention to how cookstoves are being reframed, from being a tool in development and poverty reduction programmes to a carbon offset technology. Hence, the average kitchen in the global South becomes established as a site of carbon offset production, an object to be governed through calculative practices (measurement, verification, certification, etc.), packaged in a standardised form and traded globally to 'neutralise' emissions produced elsewhere in the global North. Alongside macro-level biopolitical processes such as these, carbon markets also entail the production of certain sorts of subjectivities, creating people capable of participating in carbon markets and in the sorts of practices necessary for both carbon-entrepreneurialism and to be one's own 'carbon asset manager', a process we have elsewhere called 'My Space' (Paterson and Stripple 2010). Carbon markets can thus be understood to be one of the multiple centres of calculation and authority that shape the conduct of individuals in modern society. Descheneau and Paterson (2011) suggest that a key to the mobilisation of carbon markets is the intertwining of what they call 'desire and routine'. Carbon market actors are mobilised both by an affective process of the construction of desire (not only for specific products or services, but for the market as a whole) and by the borrowing of a variety of practices from other financial markets. The former generates an enthusiasm for carbon markets, while the latter renders the strange world of commodifying carbon familiar to financiers.

In this chapter we extend this sort of argument by borrowing from James Der Derian's idea of 'virtuous war' (Der Derian 2001). Before introducing the concept, let us first briefly introduce Der Derian and the ethico–political horizons that he challenges in his writings. Der Derian's work is situated in poststructuralist approaches in International Relations theory. Der Derian himself has, however, not spent much time on policing the borders of a particular (postmodern, poststructural, post-classical) approach to International Relations: 'whichever theorist helps me best understand the subject of my inquiry gets to the head of the class', as he once wrote (Der Derian 2000: 778). In the second edition of the book *Virtuous War* he describes the book not as a scholarly treatise on International Relations, but as a 'travelogue' through the military-industrial-media-entertainment network, pursued with a critical attitude, developed outside (geographically, culturally, educationally) of American politics and scholarship (Der Derian 2009). Jef Huysmans notes how Der Derian's interventions into the familiar worlds of world politics embodies two kinds of 'ethos of critique'. The first is the rhetoric of disruption, of defamiliarising familiar understandings of International Relations. The second is the 'exposition of dangers', where Der Derian is not looking for alternatives, new orthodoxies. His expositions are not constructed from an assumed firm external point of view, but emerge rather from the inside (Huysmans 1997). In *Virtuous War* Der Derian sheds light on why resistance does not work despite its intellectual value. Der Derian's writings arise out of a strong anti-militarist ethos, but nevertheless constitute an attempt to take seriously the internal normative logics of contemporary military practices.

In 'Virtuous war/virtual theory', Der Derian (2000) noted that at one time the two words virtual and virtuous were hardly distinguishable. Both words are derived from the medieval notion of power inherent in the supernatural, a divine being endowed with natural virtue. The two words carried a moral weight from the Greek and Roman sense of virtue, i.e. the properties and qualities of right conduct. In modern usage, the meanings of the two words diverge, with 'virtual' taking a morally neutral, more technical tone, while 'virtuous' has lost its sense of exerting influence by means of inherent qualities (Der Derian 2000: 771). Der Derian argues that the meaning of the two words now seems to have come together again through the American efforts to effect ethical change through technological and martial means. The 'virtuous war' thesis that Der Derian's book (2001) revolves around is about the belief that one can use military violence to resolve political problems. The concept of 'virtuous war' displays the affinity of a technical capability and an ethical imperative, to actualise violence 'from a distance', with no or minimal casualties. Digital information and virtual technologies bring 'there' (battlefields in Mogadishu and Baghdad) 'here' (to the US army's training centre in the Mojave Desert). The real becomes that which can be represented on the screen, a 'pixillation' of war that keeps death out of sight and mind. Der Derian argues that in virtuous wars, questions about what we *can* do and what we *should* do become mixed and the first tends to legitimate the second. These technological and representational forms of discipline and deterrence clean up the political discourse as well as the battlefield. Virtuous war projects a mythos as well as an ethos. The virtual constructs a world – not *ex nihilo* but *ex machina* – where there was none before.

It might seem far-fetched to use an argument developed to understand American hegemony and recent revolutions in military affairs to shed light on the carbon economy. What we conceptualise as 'virtuous carbon' is neither about a technological 'war on climate', nor about battlefields and deaths in distant territories. We are not interested particularly in the 'war' part of his analysis (although we accept that such an analysis could be undertaken, highlighting the violence in carbon markets or in climate change politics more broadly). Rather, we see Der Derian's argument as useful for making the connections between virtuality and virtue; the technological and the ethical in the construction of carbon markets. Drawing on this, we attempt to elaborate how this conceptualisation of 'virtuous carbon' might help us understand the dynamics of carbon market construction. We draw in this article from a diverse set of writings within governmentality studies, science and technology studies (including Actor Network Theory) and sociology of markets. Der Derian's neat insistence on 'simulation', breaking the difference between the real and the imaginary, between true and false, is particularly useful. As Huysmans notes, 'since reality is produced on a screen, an object is not physical or necessarily representational of a physical object but is made of pure data, pure information, and the information is produced through the exchange of symbols' (Huysmans 1997: 352). Similarly, there is a kind of 'pixillation of carbon', the ability to abstract carbon emissions from its complex social contexts and turn them into a digitised reality. Carbon emissions become an electronically tradeable unit on a trader's screen displaying (only) the current price and volume traded. This virtualisation of carbon simultaneously embodies the moral character, or the virtue, of the commodity being traded.

At each of the five moments we identify above, we contend, the processes involved can be understood through the double character of virtuality and virtue. It is about the imagination of a set of practices that can be brought into being (virtuality) and the ascription of a normative value to those imagined practices (virtue, what we can do, we should). A summary of how these work is elaborated in Table 5.1.

One of the important effects of this way of understanding carbon market politics is to show how the imaginations of carbon commodities are immediately normatively infused in ways that render resistance problematic. As is well known, there are major critiques made of carbon markets, opposing them variously on the grounds that they fail to reduce emissions ('climate fraud'), that they are essentially colonialist in the way they entail appropriation of

Table 5.1 Virtuality and virtue at each moment in carbon's commodification

Moment	Virtuality	Virtue
Inventing the tCO₂e	Technical capability to give carbon flows uniformity and comparability Equivalent metrics (e.g. GWP) The capital market perceiving and treating emission reductions as a commodity (e.g. carbon as an asset in Barclay's portfolio)	Imagining activities as being 'the same', no distinction between luxury and survival emissions. Blurring the line between judgement and calculation: new line drawn between that which can, and that which cannot, be calculated. Legitimacy of counterfactual reasoning
Proliferation of carbon units	Regulatory elaborations (e.g. Kyoto Protocol, Marrakesh Accords, EU ETS, REDD+). Carbon units as comparable assets, tradable for — and valued against — each other	The virtue of liberal environmentalism. Marketization as overarching ideology
Assuring the tonne is a tonne	The infrastructure of the trading system, e.g. information collection about the regulated entities. System designs, e.g. cap-and-trade, baseline-and-credit Certification standards Authority of third party verification	The virtue of carbon units understood as critical debates within science, policy and civic communities about the quality of the carbon units (HFC projects), additionality (wind power in China), baselines, carbon fraud in EU ETS, hot air in IET. Suspension of DOEs by CDM executive board, revision of allocation rules in the EU ETS
Boutique carbon	Imaginations of projects and methodologies. Displaying of the full 'production chain' Spread of carbon calculators	The virtue of particular reductions. Stories told about the benefits (poverty, development, local empowerment, gender) of the particular carbon unit. Relationships between buyer and seller
Walmart carbon	Financialised practices, e.g. derivatives (options, swaps, bonds) and arbitrage.	Pixillated carbon; a restricted range of information available on the screen. Carbon as an 'empty' unit, detached from climate mitigation as ethical duty

atmospheric space by the North and externalisation of emissions reductions via offset projects ('carbon colonialism'), that the atmosphere should simply not be commodified, that they act as a sort of 'new indulgences', with rich consumers assuaging their guilt via offsetting, or most recently that they constitute 'sub-prime carbon', with the possibility of bubble economies and collapses similar to the 2007–9 financial crisis.[1] These critiques have been widely adopted by what is often called the climate justice movement (see also Lane and Stephan, this volume).

So the virtuousness of carbon markets cannot be taken for granted. The point of the notion of 'virtuous carbon' is precisely that it describes discursive forms that circumvent or undermine these critiques. The range of commodities and business strategies around carbon markets both depend on and continually reinforce the naturalised normativity of such markets as *the* response to climate change. In the context of the 'new economy', Thrift suggests that 'effective social movements need to create background, a taken-for-granted world which, if you like, assumes the new economy's assumptions' (Thrift 2005: 117). This is redolent of Berman's more general account of modernity as ambivalence:

> This strategy [of the promoters of the 'expressway world'] was effective because, in fact, the vast majority of modern men and women do not want to resist modernity: they feel its excitement and believe in its promise, even when they find themselves in its way.
>
> (1982: 313)

Arguably, virtuous carbon attempts to reproduce this dynamic in relation to carbon markets. Virtual carbon has helped significantly to create this 'taken-for-granted' world. The performativity of carbon markets thus operates in part by conflating opposition to carbon markets with opposition to action on climate change (see more in Spash, this volume).[2] The notion of virtue/virtuality operates, as does Berman's account of the 'expressway world', at the level of affect and daily unconscious practice rather than at the level of self-conscious intellectual engagement – the virtue of carbon trading is reproduced in the daily practices of traders, policy-makers, and so on, reproducing normative identities, not through 'the force of the better argument'. If what you can do you should, then opposition to those possibilities cannot be countenanced.

Inventing the tCO_2e

Who invented the unit of a tonne of carbon dioxide equivalent (tCO_2e), and when? This history has not yet been written and what follows is our provisional take on these developments. A functionalist history of carbon markets could easily be outlined. In order to have a functioning market, so the story would go, you have to have something to buy and sell. Given that carbon markets are artifices of governments, one of the functions governments fulfil is to create the thing to be bought and sold. In the Kyoto negotiations, governments thus created the basic infrastructure of the carbon markets we now see. In particular they created the basic unit of account in (almost) all carbon markets[3] – the tonne of carbon dioxide equivalent, or tCO_2e. Such a history would, however, be misleading. The emergence of the tCO_2e as the unit of account was a much more messy and contingent process than the functional argument suggests. It starts with the notion of the Global Warming Potential (GWP). In the late 1980s, scientists felt the need to come up with a single measure by which all greenhouse gases (GHGs) could be compared to each other. They needed this in order to be able to build scenarios for future projections of climate change based on different trajectories of GHG emissions. Governments wanted these comparisons in order to be able to decide how they could respond most effectively. As they anticipated negotiations towards a treaty, they also wanted to be able to maximise freedom for manoeuvre by including as many GHGs as possible rather than just CO_2.

The scientists involved borrowed an idea from the ozone depletion issue, which had developed the notion of 'Ozone Depletion Potentials' as a means of comparing the various gases involved in that process (Smith and Wigley 2000). Rogers and Stevens (1988) first introduced the idea of a GWP and there was, in the late 1980s, a flurry of activity trying to come up with the best way to enable the commensuration. Lashof and Ahuja's article 'Relative contributions of greenhouse gas emissions to global warming' published in *Science* in 1990 became particularly influential in part because their approach was easily understandable for the policy-making world, not least because they used the term 'discount rates' instead of wrapping it in a language of physical differential equations. It became the basis for the version launched by the IPCC in its First Assessment Report (Shine *et al.* 1990: 58–61). The GWP takes carbon dioxide as its basic measure, giving it a GWP of 1, and gives other gases a GWP relative to that of carbon dioxide. While there were, and are, various controversies about the specifics of the GWP, the measure has nevertheless stabilised.

The stabilisation of the GWP, which MacKenzie (2009) eloquently dubbed as 'making things the same', became crucial for the future development of carbon markets, although this was not its aim at all. The GWP concept is a necessary abstraction from which the 'tCO_2e' could be imagined. It enables the incorporation of the six gases into the Kyoto Protocol, providing the measure of equivalence between them, and enabling all to be covered in the trading elements of Kyoto. But the GWP is still not that thing-like tangible unit, partly because GWP was construed to account for aggregate emissions. Early IPCC reports talk much more either about million tonnes of carbon or about ktC because they are talking either about global emissions and thus likely climate impacts, or national emissions to apportion responsibility. In the Kyoto Protocol, there is no mention of a single unit. All there is in the Protocol itself is a recognition that, if parties were to be allowed to trade emissions reductions through one of the Protocol's flexibility mechanisms, then two things would follow. First, the obligations to reduce those emissions needed to be expressed in terms of a right to a certain amount of emissions. This is what is expressed in Annex B of the Protocol, which details the emissions reductions obligations of industrialised countries in terms of a percentage of their 1990 emissions levels. So the EU has a right to emit 92 per cent of its 1990 levels in the 2008–2012 period, and so on. Second, those levels are referred to in the Protocol as 'assigned amounts' – the amount of emissions that each country can emit. Later in the negotiations, this becomes the basis for the Assigned Amount Unit (AAU), which is the unit traded under the emissions trading provision in the Protocol (article 6). So in legal terms, Kyoto introduces the terms 'emission reductions units', 'certified emission reductions' and 'part of an assigned amount' but does not define them, that was the work of the Marrakesh accord a few years later.

In the Kyoto Protocol, certified emissions reductions and emissions reductions units (assigned amounts never have 'unit' after them in the Protocol itself) are always spelled with lower case. Four years later in the Marrakesh Accords, we can see the unit coming into being:

> An 'emission reduction unit' or 'ERU' is a unit issued pursuant to the relevant provisions in the annex to decision -/CMP.1 (*Modalities for the accounting of assigned amounts*) and is equal to one metric tonne of carbon dioxide equivalent, calculated using global warming potentials defined by decision 2/CP.3 or as subsequently revised in accordance with Article 5.
>
> (UNFCCC 2001: 8)

A 'certified emission reduction' or 'CER' is a unit issued pursuant to Article 12 and requirements thereunder, as well as the relevant provisions in these modalities and procedures, and is equal to one metric tonne of carbon dioxide equivalent, calculated using GWPs defined by decision 2/CP.3 or as subsequently revised in accordance with Article 5 (UNFCCC 2001: 8).

The Marrakesh Accords reify the carbon units and spell them out as capitalised Certified Emissions Reductions, the CER, and Emission Reduction Units, ERUs.[4] This creation of a fictional single unit is the corollary of something that can be exchanged in a market process. If the tCO_2e is the equivalent of a hectare or acre in land, the abstract measure to be used, a CER or ERU is more like a specific plot of land that can be exchanged for another via the referent to a hectare. Each unit is given an exact value in the Marrakesh Accords, although the value is open for reformulations if GWPs change.[5] However, in the Marrakesh Accords there is no mention of 'tCO_2e' as a specific unit, instead each of the units is defined as equivalent to one tonne of carbon dioxide equivalents – spelled out. So it seems that the history is that the AAU, the ERU, the RMUs and the CER were invented before the tCO_2e. In effect, the latter is a bundling of the individual units into something more generic or abstract. If one does not put too much weight on the acronym 'tCO_2e', then there is a case for the co-invention of the tCO_2e with the Kyoto units AAU, ERU and CER since 'one metric tonne of carbon dioxide equivalent' is written out

16 times in the Accords. Compared to the GWP, the invention of the tCO_2e took the development one step further. It translated a scientific/nationalist way of accounting for carbon into one that could become fungible for market exchange. This was not an inevitable outcome but rather the product of messy negotiations about the details of the Kyoto Protocol, where many actors were still resisting the various flexibility mechanisms entailed in the Protocol.

The tCO_2e exemplifies the sort of imaginative abstraction Der Derian identifies as key to virtuous war. It concretises a whole series of imaginative moves (the GWP as scientific equivalence, the flexibility proposals in Kyoto, the desire to create a set of market processes around climate, in particular) and operates as a unit that combines each move while obscuring the messy history around each (the problematic nature of the GWP, the ethical contestation of flexibility mechanisms and potential resistance to the 'commodification of the atmosphere' and markets in nature). At the same time its abstraction operates normatively to (re)focus attention on the character of the challenge of climate change. This becomes oriented around reducing the amount of tCO_2e in the atmosphere, a task translated from the more immediate one of reducing GHGs by the way tCO_2e operates as a collectivity – it is not a single thing that can be directly measured, but has to be inferred and calculated in order to bring each tonne into existence. Those involved in the tCO_2e business gain legitimacy from their association with the signifier itself: one carbon market firm originally called itself simply CO2e.com.[6] Der Derian's notion of imaginative abstraction is similar to Callon and Muniesa's (2005) focus on 'calculation' in the construction of markets. Both sets of practices negotiate the boundary between the technological and the ethical–'the most appropriate dividing line is no longer between judgement and calculation, but between arrangements that allow calculation (either quantitative or qualitative) and those that make it impossible' (Callon and Muniesa 2005: 1232).

Proliferating carbon commodities

Once the logic of a single unit was established, it set up a sort of conceptual path dependence. Those establishing particular markets tended to adopt that unit as the basis for their specific units. More precisely, a series of carbon units have been created, and their creation is an ongoing exercise. We have a series of carbon 'asset classes' that can now be traded for and valued against each other – AAUs, CERs, EUAs, NZAs, ERUs, RGGI units, VERs, and so on. Each has a series of 'vintages', as they are only valid for a given year, but trading across these vintages (a CER 2011 for an EUA 2008, for example) is possible. Due to the transfer of financial sector techniques to carbon markets (Descheneau and Paterson 2011), many of these now also have simple derivative products – futures, options and swaps – that enable a secondary commodification process to take place. At root, these commodities are not carbon 'itself' but are either based on a commodification of a right to emit carbon (usually expressed as a tCO_2e or tonne of carbon dioxide equivalents) established by some sort of political authority (as in 'cap-and-trade' markets) or of a promise not to emit carbon (as in 'baseline and credit' markets).

The history of these units is largely the history of the scheme that creates them, and does not need elaborate retelling here (see in particular Betsill and Hoffmann 2011). The initial moment of proliferation is (again) in the negotiations just after the Kyoto Protocol was adopted. In Kyoto, negotiators had agreed three 'flexibility mechanisms' – emissions trading, Joint Implementation, and the CDM, providing different means for states to meet their obligations through investment abroad. Each was given its own unit, each of which was defined as a tonne of carbon dioxide equivalent (e.g. UNFCCC 2001). The simplest form of trading, emissions trading, was given the name coming out of the language in the Kyoto Protocol itself, that of assigned amounts, becoming the AAU. The CDM got the Certified Emissions Reduction while Joint Implementation got the Emissions Reduction Unit. Marrakesh defines each as being fungible to the other, under certain conditions. So we already at the time of Marrakesh in 2001 have an imagination of a set of differentiated forms of carbon that can be exchanged for each other. Shortly afterwards, other actors start to imagine adopting similar units (usually with the three-letter acronym) for schemes they start to develop. Table 5.2 outlines the units for those markets that are up and running.

The proliferation of carbon units has slowed recently, in particular because of stalled legislative processes in the US and Australia, and stalled negotiations on how the international climate regime will evolve post-Kyoto, but a number of other units are expected to come into being and some have been retracted. In 2013 nine additional cap-and-trade schemes started in North America, Central Asia and East Asia (five of them in China). The first compliance period under the California and Québec ETS began in January 2013 and while this is smaller than the initial plans for the Western Climate Initiative, other jurisdictions, most plausibly Ontario and British Columbia, might join in coming years. In January 2014 California and Québec linked their systems, which then became the first cap-and-trade scheme consisting of subnational jurisdictions

Table 5.2 Units in the compliance and voluntary carbon markets

Carbon market	Compliance market	Voluntary market
Cap and trade	Kyoto Emissions Trading (AAU), EU ETS (EUA) New Zealand ETS (NZA) Regional Greenhouse Gas Initiative (RGGI units) California/Québec	Chicago Climate Exchange (CCX units) Japanese Voluntary ETS (j-CER/JPA)
Baseline and credit	Clean Development Mechanism (CER) Joint Implementation (ERU)	Voluntary Carbon Offsetting (VER)

(ICAP 2014). The cap-and-trade system of the Chicago Climate Exchange (CCX) was closed down in late 2010, but CCX still operates as a carbon off-sets registry that facilitates 'carbon exchanges' based on a comprehensive set of verification protocols.[7] In the autumn of 2013, Australia's new government introduced a draft legislation to repeal their emissions trading scheme (see also Spash, this volume). In the coming years, it is possible that a trading system for Reduced Emissions from Deforestation and Degradation will emerge, as well as systems in a number of developing countries, notably South Korea in 2015, a few more in China and possibly India (see more in Lederer and Engels *et al.*, this volume). Another interesting trend is the use of an internal carbon price by companies as an incentive and strategic planning tool (CDP 2013). The 'pro-liferation moment' in the commodification of carbon entails the imaginative transformation of what might be a set of disparate systems into a single universe of commodities. These are united virtually not only through the technical commensuration of the tCO_2e, but also through their rendition into a simple set of acronyms, which give traders easy means of comparison and the ability to develop rules of thumb about their relationships. The proliferation of units produces a normalising tendency, reinforcing for traders the habit of compar-ing different units to engage in arbitrage, invention of derivative instruments, and so on. They are also united virtuously through their normative frames of reference: to trade a CER is to act virtuously in relations to climate change and, thus, to proliferate the various units is to increase the universe of possible climate change-positive actions, embedding the virtuous character of carbon trading more deeply in the daily practice of traders. But this claim to virtue rests on an underpinning of claims that relate these abstracted acronyms to the overall social purpose, an underpinning to which we now turn.

Assuring the tonne is a tonne

Having invented a commodity as an abstract, virtual entity, the central question that follows is an epistemological one. How do you know an EUA or CER when you see one? How do you know a tonne is a tonne? A central element in the invention of these carbon commodities has thus been the methods by which it is claimed that the commodity exists. Much of the literature on the construction of carbon markets has focused on this part of the process. Bumpus and Liverman (2008) discuss the emergence of certification standards, Lovell and Liverman (2010) examine audits and verification systems while Hoffman (2011) discusses the emergence of infrastructure systems (inventories, transac-tion logs and registries).

It is now conventional to distinguish carbon markets into cap-and-trade, or baseline and credit, markets. In cap-and-trade systems, the commodity is invented by fiat of a central authority (EU Commission, the CCX, the Kyoto Protocol). First, central to this creation is a system of information collection and management about the emissions of the regulated entities (states or firms, usually, although if a system of personal carbon allowances (PCA) ever takes off it will go down to the individual level). This infrastructure is crucial to the invention of the commodity in that it is the capacity to create credible claims about emissions levels that allows governors both to decide on appropriate obligations and for enforcement to take place. Second, governors have developed systems of allocation of allowances. These have tended to coalesce around the question of direct allocation versus auctioning although, in principle, many other modes of allocation are possible.

In baseline and credit markets, however, the invention of a carbon credit is much more elaborate. These markets essentially entail investment in projects that claim to reduce carbon emissions, and generate carbon credits on the basis of the reduced *projected* emissions. When this is to be done on the basis of an individual project, it entails considerable effort to demonstrate the emissions reductions. In the CDM, much of the effort has been around developing methodologies to demonstrate the 'additionality' of the project, which is a clear example of the logic of virtuous carbon. The crux of the matter, as Lohmann (2005) points out, is the establishment of a counterfactual – a claim about what the emissions would be without the project is an enormous imaginative exercise, involving the creation of a basic baseline (imagining the future emissions of a particular community without the project), and various scenarios for what will happen to emissions with the project – imagining not only the emissions from the project itself but whether there will be leakage elsewhere (if you build a windfarm somewhere to replace electricity from a coal plant, will others not just consume the coal plant's electricity?), whether there will be other displacement effects (see also Methmann and Stephan, this volume). The end result of all this imagination is the creation of a simple unit (CER, in the CDM's case) out of widely diverging socio-ecological practices that, in turn, enables market exchange of carbon commodities to take place.

The voluntary carbon market has been rapidly evolving through several different standards that assure the tonne is a tonne in slightly different ways. A couple of standards are currently dominant: the CDM Gold Standard, developed under the lead of environmental NGOs and launched in 2003, offers the most ambitious high-quality carbon standard at present. Credits certified by the Gold Standard are called GS VER and trade at prices higher than for other carbon currencies. The Verified Carbon Standard (VCS) is the outcome of an extensive consultation process within the carbon market business community. The VCS was launched in November 2007 and the certified credits are called Verified Carbon Units. The California Climate Action Registry established, in 2008, the Climate Action Reserve (CAR) standard for the United States market. The credits are called Climate Reserved Tonnes. Finally, the American Carbon Registry was established as a standard in 2008 by, for example, the Environmental Defense Fund and the credits are called Emission Reduction Tons.

As this brief overview of the technologies of carbon government has shown, each simulation, each of the exercises in imaginative abstraction is simultaneously normatively fused. Carbon market promoters express the logic of 'because we can, we should'. For example, Bayon *et al.* (2007) state that voluntary carbon markets are the means that individuals or corporations can participate in the fight against climate change (forgetting that they could simply reduce their own emissions!). The ethical tensions are perhaps most visible in the commodification of forest ecosystems. This imagination depends on the assignment of a property title over emission reductions to a named individual, group or institution, which can then dispose of the forest carbon offset as they wish and exchange for money. But the assignment of property rights over forest carbon is only possible after the tonnes of sequestered carbon have been appropriately accounted for. Since there are large uncertainties and scientific controversies around carbon exchanges in forest and soils, large work is currently invested into forging consensus and build trust among various research and policy communities with regard to standards, methods and requirements that national forest monitoring and carbon accounting systems will deploy (see Methmann and Stephan as well as Lovell and MacKenzie, this volume).

'Boutique carbon'

Our final two moments of virtuous carbon refer to the differentiation of markets into those that operate around the particularities of projects to reduce emissions and those that are 'fully commodified' in the sense that financiers understand the term. 'Boutique' carbon (in contrast to what we call below Walmart carbon) refers to those sorts of trades where the individual character of the project, or the relationship between buyer and seller, matters strongly. It is most present in offset markets, although AAU trades between states also have this sort of boutique character. In fact this boutique character of AAU trades is one reason the emissions trading part of Kyoto has not taken off particularly strongly. Relations between buyer and seller can be negative as well as positive, and given that most AAU purchases would be from Russia or Ukraine, this has in part inhibited such exchanges. In Canada, for example, who would have been the biggest buyer of AAUs if the government had not decided not to try to meet its Kyoto obligations, part of the discourse rejecting Kyoto by the incoming Conservative government in 2006 was a desire not to buy AAUs from Russia, and a preference for what the government called a 'made in Canada' solution instead.

Here, the virtualisation is more closely linked to the imagination in the methodologies, the imaginations of particular projects and the emissions reductions that are claimed for them. As Lovell and Liverman (2010) point out, offset providers routinely tell stories about their individual projects in their publicity, and focus on the relationships with project developers (almost always in developing countries). The virtualisation is thus more closely connected with the virtue of particular emissions reductions, in contrast with Walmart carbon's abstraction to the notion of a carbon price and its active disassociation of information about how the credit was produced. Boutique carbon is also connected to a virtue of face-to-face contact – between offset provider and purchaser, between provider and project developer, in particular.

This particularistic form thus permits all sorts of novel relationships to be established around particular ways of imagining the building of carbon markets. It can work through particular relationships as in the way that offset providers have built extensive links with airlines, enabling firms and individuals to offset that particular source of emissions. This relationship has a semi-arbitrary quality: while it is the case that air travel accounts for a great deal of the emissions of those who fly, and is also the source of emissions for which there is as yet no feasible technofix (so its emissions can only be offset – the alternative is not to fly), it has arguably thrived as a marketing strategy as much because of the technical ease of integrating offset purchases into flight purchases. It has also generated other sorts of infrastructural and discursive innovations. The spread of carbon calculators where individuals can calculate their own emissions has occurred, in part, as an extension of offsetting logic – if you are to offset, you need to know how much you need to offset – and offset providers now mostly have their own carbon calculator or refer clients to a specific one. Discursively, the virtue of going 'carbon neutral' has spread as the end state of a fully offset life.

'Walmart' carbon

With Walmart carbon, by contrast, the carbon is entirely abstracted from its conditions of production – the various methodologies and infrastructures discussed in the previous section. It is, in Marx's terms, fully 'fetishised' (Lohmann 2010) – abstracted from the conditions of production that make it possible. Financiers understand the various forms of the carbon commodity – EUA, CER, and so on – as fungible 'asset classes'. The original commodities are supplemented by a range of (for the moment, relatively modest) derivative instruments like futures, options and swaps. Carbon is thus imagined within a world of financial trading already familiar to those in that world. Carbon becomes virtualised through the representational device of the computer screen, with traders observing price changes and engaging in arbitrage on short-term price changes in different carbon commodities (and outside carbon markets also – observing changes in the closely related markets of electricity, natural gas, and oil). As in other financial markets, the screen conveys a closely restricted range of information – the prices of the various commodities, and a range of requests for sales or purchases by clients that the broker is trying to respond to. The logic of 'because you can, you should' is instantaneously operationalised through the response to the stimuli on the screen – to sell or buy according to the conditions, to beat a competitor to a deal.

The 'Walmart moment' also entails a virtue almost entirely abstracted from its climate change context. The universalising climate change ethical imperative becomes re-read through the universalising economic discourse – the Smithian presumption that if economic agents pursue their own interests (buying or selling carbon according to their requirements) then the global public good will be produced. This 'invisible hand' logic only operates in the moment of buying and selling – outside this moment actors are acutely conscious that carbon markets are politically driven 'fictitious commodities'. Demand and supply for carbon allowances and credits are driven by policy in various countries and by international agreements. But in the moment, the aim is to get the best deal, to drive a price down or up depending on your position, and the climate change virtue, its related normativity, arises out of the 'carbon price' that is the result of the aggregation of all this trading activity.

Conclusion

This chapter provides an overall framework for thinking about the construction of carbon markets. We draw together emerging research on how particular carbon markets get built to provide a fuller understanding of the way these various elements fit together, and thus how the 'actually existing' carbon economy has been assembled and how it becomes possible as a political imagination. We advance an analysis of how the carbon economy comes into being without insisting on the state, or the system of states, as its locus and origin. Our analysis grapples with the practices of governing themselves, e.g. in terms of forms of calculation and methodologies for categorising projects and certifying emission reductions. What is particular about the carbon economy is that it becomes an international reality by being 'virtuous', a figure of thought that we borrow from James Der Derian and his notion of 'virtuous war'. Our figure of 'virtuous carbon' highlights carbon's fictitious character (virtuality) as well as the possibility for ethical contestation at every moment of commodification (virtue). Once upon a time, virtue and virtual meant almost the same thing (the properties and qualities of doing the right thing). What you can do was what you also should do. Carbon markets, as centres of calculation and authority that shape the conduct of individuals and communities (firms, states, regions, etc.), display a reassembling of the virtual and the ethical. The carbon economy thus becomes 'the right thing', a divine being endowed with natural virtue that limits the space for critical engagement.

Acknowledgements

The authors would like to thank Harro van Asselt, Björn Badersten, Joanna Depledge and Bo Wiman for helpful suggestions on various parts of the text. We are also grateful for comments from the participants at the SGIR 7th Pan-European International Relations Conference: 9–11 September 2010. The study has been made possible through financial support from BECC, the Swedish government's strategic focus on climate change hosted by the Faculty of Science at Lund University and the Low-Carbon Energy and Transport Systems (LETS) project, financed by the Swedish Environmental Protection Agency (among others). This chapter was originally published as Paterson, M. and Stripple, J. (2012) Virtuous carbon. *Environmental Politics*, 21 (4), 563–582, but this version is revised to include recent developments in the carbon economy. We would like to thank the editors for useful comments in preparation of the revised version.

Notes

1 On this, see variously Lohmann (2006), Smith (2007), Bachram (2004), Böhm and Dabhi (2009), Gilbertson and Reyes (2009) or Chan (2009).
2 One of us (MP) had one striking conversation with a journalist working in carbon markets, who, even with an MA in International Political Economy, and very familiar and sympathetic to neo-Gramscian approaches, asked, when I was discussing anti-carbon market activism, 'Why would anyone oppose carbon markets?'
3 The current exceptions are the Alberta offset market and the Regional Greenhouse Gas Initiative (RGGI), which is based on an American or 'short' ton rather than a metric tonne.
4 There is also a third, the Removal Unit, or RMU applicable to sinks within industrialised countries. However, it has never become a significant part of carbon markets.
5 Yamin and Depledge (2004: 79) have a table showing the changes of GWPs between 1990 and 1995. For the 2008–2012 commitment period, Article 5.3 in the Protocol sets out a political decision to the 1995 values for reporting purposes, even though the IPCC can update the GWPs on the basis of new scientific knowledge.
6 The company CO2e.com was part of the Cantor Fitzgerald group. The firm was, in August 2011, acquired by BGC Partners, L.P. and became BGC Environmental Brokerage Services, L.P. Read more about them here http://www.bgcebs.com/AboutUs/
7 Even though the CCX is now defunct, we keep it in the table because it is a clear example of a voluntary cap-and-trade scheme. While the scheme in California has had a 'soft start' it will be mandatory for the covered entities. Starting in the second compliance period (2015–2017), the cap-and-trade programme will cover sources responsible for 85 per cent of California's GHG emissions (ICAP 2014).

References

Asselt, H. van and Biermann, F. (2007) European emissions trading and the international competitiveness of energy-intensive industries: a legal and political evaluation of possible supporting measures. *Energy Policy*, 35 (1), 497–506.

Bachram, H. (2004) Climate fraud and carbon colonialism: the new trade in greenhouse gases. *Capitalism, Nature, Socialism*, 15 (4), 5–20.

Barrett, S., Grubb, M., Rolland, K., Rose, A., Sandor, R. and Tietenberg, T. (1992) *Study on a Global Scheme for Tradeable Carbon Emission Entitlements. Tradeable Entitlements for Carbon Emission Abatement* (Project INT/91/A29). Geneva: UNCTAD.

Barry, A., Osborne, T. and Rose, N., eds. (1996) *Foucault and Political Reason: liberalism, neo-liberalism and rationalities of government.* London: UCL Press.

Bayon, R., Hawn, A. and Hamilton, K. (2007) *Voluntary Carbon Markets.* London: Earthscan.

Berman, M. (1982) *All That is Solid Melts into Air: the experience of modernity.* London: Verso.

Betsill, M. and Hoffmann, M. (2011) The contours of 'cap and trade': the evolution of emissions trading systems for greenhouse gases. *Review of Policy Research*, 28 (1), 83–106.

Böhm, S. and Dabhi, S., eds. (2009) *Upsetting the Offset: the political economy of carbon markets.* Colchester: Mayfly Books.

Brunnengräber, A. (2007) The political economy of the Kyoto Protocol. *Socialist Register*, 43, 224–225.

Bumpus, A. and Liverman, D. (2008) Accumulation by decarbonization and the governance of carbon offsets. *Economic Geography*, 84 (2), 127–155.

Callon, M. (2009) Civilizing markets: carbon trading between in vitro and in vivo experiments. *Accounting, Organizations and Society*, 42 (3–4), 535–548.

Callon, M. and Muniesa, F. (2005) Peripheral vision: economic markets as calculative collective device. *Organization Studies*, 26 (8): 1229–1250.

CDP, Carbon Disclosure Project (2013) Use of internal carbon price by companies as incentive and strategic planning tool. Available from: www.cdp.net/CDPResults/companies-carbon-pricing-2013.pdf [accessed 27 February 2014].

Chan, M. (2009) *Subprime Carbon? Re-thinking the world's largest new derivatives market.* Washington, DC: Friends of the Earth.

Dean, M. (2004) *Governmentality: power and rule in modern society.* London: Sage Publications.

Der Derian, J. (2000) Virtuous war/virtual theory. *International Affairs*, 76 (4), 771–788.

Der Derian, J. (2001) *Virtuous war: mapping the military-industrial-media-entertainment network.* Boulder, CO: Westview Press.

Der Derian, J. (2009) *Virtuous war: mapping the military-industrial-media-entertainment network.* 2nd Edition. New York: Routledge.

Descheneau, P. and Paterson, M. (2011) Between desire and routine: assembling environment and finance in carbon markets. *Antipode*, 43 (3), 662–681.

Ellis, K., Winkler, H., Corfee-Morlot, J. and Gagnon-Lebrun, F. (2007) CDM: taking stock and looking forward. *Energy Policy*, 35 (1), 15–28.

Gilbertson, T. and Reyes, O. (2009) Carbon trading: how it works and why it fails. *Critical Currents*, Volume 7. Uppsala: Dag Hammarskjöld Foundation.

Grubb, M. (1989) *The Greenhouse Effect: negotiating targets.* London: Royal Institute of International Affairs.

Hoffmann, M. (2011) *Climate Governance at the Crossroads: experimenting with a global response after Kyoto.* Oxford: Oxford University Press.

Huysmans, J. (1997) James Der Derian: the unbearable lightness of theory. *In:* O. Waever and I. B. Neumann, eds., *The Future of International Relations: masters in the making?* New York: Routledge, 367–358.

ICAP, International Carbon Action Partnership (2014) Emissions Trading Worldwide. *Status Report 2014.* Available from: www.icapcarbonaction.com/news-archive/209-emissions-trading-worldwide-icap-status-report-2014 [accessed 27 February 2014].

Lane, R. (2012) The promiscuous history of market efficiency: the development of early emissions trading systems. *Environmental Politics*, 21 (4), 583–603.

Lashof, D. and Ahuja, D. (1990) Relative contributions of greenhouse gas emissions to global warming. *Nature*, 344, 529–531.

Lohmann, L. (2005) Marketing and making carbon dumps: commodification, calculation and counterfactuals in climate change mitigation. *Science as Culture*, 14 (3), 203–235.

Lohmann, L. (2006) Carbon trading: a critical conversation on climate change, privatization and power. *Development Dialogue*, 48, 1–359.

Lohmann, L. (2010) Commodity Fetishism in Climate Science and Policy (in which various men with beards are enlisted to help explain why official efforts to address climate change have reached an impasse). *Presentation at Imperial College*, London, February 2010.

Lövbrand, E. and Stripple, J. (2011) Making climate change governable: accounting for carbon as sinks, credits and personal budgets. *Critical Policy Studies*, 5 (2), 187–200.

Lovell, H. (2010) Governing the carbon offset market. *Wiley Interdisciplinary Reviews: Climate Change*, 1 (3), 353–362.

Lovell, H. and Liverman, D. (2010) Understanding carbon offset technologies. *New Political Economy*, 15 (2), 255–273.

MacKenzie, D. (2009) Making things the same: gases, emission rights and the politics of carbon markets. *Accounting, Organizations and Society*, 34 (3–4), 440–455.

Meckling, J. (2011) *Carbon Coalitions: business, climate politics, and the rise of emissions trading.* Cambridge: MIT Press.

Methmann, C., Rothe, D. and Stephan, B., eds. (2013) *Interpretive Approaches to Global Climate Governance: (de)constructing the greenhouse.* London: Routledge.

Miller, P. and Rose, N. (2008) *Governing the Present.* Cambridge: Polity Press.

Newell, P. and Paterson, M. (2010) *Climate Capitalism: global warming and the transformation of the global economy.* Cambridge/New York: Cambridge University Press.

Oels, A. (2005) Rendering climate change governable: from biopower to advanced liberal government? *Journal of Environmental Policy and Planning,* 7 (3), 185–207.

Paterson, M. and Stripple, J. (2010) My Space: governing individuals' carbon emissions. *Environment and Planning D: Society and Space,* 28 (2), 341–362.

Paulsson, E. (2009) A review of the CDM literature: from fine-tuning to critical scrutiny? *International Environmental Agreements: Politics, Law and Economics,* 9 (1), 63–80.

Peeters, M. (2006) Enforcement of the EU greenhouse gas emissions trading scheme. *In:* M. Peeters and K. Deketelaere, eds., *EU Climate Change Policy: the challenge of new regulatory initiatives.* Cheltenham: Edward Elgar, 169–187.

Rabe, B. (2008) *The Complexities of Carbon Cap and Trade Policies: early lessons from the States.* Governance Studies Paper. Washington, DC: The Brookings Institute.

Reyes, O. (2011) Zombie carbon and sectoral market mechanisms. *Capitalism, Nature, Socialism,* 22 (4), 117–135.

Rogers, J. D. and Stephens, R. D. (1988) Absolute infrared intensities for F-113 and F-114 and an assessment of their greenhouse warming potential relative to other chlorofluorocarbons. *Journal of Geophysical Research,* 93, 2423–2428.

Ruggie, J. G. (1998) What makes the world hang together? Neo-utilitarianism and the social constructivist challenge. *International Organization,* 52 (4), 855–885.

Shine, K. P., Derwent, R. G., Wuebbles, D. J. and Morecrette, J.-J. (1990) Radiative forcing of climate. *In:* J. T. Houthdon, G. J. Jenkins and J. J. Ephraums, eds., *Climate Change: The IPCC Assessment.* Cambridge: Cambridge University Press for the Intergovernmental Panel on Climate Change, 41–68.

Skjærseth, J. B. and Wettestad, J. (2008) *EU Emissions Trading: initiation, decision-making and implementation.* Aldershot: Ashgate.

Smith, K. (2007) *The Carbon Neutral Myth: offset indulgences for your climate sins.* Amsterdam: Transnational Institute.

Smith, S. and Wigley, T. (2000) Global Warming Potentials: 1. Climatic implications of emissions reductions. *Climatic Change,* 44, 445–457.

Spaargaren, G. and Mol, A. (2013) Carbon flows, carbon markets, and low-carbon lifestyles: reflecting on the role of markets in climate governance. *Environmental Politics,* 22 (1), 174–193.

Stephan, B. (2012) Bringing discourse to the market: the commodification of avoided deforestation. *Environmental Politics,* 21 (4), 621–639.

Streck, C. and Lin, J. (2008) Making markets work: a review of CDM performance and the need for reform. *European Journal of International Law,* 19 (2), 409–442.

Stripple, J. and Bulkeley, H., eds. (2014) *Governing the Climate: new approaches to rationality, power, and politics.* Cambridge: Cambridge University Press.

Stripple, J. and Lövbrand, E. (2010) Carbon market governance beyond the public-private divide. *In:* F. Biermann, P. Pattberg and F. Zelli, eds., *Global Climate Governance Post 2012: architectures, agency and adaptation.* Cambridge: Cambridge University Press, 165–183.

Thrift, N. (2005) *Knowing Capitalism.* London: Sage.

UNFCCC (2001) Report of the Conference of the Parties on its Seventh Session, held at Marrakesh from 29 October to 10 November 2001, Addendum, Part Two: Action Taken by the Conference of the Parties. Bonn: United Nations Framework Convention on Climate Change.

Yamin, F. and Depledge, J. (2004) *The International Climate Change Regime: a guide to rules, institutions and procedures.* Cambridge: Cambridge University Press.

Part II

The politics of carbon

6 A neo-Gramscian account of carbon markets

The cases of the European Union Emissions Trading Scheme and the Clean Development Mechanism

Elah Matt and Chukwumerije Okereke

Introduction

Over the past decades, carbon commodification has emerged as the chief solution for addressing global climate change. To this end, a range of 'new' (Jordan *et al.* 2003, 2005), often market-based policy instruments have been deployed across many jurisdictions and hailed as credible approaches for addressing climate change (Ellerman and Harrison 2003; Victor *et al.* 2005; Wara and Victor 2008). Emissions trading schemes, Joint Implementation (JI), regulatory-compliance and voluntary carbon markets, corporate targets, rating and disclosure, are some notable examples of these market-creating policy instruments.

Despite their increasing popularity, the legitimacy and effectiveness of these instruments are vigorously challenged, and counter-measures are dotted across the climate governance landscape (Fuhr and Lederer 2009; Hoffmann 2011; Paterson 2009). Agitations over climate justice or, more broadly, the demand for just transitions to a low-carbon society, provide a platform for articulating the ineffectiveness of these market-based practices and challenging their orthodoxy (Hayward 2007; Lohmann 2008; Okereke and Dooley 2010). Across different geographies of scale, scholars and advocacy groups, as well as practical experience, highlight the inherent contradictions and inequitable outcome of carbon markets (Bachram 2004; Bond 2011; Borger 2012; Burkett 2008). At the same time, the effectiveness and economic efficiency of market approaches to climate governance, which have been their main selling point, have also been brought into question (Böhm *et al.* 2012; Lohmann 2006).

Regardless of these potent challenges and contradictions, there is little ground for optimism that states or civil society will press ahead to devise radical regimes that will lead to significant decarbonisation of the global economy in the near future. The main reason is that a commitment to 'The Good Life', defined mainly in terms of economic growth and continued capital accumulation, appears to run deeply in the consciousness of both the producers and the recipients of carbon pollution and climate impacts.

In this chapter, we offer a neo-Gramscian perspective for understanding the emergence and persistence of carbon markets as key instruments for addressing global climate change. Following neo-Gramscian insights, we argue that the introduction and prevalence of carbon markets are best understood as a compromise among competing societal and political actors. Although carbon commodification is presented as working to the benefit of all, it continues to serve the interest of the most powerful configuration of actors. Central to this explanation is the Gramscian concept of hegemony, as well as cognate notions such as war of position, passive revolution, and the historical bloc – all of which are explained in the following section.

Utilising this neo-Gramscian perspective, we then examine the European Union Emissions Trading Scheme (EU ETS) and the Kyoto Protocol's Clean Development Mechanism (CDM) as prominent case studies of the marketisation of key climate governance instruments. The EU ETS is of interest for at least two reasons. First, it is one of the highest profile climate instruments, implemented at multiple levels of governance, and the largest functioning emissions trading scheme globally (Braun 2009). Second, tracing the policy process of the scheme allows us to examine the development of EU and international climate politics for over two decades. Equally, the CDM is a representative programme, which held promise for reconciling the need for flexibility in emissions reduction approaches on the one hand, and technology transfer and sustainable development on the other. Since the EU ETS and the CDM are emblematic of broader developments in international climate governance, they offer a powerful snapshot of the difficulties in creating an effective climate regime to date. Finally, we offer a neo-Gramscian assessment of the effectiveness of these climate instruments, as well as a justice critique.

Introducing a neo-Gramscian political economy approach

Antonio Gramsci developed Marxist political theory in the early twentieth century (Mouffe 1979). He inspired generations of political theorists and International Relations scholars (Bieler and Morton 2001; Cox 1981, 1983; Cox and Sinclair 1996; Jessop 1990; Mouffe 1979; Showstack Sassoon 1982, 1987, to name a few). Drawing on Gramsci's work, most of which was written in an Italian prison and was often fragmentary and difficult to interpret, neo-Gramscian scholars provide a range of theoretical perspectives inspired by his work (Bieler and Morton 2001). The strength of these approaches lies in their ability to explain relations of power among a multitude of public and private policy actors, and across multiple spatial scales. Central to their analysis is the examination of the role of material, organisational and discursive practices in shaping societal relations of power. Particularly insightful and analytically useful are the notions of hegemony, the historical bloc, passive revolution, and war of position.

Hegemony and the historical bloc

The notion of hegemony is perhaps Gramsci's most influential conception (Cox 1981; Femia 1981; Forgacs 1988; Levy and Egan 2003; Mouffe 1979).

Gramsci used this concept to explain how economic interest groups gained and maintained dominance in modern capitalist societies, without the need for overt class struggle. Contrary to the Realists' understanding of hegemony that emphasises state-centred coercive hierarchy, and World-System approaches that emphasise material hierarchy and class domination, a neo-Gramscian account of hegemony posits a far more complex and dialectical relationship between the elite, state, and civil society, based on ideological and consensual leadership (Wittneben *et al.* 2012).

For Gramsci, hegemony is successfully established when a dominant class links its interests with those of subordinate classes, in the pursuit of a social order that reproduces its own dominant position. In other words, hegemony is effectively established when the interest of the dominant class is accepted as the universal interest of society (Cox 1983: 126–127; Gramsci 1971: 181). This implies that the state-elite need not enforce discipline through coercion. Rather, hegemonic stability is rooted in consensus and manifested in the legitimacy and universal acceptance of the core material, ideological, and social logic underpinning the polity. The result is a common social and moral language, an inter-subjective identity that is supportive of the prevailing order, and one dominant concept of reality 'informing with its spirit all modes of thought and behaviour' (Gill 2003: 58).

Thus, while hegemony is rooted in the economic sphere, it is expressed in the realms of civil society and its institutions (Anderson 1976: 18; Bates 1975: 353–357). The church, media, academia, non-governmental organisations (NGOs), trade unions and other civil society institutions all perform a crucial role in promoting and perpetuating the social order through ideological acquiescence and performances (Bates 1975: 353–357). Civil society, in Gramscian terms, is not an arena of social and industrial activity separate from political life. It is rather a state–society complex, the ideological superstructure, which, through its institutions and ideological functions, creates and diffuses dominant modes of identity and thought. This kind of consensus is made possible because many within this state–society complex have come to accept the hegemonic project as their own, even though in critical terms the project serves to reproduce the dominance of the ruling elite (Levy and Newell 2005: 50). To this end, Gramsci applied the notion of the extended state, which comprises civil society and political society (Davies 2011: 117).

Gramsci referred to the alignment of social groups and the concomitant material, organisational and discursive practices as the historical bloc. A historical bloc is configured of state authority, economic dominance and civil society legitimacy. It is more than the alliance among these groups; it is also 'the specific alignment of material, organisational, and discursive formations that stabilise and reproduce relations of production and meaning' (Levy and Newell 2005: 50).

Of particular importance in shaping the historical bloc are organic intellectuals (Bieler 2002: 581; 2006: 124; Gramsci 1971: 3). These intellectuals are 'organically linked to a specific social group'. They include politicians, scholars, journalists, industry representatives and members of NGOs (van Apeldoorn 2002: 30–31). Organic intellectuals give each social group 'homogeneity and an awareness of its own function, not only in the economic but also in social and political fields' (Gramsci 1971: 5; see also Levy and Egan 2003: 808–809). These actors 'frame transformations in a way that make[s] sense to the public at large' (Andrée 2011: 176).

Although dominance rests in the leadership and acquiescence of the state–society complex, civil society institutions simultaneously constitute the key site of political contestation, primarily because of their partial autonomy from the economic structures and bureaucratic authority of the state. From a neo-Gramscian perspective, civil society therefore has a dual role: it is a part of the 'extended state', complementing the disciplinary and universalising tendency of the capitalist state, and at the same time it is an arena for counter-hegemonic discourses and struggle (Levy and Newell 2002: 87).

Hegemony is thus contingent and accommodative. Power is neither static nor zero-sum, but resides in part in the strategies and discursive ability of constituent groups and institutional entrepreneurs. There is plenty of room for manoeuvre and reconfiguration of interests, coalitions and alliances. Ultimately, however, transformational reform possibilities are limited and firmly circumscribed by the economic superstructure and moral ideologies, favouring existing power hierarchies. A hegemonic order thus evolves through dialectical processes of contestation and compromise among competing societal groups (Bieler 2002: 581; Jessop 1982: 142; van Apeldoorn 2002: 20; van Apeldoorn et al. 2003: 36–37). Political power is concurrently maintained through compromises and alliances among these groups.

The war of position and passive revolution

Gramsci distinguished between the strategies employed by the dominant class and those employed by subordinate groups to gain influence within the historical bloc. The war of position, often employed by subordinate groups, entailed gaining influence through action within civil society. It 'constitutes a longer term strategy, coordinated across multiple bases of power, to gain influence in the cultural institutions of civil society, develop organisational capacity, and win new allies' (Levy and Newell 2005: 51). The war of position requires building alliances, organisational capacity, and germinating alternative ideologies in the institutions of civil society (Femia 1981: 52; Showstack Sassoon 1982: 113; Simon 1991: 75).

In contrast, passive revolution strategies are often deployed by the dominant class to capture, redirect or neutralise the impetus for radical change (Forgacs 1988: 224; Morton 2007: 97). Gramsci (1971: 115) described the passive revolution as 'the political form whereby social struggles find sufficiently elastic frameworks to allow the bourgeoisie to gain power without dramatic upheavals'. The concept refers to social, economic and political reforms that occur through consent rather than coercion (Adamson 1980: 186; Cox 1983: 129). It relates to the 'reorganization of economic, political, and ideological relations, often in response to a crisis that maintains the passivity of subordinate groups, and the separation of leaders and led' (Jessop 1982: 150; see also Showstack Sassoon 1982: 129). Passive revolution relies on 'extensive concessions' (Levy and Newell 2005: 51) that forestall more comprehensive challenges from other social groups, and thus serve to reproduce the dominance of the hegemonic group (Rupert 1993: 81).

The starting point for our analysis is that climate change poses a threat to the hegemony of the 'carboniferous' historical bloc (Dalby and Paterson 2009). This bloc is dominated by fossil-fuel-reliant fractions of capital, such as oil refiners, car manufacturers, and electricity producers. The operations and profitability of these groups are challenged by demands for climate-change mitigation and decarbonisation of the economy (Newell and Paterson 2010). These actors are therefore 'highly interested in the type and character of mitigation measures and strategies taken by governments' (Stephan 2011: 9). The historical bloc also comprises government allies, civil society groups and organic intellectuals; and concomitant material, organisational and discursive practices (Levy and Egan 2003: 806; Levy and Newell 2005: 50). Given this scenario, we suggest that much of climate politics across geographical scales will consist mainly of efforts, on the one hand, by progressive coalitions to implement effective climate-change policies, and on the other hand, by the 'carboniferous' bloc to de-radicalise and possibly 'tame' emergent climate governance regimes.

This perspective offers an alternative to other approaches that explain climate-change policy processes, such as those that emphasise diffusion and policy learning (Braun 2009), state-based entrepreneurial leadership (Hovi *et al.* 2003; Skjærseth and Wettestad 2010), policy innovation (Voß 2007), advocacy coalitions (Michaelowa 2008), and policy windows (Buhr 2012). While these approaches offer useful insights, their main weakness, as Stephan (2011: 3) aptly argues, is that 'they do not problematise the power structure at play', nor do they take serious 'account of the material or discursive structural context' underpinning the design and implementation of given policies (ibid.).

Our analysis complements a number of neo-Gramscian interpretations of environment and climate-change governance that have emerged in recent years (Levy and Egan 2003; Matt 2012; Newell 2008: 522; Okereke *et al.* 2009; Stephan 2011). We draw inspiration from these works and aim to further a neo-Gramscian understanding of climate-change governance. Particularly relevant is Stephan's (2011) analysis of the EU ETS, which provides an excellent starting point for our chapter. We draw on his work and extend it both theoretically, by examining in more detail the notion of the war of position, and empirically, by adding the case study of the CDM.

Carbon commodification: the EU ETS

The EU ETS was the first international carbon market, and remains the largest emissions trading scheme operating globally (Braun 2009: 470). To date, the scheme encompasses over 11,000 industrial constellations and power plants in 31 countries (DG Climate Action 2013). Although the EU pioneered the implementation of a regional carbon emissions trading scheme, it did not initially advocate this policy instrument. As Stephan (2011: 3) observes, between 1991 and 1999, the EU 'turned from a sceptic and opponent of emission trading into the biggest advocate for the policy tool' (see also Ellerman and Buchner 2007: 67).

From the emergence of climate change as a policy problem on the EU's agenda in the late 1980s and early 1990s, the EU adopted a proactive approach to climate governance, influenced by several factors. First, environmental awareness among the European public was high, following concerns regarding acid rain, the depletion of the ozone layer and long-range transboundary air pollution (Sprinz 1992). Subsequent treaties and policies resulted in growing confidence that the EU could design effective measures to manage challenging environmental problems. Second, EU-based Environmental Non Governmental Organisations (ENGOs), such as Greenpeace and Friends of the Earth, were quick to embrace and popularise the science of climate change and related potential social impacts. They were organised in their advocacy and role as policy entrepreneurs at the EU-level and internationally. These organisations played a significant role in organising some of the first wave of international conferences on climate change, such as the World Conference on the Changing Atmosphere in Toronto in June 1988, and the 1989 The Hague and Noordwijk Conferences (Paterson 1996). Third, the EU saw climate change as an opportunity to gain credentials as an international 'climate leader', while reaping economic, social and political benefits from the promotion of a low-carbon economy (Wurzel and Connelly 2011).

Initially, the EU supported a carbon tax as its instrument of choice for addressing climate change (CEC 1992; Haigh 2011; Jordan and Rayner 2010: 59). However, the tax was aborted due to a range of institutional and political barriers. Particularly, business groups were able to rely on their privileged position and access to politicians in order to mount a successful lobby against this policy instrument, which threatened their economic operations (Braun 2009: 473). Opposition to the tax was also noted among a number of Member States, and particularly the UK (MacKenzie 2009). Consequentially, the EU's attempts to introduce a carbon tax were unsuccessful, and left a void in its climate-change strategy.

Aside from lobbying against the carbon tax, European companies, mindful of public opinion and the positive intent of the European Commission, largely sought a consensual approach to climate change (Levy and Egan 2003). In contrast, the immediate response of their American counterparts was to form advocacy groups to mobilise against climate action. Perhaps the most notable example of such platforms was the Global Climate Coalition (GCC). The GCC was dominated by oil companies, car manufacturers and other fossil-fuel-dependent corporations, whose business models were threatened by climate-change mitigation efforts (Levy and Newell 2005; Stephan 2011). A cardinal tool in their strategy was to question the scientific understanding of anthropogenic climate change. In so doing, they sought to attack and destroy the basis of common belief that framed climate change as a problem caused by polluting socio-economic activities. Furthermore, they indicated that prevailing efforts to tackle climate change would lead to economic crises, poverty and scarcity, thus appealing to the deep and pervasive desire for economic prosperity among the population, and the need to ensure continued capital accumulation within the 'carboniferous' bloc (Dalby and Paterson 2009). Despite their divergent strategies, the multi-national nature of these business groups ensured that they coordinated strategies across the Atlantic (Levy and Egan 2003).

The idea of emissions trading as a policy instrument to address climate change emanated from a number of US-based ENGOs, and their close associates in academia (Dudek and LeBlanc 1991; Hahn and Stavins 1995; Stewart and Wiener 1992). Following the introduction of the SO_2 and NO_x trading scheme in the US 1990 Clean Air Act, the Environmental Defence Fund (a renowned US climate-advocacy NGO), alongside other US-based ENGOs, such as the Centre for Clean Air Policy (CCAP), promoted emissions trading as a means for addressing climate change (Braun 2009: 478; Stephan 2011: 10). This sentiment is captured by the following quote from Daniel Dudek and his colleagues at the Environmental Defence Fund, who asserted that: 'We do not need to wait for the development of grand international agreements before progress on reducing greenhouse gases can be made. The market-based approach to controlling acid rain leads the way' (Dudek *et al.* 1991: 46). Thus, the advocacy of emissions trading, which was 'deeply rooted in neo-liberal discourse' (Stephan 2011: 10), was promoted by organic intellectuals in academia and US-based ENGOs as a means of rallying support for action on climate change among adversarial political and economic actors.

The carbon trading compromise

The EU continued its efforts to introduce policy instruments to address climate change, but to little avail. In 1995, ahead of the first Conference of Parties of the United Nations Framework Convention on Climate Change in Berlin, the Commission published a working paper setting out options for a Community Climate Strategy (CEC 1995). The document showed continued support for a carbon tax, and other cost-effective measures to reduce CO_2 emissions, through which the EU would assume an international climate leadership role (Haigh 2011). The document encapsulated the ecological modernisation win-win discourse, through which climate-change mitigation efforts would encourage economic growth and political leadership (Hajer 1995; Weale 1992). In practice, however, little progress was made on introducing EU-wide climate instruments (Jordan and Rayner 2010; Wurzel and Connelly 2011).

Meanwhile, the popularity of emissions trading schemes increased globally. International organisations such as the OECD, the United Nations Conference on Trade and Development (UNCTAD) and the International Energy Agency

(IEA) supported emissions trading as their instrument of choice in addressing climate change (Braun 2009; Calel 2011: 15–16). In the US, positive experience with the SO_2 and NO_x trading scheme encouraged the government to include this instrument in the negotiations of an international climate-change treaty (Stephan 2011).

The attitude of business groups towards climate change in general, and emissions trading in particular, also shifted. As international climate efforts grew, some business groups adopted more accommodating approaches towards climate change. A milestone change occurred in May 1997, when British Petroleum (BP) publicly acknowledged the threat of climate change and joined forces with the Environmental Defence Fund in order to develop a company-wide emissions trading scheme (Levy and Egan 2003; Stephan 2011: 11). Shell similarly broke rank with the GCC and initiated an internal emissions trading scheme. The win–win ecological modernisation discourse was increasingly adopted among business groups, who also undertook organisational and material efforts to accommodate the threats of climate change to their financial operations (Levy and Egan 2003; Levy and Newell 2005).

These business strategies were important in several respects. First, they provided opportunities for companies to gain material capacities in implementing emissions trading. Second, they generated a new discourse through which companies and their products were portrayed as 'green', thus enhancing their corporate image. Third, they created new organisational capacities within companies and through alliance-building with counter-hegemonic environmental groups, as well as fostering cooperation with policy makers (Levy and Egan 2003; Levy and Newell 2005). Fourth, by investing in emissions trading, companies effectively acted to avoid the implementation of more contested policy instruments, such as command-and-control and fiscal regulations, which were potentially more harmful to their operations (Akhurst *et al.* 2003: 657; Stephan 2011: 11).

The uptake of the EU ETS and the rise of the new climate bloc

The Kyoto Conference of the Parties in 1997 marked a turning point in the EU's attitude towards emissions trading. In order to ensure US cooperation on an international climate treaty, the EU accepted emissions trading and other flexible instruments to address climate change (Braun 2009: 472). Following the uptake of emissions trading schemes by several Member States (Ellerman and Buchner 2007: 68), the Commission published a Green Paper on an EU-wide emissions trading scheme in March 2000 (CEC 2000). From the publication of the Green Paper, it took the EU institutions a relatively short period of time until the EU ETS Directive was adopted by the Council of Ministers and the European Parliament in October 2003, and the first stage of the EU ETS began in January 2005.

Oil companies were particularly instrumental in setting the agenda for emissions trading both at the EU and at Member State levels (Braun 2009: 473).

The acceptance of emissions trading schemes ensured continued political and public support for the operations of some of the most climate-harming industries, allowing these companies to largely continue business-as-usual, without requiring any major changes to their operations. Thus, the support of carbon-intensive fractions of capital for emissions trading can be understood as a Gramscian passive revolution, which served to cement the continued hegemony of the 'carboniferous' historical bloc (Stephan 2011).

ENGOs also contributed to the uptake of the EU ETS. The CCAP and other US-based ENGOs were actively engaged in the promotion and design of the EU ETS (Braun 2009: 478; Stephan 2011: 10). EU-based ENGOs, which were largely opposed to emissions trading, were left with no option but to accept carbon trading as a legitimate instrument to address climate change. In order to maintain some influence over the design and implementation of a trading scheme, they participated in the negotiations and design of the EU ETS (Braun 2009: 479). The co-optation of this subordinate group into the emerging hegemonic project is understood by Stephan (2011: 15) as a strategy to prevent ENGOs from 'organizing a successful opposition or counter-hegemonic movement to emissions trading or carbon markets'. Thus, a carbon-accommodating project emerged across horizontal and vertical levels of governance, and a coalition of economic, political and environmental actors 'became a hegemonic bloc' (Stephan 2011: 12), as discussed in more detail in Section 5. In the following section we turn to examine the CDM.

The Clean Development Mechanism: the war of positions

Although the actors and 'battle ground' are slightly different, the development of the CDM has many parallels with the EU ETS. The CDM is one of the three market mechanisms established under the Kyoto Protocol for governing climate-change mitigation, alongside emissions trading, and the JI mechanism.

The main argument of proponents of the CDM is that since it is materially irrelevant where carbon is reduced (from a global perspective), it is preferable to give developed countries flexibility in meeting their legally-binding emissions reduction obligations under the Kyoto Protocol. Thus, the CDM allows companies from developed countries (those that are subject to binding emissions reduction targets under the Kyoto agreement) to invest in emissions reduction activities in developing countries (who are not subject to legally binding reduction targets under the Kyoto Protocol). The 'additional' carbon saved through such investment is rewarded by Certified Emission Reduction units (CER), which can be subsequently monetised or used to meet industrialised countries' emissions reduction obligations, including those of the EU ETS (Newell and Paterson 2010).

With the rise in awareness of climate change in the late 1980s, developing countries were quick to point out that there was substantial asymmetry in the contribution and vulnerability to climate impact between countries (Agarwal and Narain 1991; Guha and Martinez-Alier 1997). Specifically, they argued that while developed countries were largely responsible for causing climate change, it was the less-developed countries that will bear much of the negative impacts (Dasgupta 1994). In addition, developing countries drew attention to existing problems of poverty and underdevelopment in their countries. They argued that it was unfair for them to be expected to sacrifice their development aspirations in a bid to mitigate climate change (Hayes 1993). Developing countries were adamant that any action to deal with climate change be conditional on financial and technical assistance from the industrialised countries (Dasgupta 1994). In fact, they argued that global efforts to address climate change must be used as a means of addressing wider underlying issues of global inequality (Bodansky 2001; Paterson 1996). To this effect, measures such as global carbon taxes on fuel, aviation and shipping, as well as the idea of some form of a global carbon stamp duty, were mooted (Grubb 1995).

Through these discursive and rhetorical devices, developing countries effectively framed climate change as an issue of global justice and equity. In so doing, they took a strategic stand in the emergent war of position. Early calculations suggested that several hundreds of billions of dollars will accrue to developing countries in the form of North–South financial and technology transfers (Grubb 1989; Hayes 1993). While these transfers were crucial in terms of the monetary value involved, their real significance lay in the threat they posed to the hegemonic neoliberal ideology, with its inbuilt averseness to redistribution, especially among nations (Okereke 2008). In essence, developing countries acted to promote a massive North–South economic and technology transfer as part of the international climate-change regime. However, such actions would violate the fundamental norms of neoliberal economic philosophy, enthrone the ideal of global justice and significantly disturb the configurations of powers that characterise the historical bloc. In this sense, the war of position, launched by developing countries and sympathetic voices from the North, represented a serious threat that needed to be pacified, or 'accommodated' in Gramscian terms.

Passive revolution: from carbon fund to global carbon markets

The developed countries were rattled by the coherent articulation of the global distributional implications of climate change (Parks and Roberts 2008). Arguments for North–South climate justice were presented with passion and in provoking terms by developing country politicians and their ideological demagogues. One such publication by Agarwal and Narain (1991) argued that global policies that neglect concerns for North–South distributional equity amount to nothing short of 'environmental colonialism'.

However, while recognising the intuitive appeal of international climate justice, developed countries were determined to either side-step or at least significantly dilute practical distributional responsibilities in terms of North–South transfers (Dasgputa 1994; Paterson 1996). To this effect, they argued that population growth and widespread corruption in the South was the cause for widespread poverty in developing countries. Some even suggested that developing countries were acting like 'kleptocrats', seeking to use their numerical advantage to extort money from the developed nations (Okereke 2008). Others invoked historical ignorance of the negative impacts of carbon pollution and, on that basis, argued that it was unfair, as said in philosophical parlance, 'to punish sons for the sins of their parents' (Caney 2005; Jamieson 2001; Vanderheiden 2008).

In the run up to the Kyoto agreement, Brazil tabled an elaborate 'Climate Development Fund' proposal, which called for large sums of money to be set aside by developed countries on an annual basis for the purpose of funding climate investments and the acquisition of 'clean' technology by developing countries. This proposal was widely supported by the rest of the developing countries including the G77 group and China. However, the North and the business lobby were vehemently opposed to this idea. Their main argument was that such a fund was contrary to the spirit of free market capitalism and the protection of Intellectual Property Rights (ENB 1997). The EU opposed the fund, suggesting that it offered a loophole that could render emissions targets in the North meaningless (ENB 1997: 2). A lack of agreement on this aspect proved to be one of the most important obstacles to negotiating a Kyoto Accord (ENB 1997).

When developed countries perceived the strength of the argument for climate justice and corresponding agitation over North–South financial and technology transfer, they changed tactics – from outright opposition towards a set of accommodative strategies. A major step in this strategy was the establishment of informal long-running bilateral contact between Brazil and the US to discuss the Clean Development Fund (CDF) and find common ground (Werksman 1998). In addition, several developed countries, such as South Korea and Australia, started making unilateral pledges to promote the development of clean technology and innovation centres in developing countries, especially the nations perceived to be most vulnerable to climate change (Haites and Yamin 2000). The aim was in part to break the solidarity of developing countries and encourage a more bilateral and fragmented approach to North–South climate investment, technology transfer and capacity building. After a series of exclusive talks between Brazil and the US, the proposal for the CDF was eventually changed to a Clean Development Mechanism (CDM), which was perceived to be far more flexible and business-friendly (Werksman and Cameron 2000; see also the chapter by Newell, this volume).

Since its inception in 2001, when the first CDM project was registered, to date, the CDM has issued over 1 billion CER. Interestingly, a vast proportion of these projects have been in China, and Brazil – the country that had championed the state-based Climate Development Fund (UNFCCC 2014). However, prices for CER have fallen steadily since 2012 and have been very low since January 2013. The value of CDM in reducing emissions and facilitating equity and sustainable development is questionable (Böhm *et al.* 2012; Lohmann 2008). Many feel that the transformation of the Brazilian proposal from its original focus on a North–South financial transfer to market mechanism represented a very successful move by the developed countries to take the sting off the proposal and turn it into an accumulation instrument for business and industry actors in the North (Haites and Yamin 2000; Lohmann 2006; Werksman and Cameron 2000).

Assessing climate instruments: a neo-Gramscian perspective

Ineffectiveness and contestations of justice

Both the EU ETS and the CDM have proved to be controversial measures for addressing climate change. The implementation of the EU ETS has been difficult and beset with controversies. The initial allocation of pollution permits was heavily influenced by industry lobbying, which resulted in lenient permitting conditions (Lederer 2012). To illustrate, the actual CO_2 emissions for 2005 were about 4 per cent lower than the number of pollution permits allocated that year. Consequently, carbon prices dropped drastically, reaching their lowest ebb in April 2007 (Calel 2013; Skjærseth and Wettestad 2009). The over-allocation of pollution permits, and the resulting drop in carbon prices impeded the operation and effectiveness of this instrument. Moreover, the free allocation of permits resulted in windfall profits for energy producers and other industry sectors, which passed the speculative costs on to consumers (Skjærseth and Wettestad 2010).

During the second implementation phase (2008–2012), the number of allocated permits was reduced and CO_2 emissions of EU ETS-related industries dropped. However, much of the emissions reductions made during this period can be attributed to the financial recession that began in 2008, and the resulting slowdown in polluting economic activities. Meanwhile, industry continued to make windfall profits from the scheme (Calel 2013; Okereke and McDaniels 2012). The financial crisis also fuelled a 'green growth' discourse, in which environmental innovation is perceived as a stimulus for economic growth (Runnalls 2011). Newell and Paterson (2010: 1) describe this form of 'climate capitalism' as 'a model which squares capitalism's need for continual economic growth with substantial shifts away from carbon-based industrial development'. In this view, investment in 'cleaner' technologies will stimulate the economy while promoting environmental protection and climate-change mitigation. However, at its core, this discourse does not challenge prevalent modes of production and consumption.

The recession also shaped the design and implementation of the third phase of the EU ETS (2013–2020). In December 2008, the European Council of Ministers and Parliament agreed on a revised emissions trading scheme. Wide concessions were made to industry groups, with the mediation of Member States. For example, Poland managed to secure exemptions for its coal-powered electric plants, while Germany secured concessions for its energy-intensive industries (Skjærseth and Wettestad 2010). Thus, despite the tightening of the EU ETS in its third phase, there was still an over-allocation of pollution permits (ENDS Report 459 May 2013: 27).

In April 2013, carbon prices fell to 2.63 euros, their lowest level since the 2007 carbon-market crash. The reason for this was the European Parliament's rejection of a Commission Proposal to delay the auctioning of permits – a plan known as backloading. The EU ETS was declared 'moribund as an emissions mitigation tool, if not quite dead' (ENDS Report, 459 May 2013: 27). While backloading was finally approved, the EU tries to further resuscitate the ETS. Yet obstacles to a functioning carbon market prevail. These difficulties also hinder certainty and long-term investments in 'green' technologies. 'Carboniferous' fractions of capital therefore maintain their power through making windfall profits from the EU ETS, ensuring a continued uncertain investment climate for competing low-carbon technologies, and relative regulatory freedom under the EU ETS.

The CDM has not fared any better, but has instead attracted a string of criticisms. It has been suggested that the CDM is an immoral instrument, in that it provides a cheap way for the rich West to avoid taking serious action on climate change (Lohmann 2006). Beyond this, there are strong suggestions that the CDM provided opportunities for business interests in the North to dispossess communities in the poor South and engage in primitive accumulation (Böhm *et al.* 2012; Lohmann 2006, 2009). In some instances, forests and lands that are vital for the survival of local communities in developing countries have been privatised and commoditised under the pretext of CDM (Lohmann 2009). For these reasons, the CDM has, in fact, been described as an instrument for 'carbon fraud and climate colonialism' (Bachram 2004: 5). There is virtually no evidence that the CDM has resulted in North–South technology transfer, or sustainable development, in line with the original vision of the Brazilian proposal. Furthermore, there is little ground to suggest that CDM has resulted in substantial carbon reduction. In fact, as of 1 June 2013, 57 per cent of all CERS had been issued for projects based on destroying either HFC-23 (38%) or N_2O (19%) (UNFCCC 2013). Although these are greenhouse gases, they are less central to economic growth and generally easier to eliminate than more ubiquitous carbon dioxide pollution.

By offering the means for the 'carboniferous' bloc to allocate private property rights to public goods and realise their value through the market, these instruments have more or less reinforced hierarchies and patterns of domination between the rich and the poor (see Bumpus and Liverman 2008: 144). Yet, they have been very successful in creating a sense of climate proactivity by government and industries, drawing in a large number of diverse groups, including otherwise progressive ENGOs, into a large climate coalition, shielding the 'carboniferous' bloc from societal pressure.

A neo-Gramscian critique

The uptake of these carbon-commodifying policy instruments to address climate change can fruitfully be understood in terms of a neo-Gramscian passive revolution. Through these instruments, business and capitalist governments have neutralised the impetus for radical economic, social and political transformations that were originally associated with climate-change governance across geographies. Thus, concerns regarding climate change have in effect been incorporated into the 'carboniferous' historical bloc. Incremental adjustments have been made by this hegemonic project to address climate change, resulting in some material, organisational and discursive developments within the bloc.

We term this mutated bloc the 'climate-accommodating carboniferous bloc'. On the material level, carbon markets have created favourable conditions for the continued accumulation of 'carboniferous' fractions of capital, both on regional and international levels. On the organisational level, this bloc is governed by alliances among governments and public institutions at various spatial scales, business groups, and some mainstream ENGOs, alongside an emerging group of climate professionals (Stephan 2011; Voß 2007). On the discursive level, the climate-accommodating bloc is fuelled by a green growth ideology. This paradigm evolved from the earlier ecological modernisation discourse. It provides a consensual ideological framework for continued capitalist accumulation.

The account provided in this chapter highlights the difficulties in defining, creating and maintaining a counter-hegemonic war of position. Although US-based ENGOs were ultimately successful in securing acceptance of climate change as a global policy problem, they operated within the prevailing neoliberal hegemonic order and its pro-market ideology, cooperating with a range of economic and political actors. Ultimately, the environmental interests they represented were subordinated to those of the climate-accommodating bloc. That is, demands for climate-change mitigation efforts were accommodated through the establishment of the EU ETS and concomitant carbon markets, which allowed carbon-intensive fractions of capital to continue their business operations with only minor interventions. Similarly, in accepting the substitution of direct funds with market instruments, the South was able to make the notion of North–South financial transfer politically palatable and acceptable to the North. In reality, however, the compromise has served not only to neutralise the radical content of the idea but also to present the hegemonic bloc with another opportunity for primitive accumulation. To date, these strategies have proved to be largely unsuccessful.

The challenge for counter-hegemonic social movements is to create alternative ideological narratives on which to build material and organisational capacity. This discourse could, for example, substitute existing perceptions of economic growth and prosperity (see Newell 2012: 148–151). However, challenging the prevailing hegemonic order is no simple task. The only certainty is that an alternative discourse will be met with resistance and contestation of the hegemonic bloc, requiring long-term ideational, material and organisational efforts, spearheaded by progressive organic intellectuals.

Conclusions

Using the case studies of the EU ETS and the CDM, this chapter offered a neo-Gramscian analysis of carbon-commodifying policy instruments. Particularly, we emphasised the processes of contestation and compromise that shaped climate governance across various geographical scales. We argued that both the EU ETS and the CDM can be perceived as passive revolution responses to counter-hegemonic challenges. The results of these compromises, we argued, are eminent in the carbon-accommodating historical bloc. This bloc is materially aligned with the creation of carbon markets and carbon commodification. It is steered by an alliance of economic, political and societal actors, at various spatial scales. Discursively, this bloc is fuelled by the notion of green growth, in which the prevalent capitalist mode of production can peacefully co-exist with efforts to mitigate climate change. Moreover, we illustrated that these neo-Gramscian concepts can be fruitfully used to examine and compare climate governance across national, regional and international spatial scales.

We found the concept of the war of position more elusive to apply. In both case studies, strategies that began as efforts to incorporate climate-change concerns into the hegemonic project were soon subordinated to the interests of the neoliberal hegemonic bloc. We hope that this account will stimulate organic intellectuals and social groups who are currently putting their minds together to provide alternative visions for a low-carbon future. In a neo-Gramscian view, this war of position will necessitate long-term efforts on the material, organisational and ideological levels, working outside the realms of neoliberal ideology. The question of whether Gramsci's theoretical prescription for social transformation through the war of position can work in practice remains open. Future work could explore this concept in more detail, and relate the accounts given here to broader international relations and public policy theories.

References

Adamson, W. L. (1980) *Hegemony and Revolution: a study of Antonio Gramsci's political and cultural theory.* Berkeley: University of California Press.

Agarwal, A. and Narain, S. (1991) *Global Warming in an Unequal World.* New Delhi: Centre for Science and the Environment.

Akhurst, M., Morgheim, J. and Lewis, R. (2003) Greenhouse gas emissions trading in BP. *Energy Policy,* 31(7), 657–663.

Anderson, P. (1976) The antimonies of Antonio Gramsci. *New Left Review,* 100, 5–78.

Andrée, P. (2011) Civil society and the political economy of GMO failure in Canada: a neo-Gramscian analysis. *Environmental Politics,* 20, 173–191.

Bachram, H. (2004) Climate fraud and carbon colonialism: the new trade in greenhouse gases. *Capitalism, Nature, Socialism,* 15 (4), 5–20.

Bates, T. R. (1975) Gramsci and the theory of hegemony. *Journal of the History of Ideas,* 36, 351–366.

Bieler, A. (2002) The struggle over EU enlargement: a historical materialist analysis of European integration. *Journal of European Public Policy,* 9, 575–597.

Bieler, A. (2006) Class struggle over the EU model of capitalism: neo-Gramscian perspectives and the analysis of European integration. In: A. Bieler and A. D. Morton, eds., *Images of Gramsci: connections and contentions in political theory and international relations.* Oxford: Routledge, 119–132.

Bieler, A. and Morton, A. D. (2001) Introduction: neo-Gramscian perspectives in international political economy and the relevance to European integration. *In:* A. Bieler and A. D. Morton, eds., *Social Forces in the Making of the New Europe: the restructuring of European social relations in the global political economy*. Basingstoke: Palgrave, 3–24.

Bodansky, D. (2001) The history of the global climate change regime. *In:* U. Luterbacher and D. F. Sprinz, eds., *International Relations and Global Climate Change*. Cambridge: Massachusetts Institute of Technology, 23–40.

Böhm, S., Misoczky, M. C. and Moog, S. (2012) Greening capitalism? A Marxist critique of carbon markets. *Organization Studies*, 33 (11), 1617–1638.

Bond, P. (2011) *The Politics of Climate Justice*. London: Verso/Pietermaritzburg: University of KwaZulu-Natal Press.

Borger, G. (2012) All things not being equal: aviation in the EU ETS. *Climate Law*, 3 (3), 265–281.

Braun, M. (2009) The evolution of emissions trading in the European Union: the role of policy networks, knowledge and policy entrepreneurs. *Accounting, Organizations and Society*, 34, 469–487.

Buhr, K. (2012) The inclusion of aviation in the EU Emissions Trading Scheme: temporal conditions for institutional entrepreneurship. *Organization Studies*, 33 (11), 1565–1587.

Bumpus, A. G. and Liverman, D. M. (2008) Accumulation by decarbonization and the governance of carbon offsets. *Economic Geography*, 84, 127–155.

Burkett, M. (2008) Just solutions to climate change: a climate justice proposal for a domestic Clean Development Mechanism. *Buffalo Law Review*, 56, 169.

Calel, R. (2011) Climate change and carbon markets: a panoramic history. *Working Paper No. 62*. London: Center for Climate Change Economics and Policy.

Calel, R. (2013) Carbon markets: a historical overview. *Wiley Interdisciplinary Reviews: Climate Change*, 4, 107–119.

Caney, S. (2005) Cosmopolitan justice, responsibility, and global climate change. *Journal of International Law*, 18 (4), 747–775.

CEC (1992) Proposal for a Council Directive introducing a tax on carbon dioxide emissions and energy. *COM (92)226 Final*. Brussels: Office for Official Publications of the European Union.

CEC (1995) Commission working paper on the EU climate change strategy: a set of options. *SEC(95) 288 final*. Brussels: Commission of the European Union.

CEC (2000) Green Paper on greenhouse gas emissions trading within the European Union. *COM (2000) 87 final*. Brussels: Commission of the European Union.

Cox, R. W. (1981) Social forces, states and world orders: beyond international relations theory. *In:* R. W. Cox and T. J. Sinclair, eds. (1996) *Approaches to World Order*. Cambridge: Cambridge University Press, 85–123.

Cox, R. W. (1983) Gramsci, hegemony and international relations: an essay in method. *In:* R. W. Cox and T. J. Sinclair, eds. (1996) *Approaches to World Order*. Cambridge: Cambridge University Press, 124–144.

Cox, R. W. and Sinclair, T. J. (1996) *Approaches to World Order*. Cambridge: Cambridge University Press.

Dalby, S. and Paterson, M. (2009) Over a barrel: cultural political economy and 'oil imperialism'. *In:* F. Debrix, ed., *The Geopolitics of American Insecurity: terror, power and foreign policy*. London: Routledge, 181–196.

Dasgupta, C. (1994) The climate change negotiations. *In:* J. A. Mintzer and A. J. Leonard, eds., *Negotiating Climate Change: the inside story of the Rio Convention*. Cambridge: Cambridge University Press, 129–148.

Davies, J. S. (2011) *Challenging Governance Theory: from networks to hegemony.* Bristol: Policy Press.

DG Climate Action (2013) The European Union Emission Trading Scheme. Available from: http://ec.europa.eu/clima/policies/ets/ [accessed 2 November 2013].

Dudek, D. and LeBlanc, A. (1991) *Preserving Brazil's Tropical Forests through Emissions Trading.* New York: Environmental Defense Fund.

Dudek, D., LeBlanc, A. and Miller, P. (1991) CO_2 and SO_2: consistent policy making in a greenhouse. *In:* J. C. White, W. Wagner and C. N. Beal, eds., *Global Climate Change: the economic costs of mitigation and adaptation.* New York: Elsevier, 45–72.

Ellerman, A. D. and Buchner, B. K. (2007) The European Union Emissions Trading Scheme: origins, allocation, and early results. *Review of Environmental Economics and Policy,* 1 (1), 66–88.

Ellerman, A. D. and Harrison, D. (2003) *Emissions Trading in the US: experience, lessons, and considerations for greenhouse gases.* Philadelphia/Washington, DC: Pew Centre on Global Climate Change. Available from: http://web.mit.edu/globalchange/www/PewCtr_MIT_Rpt_Ellerman.pdf [accessed 11 November 2013].

ENB (Earth Negotiation Bulletin) (1997) *12/71:* 1, 2.

ENDS Daily (various years), *daily e-mail service.* London: Environmental Data Services.

Femia, J. V. (1981) *Gramsci's Political Thought: hegemony, consciousness, and the revolutionary process.* Oxford: Clarendon Press.

Forgacs, D. (1988) *A Gramsci Reader.* London: Lawrence and Wishart Limited.

Fuhr, H. and Lederer, M. (2009) Varieties of carbon governance in newly industrializing countries. *The Journal of Environment & Development,* 18 (4), 327–345.

Gill, S. (2003) *Power and Resistance in the New World Order.* Basingstoke: Palgrave Macmillan.

Gramsci, A. (1971) *Selections from the Prison Notebooks of Antonio Gramsci.* New York: International Pubs.

Grubb, M. (1989) *The Greenhouse Effect: negotiating targets.* London: Royal Institute of International Affairs.

Grubb, M. (1995) Seeking fair weather: ethics and the international debate on climate change. *International Affairs,* 71 (3), 463–496.

Guha, R. and Martinez-Alier, J. (1997) *Varieties of Environmentalism: essays North and South.* London: Earthscan Publications Ltd.

Hahn, R. W. and Stavins, R. N. (1995) Trading in greenhouse permits: a critical examination of design and implementation issues. *In:* H. Lee, ed., *Shaping National Responses to Climate Change: a post-Rio policy guide.* Washington, DC: Island Press, 177–217.

Haigh, N. (2011) *Manual of Environmental Policy: the EC and Britain.* London/Harlow: Longman in association with the Institute for European Environmental Policy.

Haites, E. and Yamin, F. (2000) The Clean Development Mechanism: proposals for its operation and governance. *Global Environmental Change,* 10 (1), 27–45.

Hajer, M. A. (1995) *The Politics of Environmental Discourse: ecological modernization and the policy process.* Oxford: Clarendon Press.

Hayes, P. (1993) North-South transfer. *In:* P. Hayes and K. Smith, eds., *The Global Green House Regime: who pays?* London: Earthscan, 144–168.

Hayward, T. (2007) Human rights versus emissions rights: climate justice and the equitable distribution of ecological space. *Ethics & International Affairs,* 21 (4), 431–450.

Hoffmann, M. J. (2011) *Climate Governance at the Crossroads: experimenting with a global response after Kyoto.* New York: Oxford University Press Catalogue.

Hovi, J., Skodvin, T. and Andresen, S. (2003) The persistence of the Kyoto Protocol: why other Annex I countries move on without the United States. *Global Environmental Politics,* 3 (4), 1–23.

Jamieson, D. (2001) Climate change and global environmental justice. *In:* C. A. Miller and P. N. Edwards, eds., *Changing the Atmosphere: expert knowledge and global environmental governance.* Cambridge: MIT Press, 287–307.

Jessop, B. (1982) *The Capitalist State: Marxist theories and methods.* Oxford: Robertson.

Jessop, B. (1990) *State Theory: putting the capitalist state in its place.* Penn State University Press.

Jordan, A. and Rayner, T. (2010) The evolution of climate policy in the European Union: a historical overview. *In:* A. Jordan, D. Huitema, H. Van Asselt, T. Rayner and F. Berkhout, eds., *Climate Change Policy in the European Union: confronting the dilemmas of mitigation and adaptation?* Cambridge: Cambridge University Press, 52–80.

Jordan, A., Wurzel, R. K. and Zito, A. (2003) 'New' instruments of environmental governance: patterns and pathways of change. *Environmental Politics,* 12, 1–24.

Jordan, A., Wurzel, R. K. W. and Zito, A. (2005) The rise of 'new' policy instruments in comparative perspective: has governance eclipsed government? *Political Studies,* 53, 477–496.

Lederer, M. (2012) The practice of carbon markets. *Environmental Politics,* 21, 640–656.

Levy, D. L. and Egan, D. (2003) A neo-Gramscian approach to corporate political strategy: conflict and accommodation in the climate change negotiations. *Journal of Management Studies,* 40, 803–829.

Levy, D. L. and Newell, P. (2002) Business strategy and international environmental governance: toward a neo-Gramscian synthesis. *Global Environmental Politics,* 2, 84–101.

Levy, D. L. and Newell, P. (2005) A neo-Gramscian approach to business in international environmental politics: an interdisciplinary, multilevel framework. *In:* D. L. Levy and P. Newell, eds., *The Business of Global Environmental Governance.* Cambridge: MIT Press, 47–69.

Lohmann, L. (2006) Carbon trading: a critical conversation on climate change, privatization and power. *Development Dialogue* (48).

Lohmann, L. (2008) Carbon trading, climate justice and the production of ignorance: ten examples. *Development,* 51 (3), 359–365.

Lohmann, L. (2009) Climate as investment. *Development and Change,* 40 (6), 1063–1083.

MacKenzie, D. (2009) Making things the same: gases, emission rights and the politics of carbon markets. *Accounting, Organization and Society,* 34 (3–4), 440–455.

Matt, E. (2012) *The political economy of European Union environmental governance: The case of the voluntary agreement to reduce carbon dioxide emissions from new cars.* PhD Thesis. Norwich: University of East Anglia. Available from: https://ueaeprints.uea.ac.uk/39142/1/2012MattEPhD.pdf [accessed 11 November 2013].

Michaelowa, A. (2008) German climate policy between global leadership and muddling through. *In:* H. Compston and I. Bailey, eds., *Turning Down the Heat: the politics of climate policy in affluent democracies.* Basingstoke: Palgrave Macmillan, 144–163.

Morton, A. D. (2007) *Unravelling Gramsci: hegemony and passive revolution in the global political economy.* London: Pluto Press.

Mouffe, C. (1979) *Gramsci and Marxist Theory.* London: Routledge/Kegan Paul.

Newell, P. (2008) The political economy of global environmental governance. *Review of International Studies,* 34 (3), 507–529.

Newell, P. (2012) *Globalization and the Environment: capitalism, ecology and power.* Cambridge: Polity Press.

Newell, P. and Paterson, M. (2010) *Climate Capitalism: global warming and the transformation of the global economy.* Cambridge: Cambridge University Press.

Okereke, C. (2008) *Global Justice and Neoliberal Environmental Governance: ethics, sustainable development and international co-operation.* London: Routledge.

Okereke, C. and Dooley, K. (2010) Principles of justice in proposals and policy approaches to avoided deforestation: towards a post-Kyoto climate agreement. *Global Environmental Change*, 20 (1), 82–95.

Okereke, C. and McDaniels, D. (2012) To what extent are EU steel companies susceptible to competitive loss due to climate policy? *Energy Policy*, 46, 203–215.

Okereke, C., Bulkeley, H. and Schroeder, H. (2009) Conceptualizing climate governance beyond the international regime. *Global Environmental Politics*, 9 (1), 58–78.

Parks, B. C. and Roberts, T. (2008) Inequality and the global climate regime: breaking the north-south impasse. *Cambridge Review of International Affairs*, 21 (4), 621–648.

Paterson, M. (1996) *Global Warming and Global Politics*. London: Routledge.

Paterson, M. (2009) Post-hegemonic climate politics? *The British Journal of Politics & International Relations*, 11 (1), 140–158.

Runnalls, D. (2011) Environment and economy: joined at the hip or just strange bedfellows? *S.A.P.I.EN.S*, 4 (2), 1–10.

Rupert, M. (1993) Alienation, capitalism and the inter-state system: towards a Marxian/Gramscian critique. *In:* S. Gill, ed., *Gramsci, Historical Materialism and International Relations*. Cambridge: Cambridge University Press, 67–92.

Showstack Sassoon, A. (1982) *Approaches to Gramsci*. London: Writers and Readers Publishing Cooperative Society Ltd.

Showstack Sassoon, A. (1987) *Gramsci's Politics*. Minneapolis: University of Minnesota Press.

Simon, R. (1991) *Gramsci's Political Thought: an introduction*. London: Lawrence and Wishart.

Skjærseth, J. B. and Wettestad, J. (2009) The origin, evolution and consequences of the EU Emissions Trading System. *Global Environmental Politics*, 9, 101–122.

Skjærseth, J. B. and Wettestad, J. (2010) Making the EU Emissions Trading System: The European Commission as an entrepreneurial epistemic leader. *Global Environmental Change*, 20, 314–321.

Sprinz, D. F. (1992) *Why countries support international environmental agreements: the regulation of acid rain in Europe*. PhD Thesis. Maryland: The University of Maryland.

Stephan, B. (2011) The power in carbon. A neo-Gramscian explanation for the EU's adoption of emissions trading. *In:* A. Engels, ed., *Global Transformations towards a Low Carbon Society* 4 (working paper series). Hamburg: University of Hamburg/Klima Campus.

Stewart, R. B. and Wiener, J. B. (1992) Comprehensive approach to global climate policy: issues of design and practicality. *Arizona Journal of International & Comparative Law*, 9, 83.

UNFCCC (2013) Clean Development Mechanism. Available from: http://cdm.unfccc.int/ [accessed 1 June 2013].

UNFCCC (2014) CDM projects interactive map. Available from: http://cdm.unfccc.int/ Projects/MapApp/index.html [accessed 2 February 2014].

Van Apeldoorn, B. (2002) *Transnational Capitalism and the Struggle over European Integration*. London: Routledge.

Van Apeldoorn, B., Overbeek, H. and Ryner, M. (2003) Theories of European integration: a critique. *In:* A. W. Cafruny and M. Ryner, eds., *A Ruined Fortress? Neoliberal hegemony and transformation in Europe*. Lanham: Rowman & Littlefield Publishers, 17–45.

Vanderheiden, S. (2008) *Atmospheric Justice: a political theory of climate change*. Oxford: Oxford University Press.

Victor, D. G., House, J. C. and Joy, S. (2005) A Madisonian approach to climate policy. *Science*, 5742, 1820.

Voß, J.-P. (2007) Innovation processes in governance: the development of emissions trading as a new policy instrument. *Science and Public Policy*, 34 (5), 329–343.

Wara, M. W. and Victor, D. G. (2008) A realistic policy on international carbon offsets. *Working Paper for the Program on Energy and Sustainable Development*, Stanford University, 74.

Weale, A. (1992) *The New Politics of Pollution*. Manchester: Manchester University Press.

Werksman, J. (1998) The Clean Development Mechanism: unwrapping the 'Kyoto Surprise'. *Review of European Community & International Environmental Law*, 7 (2), 147–158.

Werksman, J. and Cameron, J. (2000) *The Clean Development Mechanism: the Kyoto surprise*. Brazil/US Aspen: Global Forum.

Wittneben, B. B., Okereke, C., Banerjee, S. B. and Levy, D. L. (2012) Climate change and the emergence of new organizational landscapes. *Organization Studies*, 33 (11), 1431–1450.

Wurzel, R. and Connelly, J. (2011) *The European Union as a Leader in International Climate Change Politics*. London: Routledge.

7 The politics of carbon markets in the global South

Markus Lederer

Introduction

Carbon markets are currently in poor shape: prices are low, the combined value of allowances has dropped dramatically within the last three years, investors are reluctant, politicians are doubtful, criminal activities have surfaced and civil society groups are increasingly openly hostile towards carbon trading (for a good summary of the recent critiques, see Böhm 2013). Many thus argue that the diffusion of market mechanisms resembles 'flogging a dead horse' or, in the words of Oscar Reyes, watching 'Carbon Market zombies stumble on' (Reyes 2012: 28). These problems are not only confined to the largest existing carbon market, the EU emission trading system (ETS), or the most controversial one, the Clean Development Mechanism (CDM), but are apparent in all Northern carbon markets. The Western Climate Initiative (WCI), for example, was once expected to become the second-largest cap-and-trade system in the world, but six of its US members (Montana, Utah, Oregon, Washington, New Mexico and Arizona) have declared that they will not join the cap-and-trade system and some states have even taken any climate policy off their agenda (Tuerk *et al.* 2013: 5). On the federal level, the US has, at least for the moment, completely given up on establishing an ETS and the government is using regulatory approaches via the Environmental Protection Agency (EPA) (MacNeil and Paterson 2012; Townshend *et al.* 2013: 423f.). It thus seems that those who have pointed out that capitalism cannot be reformed, that it will always be crisis prone and detrimental to nature and thus will have to be overcome, have quite a few arguments in their favor (Böhm *et al.* 2012). The notion of Newell and Paterson (2010) that climate capitalism could evolve in various varieties, potentially including one based on Keynesian macroeconomic policies, thus seems to be a rather distant policy today.

But maybe the politics of carbon markets are different in the global South? While carbon markets are in deep trouble in the North, many in the South are taking up the idea of not only pricing carbon but also allowing the trading of carbon allowances. While various studies have described the different initiatives that are ongoing in the South (Chestney 2012; IETA 2012; Mehling 2012; Sterk and Mersmann 2011; The Partnership for Market Readiness 2013; Townshend *et al.* 2013; Tuerk *et al.* 2013), few have given an explanation of why market-based approaches are still popular in the South despite the well-documented problems in the North. Those who have taken up the issue of why a potentially highly problematic policy instrument is moving into the developing world and into emerging economies have mostly made an argument claiming, in either a (neo-) Marxist or a neo-Gramscian fashion, that an ideological hegemony is in existence, being played out by Southern elites in their respective countries (e.g. Lohmann and Böhm 2012).

While this interpretation has its merits, it might underestimate how carbon markets fit nicely with the dominant practice of many governments in developing and emerging economies in actually pursuing industrial and ecological policies. Thus, the politics of carbon markets in the South might be substantially different from that of the North. This does not automatically imply that carbon markets will work more effectively in the global South or that that they will actually deliver what they promise, but the following analysis takes Southern agents and their evolving practices seriously in moving beyond some of the neo-Marxist/neo-Gramscian argumentation mentioned above.

In the first step, this chapter, therefore, shows that carbon markets have indeed diffused on a considerable scale into the global South. While primarily emerging markets are involved, the mapping of the different approaches indicates that a great variety is unfolding in all three continents of the South. This mapping exercise will primarily focus on carbon markets in the sense of setting up an ETS that includes a cap and the possibility to trade (= cap-and-trade). Baseline systems (e.g. the CDM), carbon taxes or other market-based mechanisms (e.g. feed-in tariffs that support renewable energy) will be of less interest but will also be considered wherever appropriate (for an overview of all forms of 'carbon pricing', see Aldy and Stavins 2012).

In the second step, the chapter will analyze various reasons why carbon markets are still so popular in the South. First, and making a functionalist argument, it will be shown that carbon markets in the South are perceived as just one instrument among many to counter rising CO_2 emissions in the global South. Second, carbon market activities have generated actors with strong interests in carbon trading on the international level (e.g. the World Bank), but also in Southern countries (e.g. designated operational entities (DOEs) in China or carbon financiers in India). Third and most importantly, carbon markets in the South are, mostly, part of a state-led industrial policy program (e.g. in China and South Korea) and thus depend for their success much less on private entrepreneurs or on civil society support than Northern carbon markets. This argument builds on a statist tradition within International Political Economy (IPE) that, particularly for the South, has always argued that governments have much more control of economic processes, including financial regulation, than is commonly perceived. This is an idea that is taken up in some of the more historical institutionalist arguments about the current forms of governance (for an overview of these approaches and how they might allow the analysis of carbon markets in the North as well, see Lederer 2012). The conclusion summarizes the main arguments of the chapter and briefly reflects on the implications that the politics of carbon markets in the global South might have for the future of carbon capitalism overall.

Diffusion of carbon market activities into the global South

This section provides an overview of carbon markets as they are currently either being set up or planned in Asia, in Latin America and in Africa. The following thus only focuses on domestic attempts to set up markets and does not deal with how the South has been included in global or Northern carbon market activities like the CDM, the voluntary market or the current plans to initiate new market mechanisms such as NAMA (nationally appropriate mitigation actions) crediting (for an overview of these activities and how they relate to the global South, see Lederer 2013).

Carbon markets in Asia

South Korea will only be described briefly as it is questionable whether it should still be perceived as a state of the global South. Its importance is due to the fact that, similar to that of Japan, its environmental policy is closely watched by other Asian countries, particularly China, to see what works and what does not. South Korea will start emission trading in 2015 (the legislation was approved in May 2012) and the current plans envision that about 500 companies and thus 60 percent of the national greenhouse gas (GHG) emissions will be covered by the scheme (including buildings and livestock farms). Thus, the Korean ETS will be about the same size as the Californian one (Tuerk *et al.* 2013: 19). The Korean scheme also aims for a slow introduction of auctioning and the use of domestic offsets. Interestingly, the legislation passed the national assembly with only three abstentions and no opposing votes, reflecting the strong consensus that carbon trading is a major element of South Korea's low-carbon development strategy fostering new green business developments (Townshend *et al.* 2013: 385f.). South Korea also proposed that NAMAs should be credited internationally and thus become part of a global carbon market (Röser *et al.* 2012). South Korea is thus one of the few countries of the South that is thinking about how its national market could be linked to global schemes.

Without doubt, *China*'s climate-change policy is the most important in the global South (see also Engels *et al.*, this volume). The country's leadership is highly aware of China's environmental problems and the urgent need to take action. For example, the massive air pollution in Beijing in early 2013 has again raised the issue at the highest political level and seems to be leading to China taking climate change even more seriously than before. In 2011, in its 12th Five Year Plan (2011–2015), the Chinese government endorsed pilot ETSs, and the National Development and Reform Commission initiated the setting up of carbon trading in four municipalities (Beijing, Chongqing, Shanghai and Tianjin), two provinces (Guangdong and Hubei) and one local-level city (Shenzhen in Guangdong).

These seven polities are partially building on voluntary carbon trading schemes that have been in place since 2008 (Huang 2013) and would 'account for 27.4% of China's national GDP and 18.4% of its population' (Lo 2013b). This has been characterized as a 'nation-wide bottom-up approach' (Tuerk *et al.* 2013: 18) and an example of China's determination to start using market-based instruments (Han *et al.* 2012). If China were to set an absolute national cap at one point and no longer pledge 'only' energy or carbon intensity reductions, this would eventually lead to the second-largest ETS in the world. The design of the system has partially been influenced by experiences from the EU ETS, but lessons learned from the Western Climate Initiative (WCI), Australia and particularly Tokyo, as the first urban cap-and-trade system, have been important as well (Tuerk *et al.* 2013: 22). The Chinese schemes started in 2013 and since they were designed as learning experiments they have set different objectives (e.g. Guangdong has an absolute emission cap; Tianjin and Beijing are using energy saving credits) (Sterk and Mersmann 2011: 10). The participating jurisdictions apparently see it as a privilege to be the first to conduct emission trading and seem to hope that the early experiences will generate first-mover advantages once a national ETS is established. No caps have yet been set but a large range of emitters is supposed to be included (in Shanghai even aviation, shipping and the hotel sector are envisioned to participate if they are of a certain size). Up to now (January 2014) five trading zones have gone operational. So far the system can be described as an experiment rather than a functioning market as much still has to be achieved in order to make the systems operational – particularly when it comes to administrative capacities, legal certainty and accurate emission measurements (Han *et al.* 2012: 39). Overall, the different systems vary quite strongly regarding their scope and, most likely, also their caps. A major problem for the provincial trading schemes is that leakage might occur as companies might move to provinces that have no caps and which, due to lower levels of development, have lower wages. A national-level ETS is also planned, but no date has been set for its start, and it will be interesting to see what kind of coordinating role the federal level will play vis-à-vis the regional and municipal ETSs. These pilot projects are complemented by voluntary emission trading activities that are centrally supported but locally administered. There are also plans for sectoral carbon ETSs, for example in the electricity and the building sector (Han *et al.* 2012), but so far it is not evident how they will work.

India plans to establish a carbon tax on coal, with the revenue being earmarked for clean energy research (Townshend *et al.* 2013: 22), and it has set up a pilot carbon trading scheme in three states, namely Gujarat, Maharashtra and Tamil Nadu (Lo 2013b). The main objective of the scheme is the reduction of particulate matter and thus the mitigation of climate change is, at best, a possible positive side effect of the scheme (Rosha and Freestone 2012: 352). India already has a working mandatory energy efficiency trading scheme ('Perform, Achieve and Trade') covering eight sectors, and the emission intensity should be reduced by about 20 to 25 percent (Han *et al.* 2012: 11). Here, no emissions are being traded and currently various industry sectors are raising complaints that technology that is supposed to increase energy efficiency might become more expensive and that those companies that have invested in energy efficiency in the past are now being disadvantaged as they have to fulfill their quotas independently of their past efforts (Rosha and Freestone 2012: 346). Finally, some trading is already ongoing in the field of renewable energy certificates (RECs), in which utilities can buy and sell credits to fulfill quotas, but so far most of the self-set objectives have not been met due to inconsistencies between federal-level and state-level policies (Shrimali *et al.* 2012).

Besides South Korea, China and India, some other Asian countries are also thinking about introducing carbon markets, but here the ideas are still at an early stage. *Taiwan* plans to establish a carbon offset scheme through which CO_2 levels should be stabilized at the 2005 level by 2020. A total of 270 companies have therefore agreed to supply emissions data to the government in order to kick off the scheme (Chestney 2012). *Vietnam* is considering market mechanisms within its participation in the World Bank's Partnership for Market Readiness initiative (see below for details on the PMR) covering steel, solid waste and power sectors within its green growth strategy. A carbon tax is also being considered. In January 2013, *Kazakhstan* started its first pilot phase and the government set a nationwide cap, being the first Asian state to do so (Sopher and Mansell 2013). Potentially, a regional market with Russia, Belarus and Ukraine is possible, but at this moment it is a long shot. Finally, *Thailand* also has an interest in cap-and-trade and even *Jordan* is considering whether market instruments could be part of its climate-change strategies (The Partnership for Market Readiness 2013).

Carbon markets in Latin America

In Latin America, *Brazil* is the most interesting case. The National Plan on Climate Action of 2008 stated that a cap-and-trade system should be set up, but the legislation of 2010 (National Policy on Climate Change) did not specify how an ETS should be implemented in detail. Nevertheless, the national parliament has authorized the Brazilian Securities and Exchange Commission (CVM) to set up provisions for over-the-counter (OTC) trading as well as trading on stock exchanges. Currently, the state of São Paulo is the most advanced in planning the details for market mechanisms (Townshend *et al.* 2013: 64). So far, it seems that primarily private actors are engaged in the emerging voluntary market, but the Brazilian Climate Secretary has confirmed public support for a market-based approach (Sterk and Mersmann 2011: 8).

Mexico is also of interest in that its *General Law on Climate Change* (which came into effect in 2012) incorporates an ETS that should start on the national level in 2015 as a voluntary market. In its Reducing Emissions from Deforestation and Degradation (REDD+) legislation, it has also promoted the fact that 'all economic instruments' can be used to pay the service providers (Townshend *et al.* 2013: 282). Sterk and Mersmann (2011: 13) interpret Mexico's recent communications with the World Bank in the sense that Mexico is more favorable towards crediting NAMAs (see below) than setting up a traditional ETS.

Chile also has an interest in cap-and-trade and it seems that internally the decision to push for an ETS has been set and there is no more interest in pursuing a carbon tax (Fernandez 2012). The government in Chile is also advocating a sectoral market mechanism and would most likely benefit from such an approach (Climate Focus 2012). Finally, *Peru* is exploring an ETS within the context of the PMR (The Partnership for Market Readiness 2013).

Carbon markets in Africa

Domestic attempts to set up carbon trading are certainly the least prevalent on the African continent. The only exception is *South Africa*, which in its budget for 2012 and 2013 planned a carbon tax covering all sectors and using an emission threshold of 60 percent, above which taxes will have to be paid. Furthermore, in its *National Climate Change Response Policy* of 2011, the objective of pricing carbon was stated and the current discussions focus on whether emission trading should be employed or not (Townshend *et al.* 2013: 375). *Morocco* is also thinking about carbon markets but it is at a very early stage in the process (The Partnership for Market Readiness 2013).

Various countries of the global South are thus actually moving forward with emission trading and various others are contemplating the possibility. Four overall issues are of importance for the following analysis. First, and not surprisingly, the emerging economies and again China have been much more involved than other developed countries. Furthermore, the currently emerging systems have a higher level of diversity (e.g. when it comes to coverage of sectors) mirroring the national economic circumstances and government intentions (something that has already become visible in the context of the CDM, see Fuhr and Lederer 2009). This also implies the involvement of a wider range of private actors that have to account for their emissions. Second, particularly the emerging carbon markets in Asia are seen as critical for the future of carbon trading overall, although these markets will most likely not be linked to a large extent for the foreseeable future. This, however, does not exclude the possibility that some of the Asian markets might link up with each other or with the schemes of Japan, Australia or New Zealand. Third, when we consider carbon markets in the South, we have to take into account that we are no longer just watching diffusion from the EU to the South but that also lessons learned from US or other Southern markets are being taken up (Tuerk *et al.* 2013). Finally, and most importantly, carbon trading has not yet received a major blow from the experiences of the North, but the domestic trading systems are being set up rather independently from the developments in the North. It thus seems to be worth explaining why Southern governments are so keen to instigate such schemes.

Reasons for the diffusion of carbon markets into the global South

Various reasons can be found to explain why carbon markets are being set up in the global South. In the following section, the chapter will argue that market mechanisms are simply one institution that provides some hope for emerging economies to curb their growing emissions. Second, an interest-based explanation shows that there are external as well as internal players that push and pull for carbon market activities. Third, and most importantly, the special characteristics of carbon markets fit the way in which all market activities are being planned and developed in most Southern countries. Unfortunately, it will not be possible to do justice to all the countries mentioned above as, for many, there are simply no data available at this point, and thus the following has a strong bias towards the emerging economies, particularly China, as this case is the one for which most research has been conducted.

Carbon market experiments – one among many

A good starting point for explaining why particularly emerging economies are experimenting with carbon markets is the simple fact that these countries are now emitting drastic amounts of CO_2. A couple of years ago, Wheeler and Ummel already spoke about 'another inconvenient truth', highlighting that the South is nowadays responsible for almost as many emissions as the North (Wheeler and Ummel 2007). Today, it is a well-known fact that China is not only the largest emitter in absolute terms but has also overtaken countries like France or Italy in per capita emissions, even if one discounts those emissions generated through the production of export goods (Olivier *et al.* 2011: 14). The diffusion of carbon market activities is thus, to some extent, simply due to the need to find instruments to curb the ever-growing emissions. Of course, none of these countries has an obligation under international law to reduce its emissions, but within the domestic context the need for mitigation – or at least a slowdown of the growth of carbon emissions – is generally acknowledged. This is partially due to the fact that Southern countries are also much more sensitive to the ramifications of climate change, e.g. owing to a higher proportion of people living in areas with a high level of extreme weather events. In all the countries mentioned above, market mechanisms are thus considered as one possibility to find a solution and they appear to be cost-efficient and effective ways of curbing the growing emissions (Mehling 2012: 287; for a critical evaluation of the notion of efficiency, see Lane 2012). Furthermore, elites in these countries hope to use market-based instruments as a means to curb low-carbon growth (Ward *et al.* 2012).

The schemes that are being set up are always embedded in command-and-control systems as well as information campaigns, and are accompanied by the theoretical knowledge that restructuring of the energy systems has to be undertaken in the future through various methods. For some countries, particularly China, some frustration with pure command-and-control mechanisms might also have led to the notion of experimenting with carbon markets as, for example, the simple closure of factories that was employed immediately before the 2008 Olympics cannot be repeated without serious negative economic externalities (Yu and Elsworth 2012: 15). Similarly, in India, regulators as well as some parts of the public are showing a high level of frustration with the compliance with existing command-and-control measures in the field of air pollution and are thus open to new approaches (Rosha and Freestone 2012: 350). Carbon markets are also perceived positively because they could increase the financial flows from the North to the South as Northern entities might in the future be interested in domestically generated credits from the South beyond Certified Emission Reductions (CERs) from the CDM (Mehling 2012: 288). Furthermore, there are strategic reasons for advancing domestic carbon markets, as their setting up would most likely allow the forgoing of any possible punishments in the form of carbon-based border tax adjustments that the EU is contemplating (for a good introduction to the problems related to the issue, see Dröge 2011).[1] Thus, carbon taxes and carbon markets are now the dominant alternative (Huang 2013). While the trading of carbon should not be perceived as the only instrument that governments of the global South should explore, there are discernable functional reasons why this is on the agenda.

Old and new carbon market constituencies

A catalytic role in advancing the idea of carbon market activities has to be given to international institutions, particularly the World Bank.[2] The World Bank has already been instrumental in initiating the CDM through its Prototype Carbon Fund, which dominated the early CDM market (Benecke *et al.* 2008). Pushing for domestic trading systems, the World Bank Partnership for Market Readiness (PMR) was announced at the Cancun Confernce of the Parties (COP) in 2010 and has since handed out USD 350,000 each to Brazil, Chile, China, Columbia, Costa Rica, India, Indonesia, Jordan, Mexico, Morocco, South Africa, Thailand, Turkey, Ukraine and Vietnam to set up a *Market Readiness Proposal* that is primarily supposed to identify capacity-building gaps in the field of carbon markets (The Partnership for Market Readiness 2013). These are, however, not all intended for national ETSs but also for NAMA crediting (Sterk and Mersmann 2011: 3). The PMR has held four technical workshops that have served as a basis for sharing technical expertise. The money can be upscaled to USD 8 million per country during the implementation phase – the PMR has an overall pledged budget of USD 100 million – in external technical support. From an organizational point of view, the PMR is supported by the World Bank's Carbon Finance Unit and is thus part of the larger carbon governance structure of the World Bank.[3] The World Bank's influence can thus be taken as a given and, particularly in smaller and less developed countries, it might be the major driver of carbon market activities; but it is definitely not the only one.

Besides the influence of the World Bank, the private sector in the North as well as in the South has a tremendous interest in setting up carbon market activities, although not much research on this is currently available. In a recent article, Mehling (2012: 288) points out that one advantage of carbon markets in Southern contexts is that they are easier to set up than other instruments as coalitions between economic and political actors are quite strong, something that was also visible in the early years of the EU ETS (Meckling 2011) and that is now, to some extent, also visible in countries like South Korea or China (see below for more details). It thus seems possible that a rather strong 'policy instrument constituency' (see Simons and Voß, this volume) is also developing in the global South.

The influence of domestic companies has already become visible in the CDM, whereby in China as well as in India, various domestic entrepreneurs have turned out as project developers or project financiers (Benecke 2009; Schröder 2009). Similarly, Brazilian companies, for example in the sugar industry, have been advocating the use of market instruments, as they perceive these markets as opportunities (Hultman *et al.* 2012). In addition, domestic private sector verifiers (in the lingo of the CDM: market designated operational entities) from China, Malaysia, Columbia or South Korea have become established players and are thus interested in initiating domestic markets.[4] There are also various reports from carbon asset management companies that are transnational in nature, but at least in China very often have a Chinese branch that buys and sells carbon credits. These companies have, however, been involved in only a few trades and, compared with the state-owned exchanges, these companies have not been particularly important to the market up to now (Huang 2013: 76). Finally, private sector entities from emerging economies have also been very active at side events of the climate-change conferences or at the various carbon trade expos. Overall, we still do not know enough about the political economy of these markets on the ground but there are some indicators that, at least in the emerging economies, some companies are pushing for carbon market developments and there is thus a domestic constituency visible that is putting forward the idea of carbon trading.

Carbon markets as state-led economic policy making

The developments mentioned so far (the necessity to act and the role of carbon market constituencies) do not, by themselves, explain the politics of carbon markets in the global South. The following thus argues that one has to pay close attention to how carbon market practices correspond to state-led economic policy making in most Southern countries (for an evaluation of the role of the state in the carbon markets of developed countries, see Lederer 2012).

The role of the state and how its economic constitution influences environmental policy making have, for a long time, been neglected in the literature. Only more recently has the ecological transformation literature started to discuss the 'greening of the state' as being part of the solution and not only contributing to the problem (e.g. Barry 2008; Eckersley 2004; Gough and Meadowcroft 2011; Meadowcroft 2012). Thus, the state has been highlighted as an important change agent in steering political processes in the environmental realm in OECD countries. So far, though, research on the role of the state in environmental transformations in the global South is scarce, except when it comes to the exploration of the effectiveness of environmental authoritarianism as an approach to achieve environmental goals, particularly in China (Beeson 2010; Gilley 2012; Schreurs 2011).

There is, however, a strong conceptual statist tradition dating back to the idea of the 'developmental state' that has been used to investigate the role of bureaucracies in the steering processes of economic development in the global South. Johnson introduced the concept of the 'developmental state' when he investigated the role of the Ministry of Planning in Japan (Johnson 1982). Others took up the term, analyzing different countries of South East Asia (Amsden 1989; Evans 1995; Johnson 1982; Wade 2004) and the following characteristics are usually said to be of importance for the existence of developmental states (this list draws primarily on Leftwich 1995: 405f.; Routley 2012; and Taylor 2012: 466):

- An elite – represented in the government of the respective state – that is pushing for development and a general rise in welfare, thus is not only interested in enriching itself.
- A weak civil society, but nevertheless the government possesses a high degree of legitimacy within the population.
- Effective management of non-state economic interests and access of economic players, particularly key capitalists, to the state without complete capture.
- A bureaucracy that is competent, powerful and above all autonomous from domestic pressures. Evans thus refers to the 'embedded autonomy' of the bureaucrats who are capable of steering growth-enhancing policy interventions, e.g. as 'custodians' or 'midwives' (Evans 1995), resulting in state-led industrial policies.

It is evident that not all countries of the South are 'developmental states' (traditionally only the East Asian tigers were said to have these characteristics) and certainly many of those mentioned above hardly fit this categorization, but it seems fair to summarize most developing and emerging countries mentioned above as late industrializers that share at least some of these characteristics. Using the conceptual frame of state-led development as a broad heuristic perspective, one might thus gain some insights into the politics of carbon markets in these states.

First, it is evident that the introduction of carbon markets has been sought by the elites of Southern countries. The initiation of carbon markets all over the world has never been a bottom-up process (see Matt and Okereke in this volume for the EU case) and carbon trading has had strong government backing, for example in China, from its very beginning (Han *et al.* 2012: 18), and it has been actively devised in the form of top-down policy experiments (Huang 2013). A similar story is visible in most Southern states. Governments in the South thus often perceive carbon markets as part of a larger industrial policy trying to become more independent from fossil-fuel imports (e.g. in South

Korea), to solve local air pollution (e.g. in India) or as an attempt to bring in new technologies (e.g. in the production of renewable energy technology in China (Lo 2013a)).

Second, and mirroring the first point, in most Southern countries where domestic carbon trading is planned, no strong opposition from civil society can yet be detected. The recent calls for more civil society authority in the running of Northern carbon markets (Spaargaren and Mol 2013: 185) is, at least for the moment, completely unrealistic for the global South. There are, of course, various highly critical opposition movements, particularly in many of the poorest CDM host countries (for details, see e.g. http://www.carbontradewatch.org/), but regarding domestic trading systems the opposition is much less and much less systematic than in the North, where civil society actors are often highly skeptical of cap-and-trade systems overall (e.g. in California) or lobby very hard for excluding offsets (e.g. NGOs via the European Parliament regarding the future inclusion of CERs in the third trading period). Whether opposition will rise, particularly in Latin American countries, where, for example, many NGOs are against the inclusion of REDD+ in carbon trading schemes, is of course an open question. For now, however, state-led policy is dominating the politics of carbon markets in the South.

Third, in all Southern countries, the setting up of carbon markets has been strongly influenced by key industrial players. In South Korea, industry representatives strongly influenced the timing of the initiation of the trading periods, but they did not succeed in stopping the government introducing an ETS. In China, industry representatives are involved in close discussion on how carbon trading will not lead to unfair domestic competition from companies in provinces that are not yet covered.

Finally, and most important for the overall argument, carbon markets – probably even more than any other market – depend on sound bureaucratic practices and the embeddedness Evans (1995) described so well. Various indicators show that bureaucracies in many countries of the global South are indeed guiding carbon market activities quite strongly:

• As the demand is completely artificial in all carbon markets, much depends on the government issuing a cap. Experiences from the first phase of the EU ETS have nicely shown that the 'invisible hand' of the price will simply remain complacent when the visible hand of the government has overallocated allowances. Thus, particularly in China, the current discussions within the bureaucracy are focusing on how the cap should be set in the seven provinces so that any reduction, and thus most likely also trading, does indeed occur. For the Beijing pilot, the draft design suggests that the government sets aside a certain amount of allowances that are only released if certain criteria are met (Yu and Elsworth 2012: 23), a measure that would eventually lead to the fine-tuning of demand – an instrument that many supporters of the EU ETS would wish the Commission had established at the initiation of carbon trading in Europe as well.

- Governments have realized that the uncertainty of the price and the potentially high volatility are detrimental to the market sending the right long-term signals to investors. Thus, more and more systems are adopting price ceilings or floors as 'cost containment.' In China, the credits from the CDM have always had a floor price of 8–9 euros, and it is very likely that the new systems will also use price floors and caps. In Mexico, a capped price is envisioned to avoid a loss of competitiveness of the economy. India has also established floor and price ceilings for its renewable energy trading system (Rosha and Freestone 2012: 349).

- All the existing cap-and-trade systems permit banking and borrowing of allowances over the end of specific trading periods, as would all new planned carbon markets. These allow emitters or potential investors to plan over a longer period of time and theoretically bring in a form of certainty that other markets usually do not enjoy to the same extent and this long-term perspective is much more conducive to state-led economies than to laissez-faire ones.

- Contrary to the image one receives when only following the international level, Southern governments and their bureaucracies have realized well that strong monitoring, reporting and verification (MRV) systems are highly necessary for carbon markets to work, an experience they also made with the CDM where MRV systems have rather successfully been employed in all emerging economies (Fuhr and Lederer 2009). This again follows a tradition of planning that is much more prevalent in the South than in the North.

The rather strong guidance of carbon markets by state actors is to some degree a result of more state involvement in finance overall (for a classical perspective that is still of interest today, see Haggard *et al.* 1993). Emerging and developing countries alike have liberalized their financial markets less than extreme visions of the 'Washington Consensus' would have suggested. Thus, the overall degree of financialization has been much smaller. Although it is debatable whether the South has been hit less severely by the current crisis (for an overview of the debate, see Coulibaly 2012; Didier *et al.* 2012; Nissanke 2012), the conclusion of governments in the South has not been to give up market approaches but rather to supervise and control them much more. The financial crisis has thus strengthened, not weakened, state-led development policies that are pro-market but not market-liberal.

All these elements do not guarantee that carbon markets will eventually be successful in those Southern countries where they will be employed in the sense that they reduce carbon emissions efficiently and effectively. However, taking all the indicators together, carbon markets seem to have at least a chance of making an impact. Thus, contrary to Han *et al.* (2012: 51f.) or Lo (2013a), who argue that the fact that China is not a classical free market economy is one of the dangers for using carbon markets in the country, as markets rely on free prices, proper accounting, exact measurement and legal certainty, this chapter reaches a different conclusion. In the end, the Chinese approach to steering and controlling markets might be particularly well suited to carbon trading, and, not only in China but in many countries of the global South, the political environment for employing carbon markets might be much more conducive than is generally appreciated. In a nutshell, carbon markets in the global South will not look like the classical free-trade arrangement market enthusiasts as much as some critics imagine, but they might resemble much more a hybrid mixing carbon taxes (the prices of which are controlled politically) and a market approach (whereby some trading is allowed for economic as well as political reasons) and thus they might in the end work better than appreciated at first glance.

Conclusion

This chapter has argued that we have not yet seen the end of carbon markets. Although they are indeed in poor shape and need urgent restructuring, many Southern countries are advancing plans to use carbon trading and, particularly in Asia, are also slowly implementing them. This cannot only be explained by the dominance of a neo-colonial discourse that Northern elites have successfully transported to the South. Rather, it can be explained by three interlinked arguments. First, Southern governments are experimenting with carbon markets due to the sheer pressure they face regarding their domestic emissions. Second, old and, more importantly, new constituencies seem to perceive carbon markets to be in their interest and are thus pushing for them in the political arena. Third, and of most relevance, the setting up of carbon markets corresponds very well to the tradition of state-led development that, to various degrees, is the way Southern governments and bureaucracies steer and oversee their economies. What broader lessons can be derived from this argumentation?

First, there might be some lessons for already-existing carbon markets in the North. This starts with ambition: Asian markets, in particular, are planning to include broader segments of the market (e.g. involving smaller facilities, buildings or agriculture, like in South Korea or in Shanghai). This should revive discussions in the EU and North American ETSs on how to broaden the scope of carbon trading, and could potentially increase not only the effectiveness but also the legitimacy of the instrument. Obviously, Northern regulators should also keep a close eye on the rather tight controls by which most of the markets under discussion will have to abide, and it might well be possible that some of these instruments (floor and ceiling prices and the concurrent move towards more hybrid systems mixing quantitative caps and price bands) could be re-imported into Northern trading schemes at a later point.

Second, a focus on domestic carbon market activities provides much-needed balance to the rather sad picture that climate governance displays at the international level. The rather ambitious attempts to set up carbon markets in the South show that non-Annex I governments are undertaking more at home than agreed upon internationally. This is to some extent due to the deep mistrust in international MRV systems as they might infringe sovereignty. Whether such a new bottom-up approach of national state-led initiatives will be enough to revitalize global carbon politics is, of course, highly doubtful, but at least not all hope is lost. Particularly, the outcome of the Chinese attempt to set up carbon markets will be of importance as many other emerging as well as developing countries might emulate the Chinese approach rather than the European or North American ones, strengthening the ecological dimension of the developing 'Beijing consensus' that asserts that more state-led and more autocratic steering is to be favored in contrast to the Western-led and market-oriented schemes. Lo is, therefore, right in stating that the Chinese ETS is the 'biggest test ever for the theory of carbon trading' and that the Chinese ETS 'will allow us to see whether or not a liberal market is necessary' (Lo 2013a: 73). If successful, this more authoritarian environmentalism would certainly have negative implications for the role of civil society and democratic governance, an issue that will certainly become much more dominant within the next couple of years (for an overview, see Gilley 2012).

Finally, whether the politics of carbon markets in the South might avoid becoming 'mechanisms as part of a new phase of intensified ecological exploitation and uneven capitalist growth' (Böhm *et al.* 2012: 1624), or whether we will instead potentially see a transformation of climate capitalism through the state-led politics of carbon markets in the global South seems to be a much more open question than the critical voices of carbon trading have acknowledged so far. Theoretically, the politics of carbon markets are thus a good example of the fact that governments should not be underestimated as key players. This is a lesson that mainstream IPE has tried to drive home for the last twenty years (e.g. Helleiner 1994; Lederer 2003) and it is now also being taken up by various scholars critical of neoliberalism who also see a much stronger role of the state in the North in bringing about financialization (Mirowski 2013; Panitch and Gindin 2012).[5] The open question that remains, nevertheless, is whether the much more explicit entanglement of Southern governments will, in the end, allow the many pitfalls that Northern (carbon) capitalism went through to be avoided, be that in the broad field of financial regulation or in the context of the politics of carbon markets.

Acknowledgments

I am very grateful to Anita Engels, Richard Lane, Elah Matt and Benjamin Stephan for extremely helpful comments on earlier versions of this chapter.

Notes

1 I am thankful to Wolfgang Sterk from Wuppertal Institute for introducing me to the issue of carbon-based border tax adjustments.
2 The regional development banks are also of importance, particularly the Asian Development Bank, which, for example, approved a loan of USD 750,000 for the Tianjin ETS in China in 2011 (Yu and Elsworth 2012). Similarly, national donors like the German International

Development Cooperation (GIZ) or the British Department for International Development (DFID) are heavily involved not only in China but in most of the countries mentioned.

3 Another external player is the International Carbon Action Partnership (ICAP), which was set up by fifteen governments in 2007 in order to share experience and evaluate best practices. Its ultimate objective is to initiate a global cap-and-trade system and its main interest has thus been in compatibility and design issues. Its members are, however, only OECD countries and no classical members of the South participate in the meetings (South Korea, Kazakhstan, Ukraine and Japan are observers).

4 The full list of DOEs is available at http://cdm.unfccc.int/DOE/list/index.html (last accessed 15 March 2013).

5 I am thankful to Richard Lane for pointing out this parallel.

References

Aldy, J. E. and Stavins, R. N. (2012) The promise and problems of pricing carbon: theory and experience. *Journal of Environment and Development*, 2 (2), 152–180.

Amsden, A. (1989) *Asia's Next Giant: South Korea and late industrialization*. New York: Oxford University Press.

Barry, J. (2008) Towards a green republicanism: constitutionalism, political economy, and the green state. *The Good Society*, 17 (2), 3–11.

Beeson, M. (2010) The coming of environmental authoritarianism. *Environmental Politics*, 19 (2), 276–294.

Benecke, G. (2009) Varieties of carbon governance: taking stock of the local carbon market in India. *Journal of Environment and Development*, 18 (4), 346–370.

Benecke, G., *et al.* (2008) From public-private partnership to market. The Clean Development Mechanism (CDM) as a new form of governance in climate protection. *SFB Governance Working Paper Series 10*. Berlin: SFB 700.

Böhm, S. (2013) Why are carbon markets failing? *Guardian Sustainable Business Blog*. Available from: http://www.guardian.co.uk/sustainable-business/blog/why-are-carbon-markets-failing [accessed 14 January 2014].

Böhm, S., Misoczky, M. C. and Moog, S. (2012) Greening capitalism? A Marxist critique of carbon markets. *Organization Studies*, 33 (11), 1617–1638.

Chestney, N. (2012) *Factbox: Carbon trading schemes around the world*. Reuters. Available from: http://www.reuters.com/article/2011/07/11/us-carbon-schemes-idUSTRE76A2GJ20110711 [accessed 26 September 2012].

Climate Focus (2012) *Scaled-up Crediting Mechanism: options and recommendations for Chile*. Amsterdam: Climate Focus.

Coulibaly, B. (2012) *Monetary Policy in Emerging Market Economies: what lessons from the global financial crisis?* Washington, DC: Board of Governors of the Federal Reserve System.

Didier, T., Hevia, C. and Schmukler, S. L. (2012) How resilient and countercyclical were emerging economies during the global financial crisis? *Journal of International Money and Finance*, 31 (8), 2052–2077.

Dröge, S. (2011) Using border measures to address carbon flows. *Climate Policy*, 11 (5), 1191–1201.

Eckersley, R. (2004) *The Green State: rethinking democracy and sovereignty*. London: MIT Press.

Evans, P. (1995) *Embedded Autonomy: state and industrial transformation*. Princeton: Princeton University Press.

Fernandez, I. (2012) *Decision Making and Policy Choice. Why an Emissions Trading Scheme*. Partnership for Market Readiness Technical Workshop Domestic Emissions Trading: Ministry of Energy, Chile.

Fuhr, H. and Lederer, M. (2009) Varieties of carbon governance in newly industrializing countries. *Journal of Environment and Development*, 18 (4), 327–345.

Gilley, B. (2012) Authoritarian environmentalism and China's response to climate change. *Environmental Politics*, 21 (2), 287–307.

Gough, I. and Meadowcroft, J. (2011) Decarbonizing the welfare state. In: J. S. Dryzek, R. B. Norgaard and D. Schlosberg, eds., *The Oxford Handbook of Climate Change and Society*. Oxford: Oxford University Press, 490–503.

Haggard, S., Lee, C. H. and Maxfield, S., eds. (1993) *The Politics of Finance in Developing Countries*. Ithaca: Cornell University Press.

Han, G. et al. (2012) *China's Carbon Emission Trading: an overview of current development*. Stockholm: SEI.

Helleiner, E. (1994) *States and the Reemergence of Global Finance: from Bretton Woods to the 1990s*. Ithaca: Cornell University Press.

Huang, Y. (2013) Policy experimentation and the emergence of domestic voluntary carbon trading in China. *East Asia*, 30, 67–89.

Hultman, N. E. et al. (2012) Carbon market risks and rewards: firm perceptions of CDM investment decisions in Brazil and India. *Energy Policy*, 40, 90–102.

IETA (2012) *Greenhouse Gas Market 2012: new markets, new mechanisms, new opportunities*. Geneva: IETA.

Johnson, C. (1982) *MITI and the Japanese Miracle: the growth of industry policy 1925–1975*. Stanford: Stanford University Press.

Lane, R. (2012) The promiscuous history of market efficiency: the development of early emissions trading systems. *Environmental Politics*, 21 (4), 583–603.

Lederer, M. (2003) *Exchange and Regulation in European Capital Markets*. Münster: Lit.

Lederer, M. (2012) Market making via regulation: the role of the state in carbon markets. *Regulation & Governance*, 6, 524–544.

Lederer, M. (2013) The future of carbon markets: carbon trading, the Clean Development Mechanism and beyond. In: U. Frauke and J. Nordensvärd, eds., *Low Carbon Development: key issues*. London: Routledge/Earthscan, 94–106.

Leftwich, A. (1995) Bringing politics back in: towards a model of the developmental state. *The Journal of Development Studies*, 31 (3), 400–427.

Lo, A. (2013a) Carbon trading in a socialist market economy: can China make a difference? *Ecological Economics*, 87, 72–74.

Lo, A. (2013b) Carbon trading in the Asian century: China's ETS on track. *The Conversation*. Available from: http://theconversation.com/carbon-trading-in-the-asian-century-chinas-ets-on-track-11438 [accessed 15 January 2013].

Lohmann, L. and Böhm, S. (2012) Critiquing carbon markets: a conversation. *ephemera: theory & politics in organization*, 12 (1/2), 81–96.

MacNeil, R. and Paterson, M. (2012) Neoliberal climate policy: from market fetishism to the developmental state. *Environmental Politics*, 21 (2), 230–247.

Meadowcroft, J. (2012) Greening the state? In: P. F. Steinberg and S. D. VanDeveer, eds., *Comparative Environmental Politics: theory, practice, and prospects*. Cambridge: MIT Press, 63–87.

Meckling, J. (2011) *Carbon Coalitions: business, climate politics, and the rise of emissions trading*. Cambridge: MIT Press.

Mehling, M. (2012) Between twilight and renaissance: changing prospects for the carbon market. *Carbon & Climate Law Review*, 4, 277–290.

Mirowski, P. (2013) *Never Let a Serious Crisis Go to Waste: how neoliberalism survived the financial meltdown*. London: Verso.

Newell, P. and Paterson, M. (2010) *Climate Capitalism: global warming and the transformation of the global economy*. Cambridge: Cambridge University Press.

Nissanke, M. (2012) Introduction: transmission mechanisms and impacts of the global financial crisis on the developing world. *The Journal of Development Studies*, 48 (6), 691–694.

Olivier, J. G. J. *et al.* (2011) *Long-term trend in global CO₂ emissions. 2011 report*. The Hague: PBL Netherlands Environmental Assessment Agency.

Panitch, L. and Gindin, S. (2012) *The Making of Global Capitalism: the political economy of American empire*. London: Verso.

Reyes, O. (2012) Carbon markets after Durban. *ephemera: theory & politics in organization*, 12 (1/2), 19–32.

Röser, F. *et al.* (2012) Nationally appropriate mitigation actions (NAMAs) and carbon markets. *Policy Update, issue IV*. Utrecht: Ecofys.

Rosha, A. and Freestone, D. (2012) A green emerging market: India's experiments with market based mechanisms for climate mitigation. *Carbon & Climate Law Review*, 4, 342–353.

Routley, L. (2012) *Developmental States: a review of the literature*. Manchester: Effective States and Inclusive Development Research Centre (ESID).

Schreurs, M. A. (2011) Climate change politics in an authoritarian state: the ambivalent case of China. *In:* J. S. Dryzek, R. B. Norgaard and D. Schlosberg, eds., *The Oxford Handbook of Climate Change and Society*. New York: Oxford University Press, 449–463.

Schröder, M. (2009) Varieties of carbon governance: utilising the CDM for Chinese priorities. *Journal of Environment and Development*, 18 (4), 371–394.

Shrimali, G., Tirumalachetty, S. and Nelson, D. (2012) *Falling Short: an evaluation of the Indian Renewable Certificate Market*. San Francisco: Climate Policy Initiative.

Sopher, P. and Mansell, A. (2013) *Kazakhstan: the world's carbon markets: a case study guide to emissions trading*. Washington, DC: EDF and IETA.

Spaargaren, G. and Mol, A. P. J. (2013) Carbon flows, carbon markets, and low-carbon lifestyles: reflecting on the role of markets in climate governance. *Environmental Politics*, 22 (1), 174–193.

Sterk, W. and Mersmann, F. (2011) Domestic emission trading systems in developing countries: state of play and future prospects. *JIKO Policy Paper 2/2011*. Wuppertal: Wuppertal Institute for Climate, Environment and Energy.

Taylor, I. (2012) Botswana as a 'development-oriented gate-keeping state': a response. *African Affairs*, 111 (444), 466–476.

The Partnership for Market Readiness (2013) *Shaping the Next Generation of Carbon Markets*. Washington, DC: World Bank.

Townshend, T. *et al.* (2013) *Climate Legislation Study: a review of climate change legislation in 33 countries*. 3rd ed. London: World Summit of Legislators.

Tuerk, A. *et al.* (2013) *Emerging Carbon Markets: experiences, trends, and challenges*. London: Climate Strategies.

Wade, R. H. (2004) *Governing the Market: economic theory and the role of government in East Asian industrialization*. 2nd ed. Princeton: Princeton University Press.

Ward, J. *et al.* (2012) Self-interested low-carbon growth in Brazil, China, and India. *Global Journal of Emerging Market Economies*, 4 (3), 291–318.

Wheeler, D. and Ummel, K. (2007) Another inconvenient truth: a carbon-intensive South faces environmental disaster, no matter what the North does. *Working Paper 134*. Washington, DC: Center for Global Development.

Yu, G. and Elsworth, R. (2012) *Turning the Tanker: China's changing economic imperatives and its tentative look to emissions trading*. London: Sandbag.

8 Carbon governance in China by the creation of a carbon market

Anita Engels, Tianbao Qin and Eva Sternfeld

Introduction

Anyone interested in global CO_2 emission levels or the emergence of global carbon markets must eventually consider China and its rapidly growing CO_2 emissions. In an attempt to reconcile energy security needs with environmental goals, the Chinese government has voluntarily announced that it intends to reduce the CO_2 intensity per unit GDP of its industrial production; however, its total emissions are nevertheless expected to grow until 2030. In 2013, seven pilot Emission Trading Schemes (ETSs) were introduced to experiment with this policy tool, in addition to other instruments that might help the government to achieve its goals. However, China's experiments with ETSs face poorly developed carbon markets, and the effectiveness of this approach is a matter of fierce debate. While some emphasise that carbon markets provide incentives for developing countries without the imposition of binding emission targets, which allows them to implement voluntary mitigation policies (Carbon Finance at the World Bank/Ecofys 2013: 28), critics doubt that carbon markets will contribute to the transition towards a low-carbon economy. Instead, they fear that these markets will lead to a variety of negative side effects such as increased land-grabbing in developing countries, rising inequality and a further decrease in the democratic quality of the international carbon regime (Alianca Redes de Cooperacao Comunitaria Sem Fronteiros Asociacion Ambiente y Sociedad *et al.* 2013). The weakness of the European Union ETS (EU ETS) prompts another caveat: if regulators cannot prevent the market from becoming oversaturated with unused emission allowances, the market might function as a price setting device (as a low price reflects the supply and demand relationship), but it would not deliver any environmental improvements.

The current state of the debate on carbon markets calls for two types of contributions. The first would be to improve our understanding of how the politics of carbon markets lead to specific market structures in various political contexts, how these markets function and whether and how these markets serve as an incentive to reduce carbon emissions. The second would be to engage in a more fundamental critical discourse on the potentially far-reaching societal consequences of ongoing marketisation and to search for viable alternatives to carbon markets. In this chapter, we confine our efforts to the former perspective, while our future research will also delve into the latter perspective.

The goals of this chapter are, therefore, to describe and analyse the basic features of the pilot ETSs in China and how they are embedded in the broader approach to carbon governance (see also Lederer in this volume). To do so, we adopt a theoretical perspective on the politics of carbon markets informed by economic sociology and law and society studies (Ellickson 1991; Fourcade 2007; Friedman 1975). The following sections are structured as follows: we begin with an outline of our theoretical framework, from which we derive a set of specific questions that we wish to address. This is followed by a description of the process by which ETSs were introduced in China. The next section presents a deeper analysis using our theoretical framework, followed by some preliminary answers to our research questions.

Carbon governance: policy goals, lawmaking and the creation of markets

We begin our analysis using the concept of carbon governance. Various agreements and programmes worldwide highlight a growing desire to reduce CO_2 and other greenhouse gas (GHG) emissions in the long run, even if this regime is not yet binding for most states. Carbon governance, in this sense, refers to the definition of rules and institutional settings for legitimate wealth accumulation under the conditions of a carbon-constrained future. The term 'governance' encompasses various policy instruments that can be applied individually or in combination. In recent years, the discussion has focused on the diffusion of market-based policy instruments (Jordan *et al.* 2003). In recognition of the changing role of states and the limited role of governments as rule makers in any international regime (Leibfried and Zürn 2005), governance comprises both state and non-state rules and institutions, established by various actors to regulate the carbon externalities of economic activities. Governance is thus not regulation without government, but instead refers to the specific combinations of various (state and non-state) sources of power and enforcement. This is of particular importance for analysing governance in China, as the relationship between the state and the ruling party, and that between the legal framework and the policies implemented by the ruling party, are crucial for understanding the institutional preconditions for carbon markets. We build on the work of Fuhr and Lederer (2009), who employed the concept of varieties of carbon governance to examine the myriad ways in which the Clean Development Mechanism has been introduced in newly industrialised countries such as Brazil, China and India (Benecke 2009; Friberg 2009; Schroeder 2009). We apply the concept to ETS with an emphasis on policy goals and lawmaking.

If carbon governance includes the creation of markets for tradeable emission rights, it is useful to draw on insights from economic sociology to analyse the specific dynamics of these emerging carbon markets. Economic sociology – and in particular the sociology of markets – provides insights into how markets are created and maintained and how economic actors make sense of markets (Fourcade 2007). This perspective emphasises the institutional work required to create and maintain markets (Levin and Espeland 2002) and the political nature of this process (Fligstein 2001). Empirical markets nearly always differ from the abstract, ideal market of classical economic theory, and economic sociology has adopted a pluralistic perspective on markets, with numerous empirical studies demonstrating that each concrete market operates under very specific conditions (Callon 1998). Markets are institutionally embedded in the broader societal contexts in which they exist. This theoretical perspective would therefore lead us to expect strong evidence of institutional imprinting in the creation of new carbon markets. A comparative analysis of carbon markets in China, the US and Australia, for example, would reveal a variety of outcomes in terms of the specific market design characteristics and the effects of these designs on carbon pricing and the strategies of market participants. Prior studies have demonstrated such variation in the implementation of the EU ETS in EU member states, which differ in their macro-economic institutional settings and regulatory traditions (Engels *et al.* 2008).

To account for the specific conditions of carbon governance in China, we employ the term adaptive governance. This concept was introduced to describe the ability of public policy and economics to respond to the challenges of global changes, especially in the fields of natural resource policy management and environmental policy (Hatfield-Dodds *et al.* 2007). Heilmann and Perry (2011) employed the term to characterise the Chinese political system's resilience and ability to adapt to the complexity and uncertainty of an emerging economy. Therefore, this concept is well suited to investigate the introduction of markets for tradeable emission rights as an active policy to improve energy efficiency and urban quality and control GHG emissions in China. The ongoing, gradual process of transformation from a system of centralised planning to a market-based system is crucial for our investigation of the Chinese ETSs. In November 2013, the Third Plenary session of the 18th Party Congress agreed on further economic reforms, including the liberalisation of the financial sector, opening up of economic sectors (including energy and logistics) to private and international investors and a strengthening of market mechanisms. It will be interesting to examine the extent to which economic liberalisation and relaxed state control over key industries provide an improved environment for the implementation of ETSs and how this relates to the traditionally strong influence of the authoritarian developmental state (see Lederer in this volume). *What institutional imprinting can be observed in the creation of the pilot ETSs, and how does this influence the pricing of carbon emissions?* The use of a policy of cautious experimentation is another feature of adaptive governance. The introduction of ETSs in selected pilot regions and cities can be regarded as an example for this type of policy. *It will be important to analyse the type of incentive structure this creates for market participants and other interested actors. What lessons can be drawn from this policy for the introduction of a national scheme?*

While other commentators have focused on problems with data collection, accountability and a shortage of technical knowledge, human resources and market-oriented behaviour (Han *et al.* 2012; Wang 2013), our focus is on the relationship between policies and instruments, on the one hand, and the creation of a legal framework on the other. The pilot ETSs were introduced as a social experiment in the field of carbon governance in China. The extent to which this approach will actually achieve transparent and reliable market conditions remains an open question. The Chinese lawmaking tradition can be characterised by a preference for a rather vague and general legal framework, even for addressing complex problems, such as with ETSs, which require highly specific regulations to become workable.[1] This typically leads to weak law enforcement and creates the potential for ad hoc local influences. The specific relations between the central and regional or local government levels shape the practical implementation of ETSs. *We will examine these relations and assess the extent to which transparent legislation and efficient law enforcement have been achieved in the ETSs. What can we determine about the relationship among policy goals, the general legal framework and specific ETS regulations?*

At present, little systematic data is available on the strategies and market behaviours of companies involved in the pilot ETSs. However, the questions generated in this section will guide our description of the empirical experiences with the ETSs, and we will focus on the question of how institutional imprinting in the pilot ETSs and the creation of the legal framework affect the potential outcomes of carbon markets in China.

Experimenting with ETSs in China

From energy governance to carbon governance: experimenting with ETSs as an additional tool to achieve CO_2 intensity targets

China is not subject to a binding emission target in the first Kyoto period. However, since the early 2000s, China has experienced economic growth accompanied by emissions increases that substantially outpaced any scenario foreseen during the Kyoto Protocol negotiations.[2] Chinese CO_2 emissions increased by more than 150 per cent over the last decade, and by 2012 China contributed 29 per cent of the total global CO_2 emissions (PBL 2012: 6, 11). This makes it increasingly difficult for Chinese UNFCCC negotiators to maintain the principle of 'common but differentiated responsibilities'. Instead, the country finds itself under pressure to play a more active role in the future. China's mitigation efforts will be crucial for the success or failure of global carbon governance.

Moreover, China faces domestic challenges. Its economic growth led hundreds of millions of its people out of poverty, but this came at a high cost. The massive increase in fossil fuel consumption was accompanied by a worsening energy and environmental crisis. China's current carbon governance efforts are therefore closely linked to policies promoting energy efficiency and clean energy sources to slow growth in the demand for fossil fuels.

In the first decade of this century, Chinese carbon governance was essentially a by-product of measures targeting energy security. The 11th Five Year Plan (FYP) (2006–2010) directly addressed the energy crisis by establishing a goal of improving energy efficiency by 20 per cent per unit GDP by 2010.[3] The critical year for the implementation of policies explicitly addressing climate change issues was 2007. At that time, the term *low-carbon economy* (LCE 低碳经济) appeared on the political agenda (Hofem and Heilmann 2013).[4] The first *National Climate Change Assessment Report* (MoST 2007) demonstrated that China is extremely vulnerable to climate change and more active attitudes concerning domestic climate issues were required.[5] The National Development and Reform Commission (NDRC) published *China's National Climate Change Programme* (NDRC 2007). This policy paper reads as a reinterpretation of energy security policies in terms of their relevance for climate protection, but it lacks detailed and specific policies to address increasing GHG emissions. 'Restructuring the economy, promoting technology advancement and improving energy efficiency' (NDRC 2007) are mentioned as the basic strategies for mitigating GHG emissions. The programme includes calculations of the amount of CO_2 emissions saved by improved energy efficiency.

Since 2008, China has issued an annual White Paper 'China's policies and actions for addressing climate change' (State Council 2008, 2009, 2011a, 2012, 2013). These policy papers reveal that China's carbon governance increasingly became defined in terms of policy instruments and measurable goals. Between 2008 and 2010, new institutions were established, including the China Clean Development Mechanism (CDM) Fund and Management Centre, and several exchanges specialising in trading environmental and energy-related products.

The establishment of an infrastructure for a carbon trading scheme is an example of the pragmatism and curtailed experiments typical of the Chinese politico-economic system. In 2010, the NDRC announced that five provinces (Guangdong, Liaoning, Hubei, Shanxi and Yunnan) and eight cities (Tianjin, Shenzhen, Chongqing, Xiamen, Hangzhou, Nanchang, Guiyang and Baoding) had been designated as low-carbon experimental zones. These provinces and cities, which represent a spectrum of regions with different economic and industrial development levels, are required to develop strategies for low-carbon development, promote low-carbon industries and establish a monitoring and management system (NDRC 2010).

Strategies for low-carbon development and measurable targets for CO_2 intensity were included in the 12th FYP (2011–2015), which was the first to explicitly address climate protection and low-carbon development. The plan aims to increase the share of non-fossil energy sources in the mix of primary energy sources to 11.4 per cent.[6] In addition to the target of improving energy efficiency by 16 per cent per unit GDP, it also establishes a target of reducing CO_2 intensity by 17 per cent per unit GDP. The carbon intensity cap is an important step towards the introduction of an ETS. *The Work Plan on Controlling GHG Emissions during the 12th FYP*, which was released by the NDRC in 2011, provides details on the low-carbon policies included in the 12th FYP (State Council 2011b). Similar to EU policies, it includes a 'burden sharing' principle; that is, the differentiation of energy efficiency and carbon intensity targets according to provinces' development status.[7] The targets are highly ambitious, although – in contrast to an absolute carbon cap – they still allow for additional emissions growth.

The 12th FYP also includes market-oriented instruments such as environmental taxation and the gradual introduction of ETSs for selected 'low carbon demonstration sites'.[8]

The creation of carbon markets

Experimenting with different forms of carbon markets and market-based policies

The Chinese roadmap for implementing a national ETS system during the next FYP (2016–2020) reveals that the national scheme will be based on several types of experiences with market-based approaches (Figure 8.1). Experiences with market-based mechanisms in China began with the CDM in 2005, continued with voluntary emission reductions (VER) trading by domestic enterprises, and included experiments with SO_2 emissions trading during the 11th FYP.

China's first entry into carbon markets occurred when it endorsed the CDM in which it rapidly emerged as the major global supplier of certificates for emission reduction (CERs) (UNFCC 2013) and became the largest beneficiary of the Flexible Mechanisms defined by the Kyoto Protocol. Although the CDM targeted international buyers from Annex 1 countries and hence had no direct impact on the GHG emissions of domestic enterprises, it provided some experience and capacities for establishing a domestic trading scheme. The mode of governance was shaped by the substantial involvement of the central government. The state promoted CDM projects and actively interfered in negotiations between Chinese companies and foreign project developers and

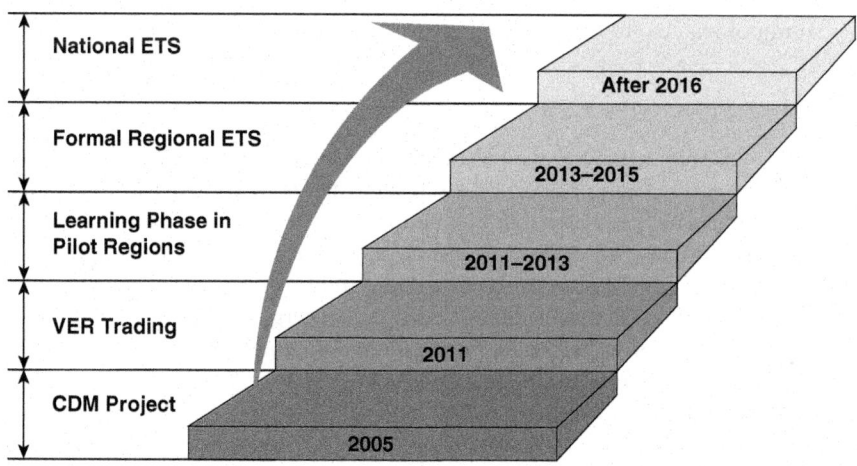

Figure 8.1 Overall layout – roadmap for China's ETS

Source: T. Qin's elaboration

buyers of CERs. Through these interventions, a price floor was established that, due to the sheer size of the Chinese CDM market, automatically established a global price floor for CERs (Lewis 2010; Schroeder 2009). China introduced a carbon emission trading infrastructure led by the China Beijing Environmental Exchange, Shanghai Environment Energy Exchange and Tianjin Climate Exchange, which had all been established in 2009. These exchanges are registered as corporations but enjoy strong government support and are linked with international exchanges such as the Chicago Climate Exchange and the (now defunct) European BlueNext (Han *et al.* 2012: 16). Thus far, they have primarily handled CDM and so-called VER (Han *et al.* 2012: 20). Public trading of carbon emission credits on these exchanges was another approach meant to help China set the prices in the global carbon emissions market. However, these exchanges only served as platforms for individual, small-scale deals.[9] After two years of preparation, the *Interim Measures for the Management of Voluntary Greenhouse Gases Emission Reduction Trading in China* were issued in June 2012 (The Climate Group 2012). The 11th FYP (2006–2010) established a mandatory national SO_2 emission reduction target of 10 per cent. The market-based SO_2 trading system was introduced to complement command-and-control instruments.[10] The success of experiments in pilot SO_2 trading projects at the city and provincial levels was limited for a variety of reasons. For example, SO_2 trading schemes lacked legislation concerning emission rights, and financial penalties were not sufficiently high to serve a prohibitive function. Finally, local authorities were interested in economic growth rather than environmental protection and not particularly active in ensuring compliance (Qin 2012). Despite these shortcomings, experiences with SO_2 trading provided valuable lessons for developing a carbon trading scheme. The introduction of market-based instruments and ETSs for selected pilot municipalities and provinces (Beijing, Tianjin, Shanghai, Chongqing, Shenzhen, Hubei and Guangdong) are planned as an important step towards China's introduction of a national ETS.

Description of the pilot ETSs

We now examine in greater detail three of the seven pilot cities and regions that officially began trading in 2013. We selected the Shenzhen Special Economic Zone (SEZ), which is situated in Guangdong, Hubei Province and the City of Shanghai to discuss the current situation of pilot ETSs in China (Table 8.1). They are interesting examples because they differ with respect to economic strength, the industrial composition and related energy demands.

With 105 million inhabitants, Guangdong is China's most populous province. With a per capita GDP of US$ 8,500 (2012) and representing 11 per cent of the country's GDP, it is also one of China's most productive economic regions and home to many export-oriented industries (National Bureau of Statistics 2013). Moreover, it has high energy consumption and accordingly high GHG emissions. Per capita CO_2 emissions are higher than the global average.[11] Guangdong was among the provinces that substantially failed to achieve their energy efficiency targets in the 11th FYP. According to the 12th (current) FYP, the province is now required to reduce CO_2 emissions per unit GDP by 35 per cent by 2015 (relative to 2005) and by 45 per cent by 2020.

At the sub-provincial division level, Shenzhen enjoys the same type of autonomy as any provincial capital city in terms of policy making in economics and law; that is, it has the authority to develop local economic and legal policies, at least in some well-defined respects. In addition, the legacy of being an SEZ provides Shenzhen with greater flexibility to experiment with nascent policies, given its achievements in economic development over the last 30 years under more progressive approaches. Thus far, the China Shenzhen Emission Rights Exchange has accumulated experience in voluntary emission reduction, in preparation for the forthcoming mandatory scheme in the city. On 18 June 2013, with the implementation of the carbon emissions trading platform, Shenzhen became the first official carbon trading pilot city and mandatory carbon market in China. The prices established in the initial rounds of trading were below four euros per one-ton CO_2 allowance.

Hubei, with 57 million inhabitants and a per capita GDP of approximately US\$ 2,800 (2012), ranks near the middle of Chinese provinces (National Bureau of Statistics 2013). Its pilot ETS was to be officially launched by the end of 2013; specific trading rules and supporting programmes are currently being developed. According to the plan, 150 companies, including private and state-owned companies, with annual energy consumption exceeding 60,000 tons of standard coal will be subject to mandatory carbon trading. Their total volume of CO_2 emissions accounts for 35 per cent of total carbon emissions in Hubei and 52 per cent of Hubei's industry in 2011.[12] As the only pilot province located in central China, Hubei is seeking a feasible approach to an ETS appropriate for the economic and social characteristics of the central region.

Guangdong initiated its pilot ETS in September 2012 and Shenzhen followed in May 2013. The pilot ETS in Hubei remains in the planning stage. Thus, it is too early to draw clear conclusions from the pilot ETSs. However, the central government established a precondition for the implementation of the seven pilot regions' ETSs that their *Working Plan for Regional Pilot ETS* must be approved by the NDRC. The working plan plays a crucial role in the pilot ETSs. However, the draft procedures and decision-making process for the working plan are not available to the public. For example, while Shenzhen has initiated its pilot ETS, its working plan has not been made available to the public.

The roadmap for the creation of a national Chinese ETS can be summarised along five dimensions: from voluntary to compulsory; from targeting specific industries to the entire economy; from regional pilots to the national market; from the spot market level to secondary financial markets; and from the domestic level to the international level. However, the differences between the pilot regions make it difficult to imagine how an integrated national scheme can be developed on the basis of this pilot phase. Alternatively, one could imagine a differentiated approach in which rules and regulations vary according to the

Table 8.1 Comparison of three ETS pilots

Allowance Allocation and Management of ETSs in three pilot areas

	Hubei	Guangdong	Shenzhen
Setting the cap (cap and trade model)	based on national and provincial carbon emissions control targets (top-down approach)	based on national and provincial carbon emissions control targets (top-down approach)	based on national and provincial carbon emissions control targets (top-down approach)
Mandatory carbon intensity targets (to be achieved by 2015)	17% (set by the central government)	19.5% (set by the central government)	15% (set by the central government); increased by Shenzhen's local government to 21%
Determining which emitters are subject to the regulation	Annual comprehensive energy consumption exceeds 60,000 tons of standard coal in either 2010 or 2011; approximately 150 companies	Annual comprehensive energy consumption exceeds 10,000 tons of standard coal or annual total CO_2 emissions volume exceeds 20,000 tons in each year from 2010–2014; approximately 202 companies	List of companies/emitters: 635 companies and 197 government buildings. Their total CO_2 emissions volume is 31,730,000 tons, accounting for 38% of total carbon emissions in the city in 2010
Allocation of allowances	Free allocation Gradual increase through auctions	Free allocation Gradual increase through auctions	Free allocation Gradual increase through auctions
Regulation concerning the use of carbon offsets and CCER (China CER)	2012 Interim Regulations of Voluntary Greenhouse Gases Emission Trading in China. Derived from the projects within the Hubei administrative area ≤10%	2012 Interim Regulations of Voluntary Greenhouse Gases Emission Trading in China. The Provincial Development and Reform Commission (PDRC) of Guangdong will draft the regional measures for the management of CCER. The CCER of the project in Guangdong concerning the national or regional measure could fall under the ETS	– unknown –
Regional interim ETS measures of (legal basis)	In draft	In draft The working plan for the Guangdong ETS pilot was issued in 2012	In draft Several Provisions on Carbon Emissions Management for Shenzhen Special Economic Zone was issued in 2012

Source: Hubei Interim Measures for the Management of ETS (un-published draft); the working plan for ETS pilot of Guangdong: http://zwgk.gd.gov. cn/006939748/201209/t20120914_343489.html; http://www.cerx.cn/cn/trade_details.aspx?ArticleID=274

specific demands of each province. In particular, the selection of emitters subject to the regulation varies, as do the interim regulations for the management of the ETS and the determination of the extent to which emitters can employ carbon offsets and CERs to comply with the ETS. It is unclear whether the central government seeks to develop a harmonised approach (a combination of the different schemes) or whether one of the schemes will set the standard. Moreover, inter-regional carbon trading is not allowed during the pilot phase.

Creating the legal basis for an ETS in China

The pilot ETSs were introduced as a social experiment in the field of Chinese carbon governance. A transparent and reliable legal basis would be a critical element for establishing sound ETSs in China, especially in the pilot phase. At the central government level, China frequently employs command-and-control measures to achieve social, economic and political goals. It is therefore unsurprising that the 12th FYP, including its *Working Plan on Controlling GHG Emissions during the 12th FYP*, includes numerous mandatory intensity and efficiency targets to achieve its commitment. China intends to achieve these targets through innovation: not only do provinces have greater autonomy in deciding how they are achieved, but market mechanisms will be used to achieve them. The introduction of market mechanisms is based on an understanding that command-and-control measures alone are not sufficient to reach carbon intensity targets. Regardless of the substantial differences in regional social and economic bases, the complexity of an ETS at the national level and the disappointing performance of the SO_2 ETS, the central government decided to develop a common national framework via a bottom-up approach. The central government simply established targets at the provincial and city levels, leaving the specific details to the seven pilot ETSs.

In 2011, the NDRC issued the *Notice for Launching Pilot ETSs*, which establishes a basis for the validity of the ETSs in the seven pilot areas. The NDRC determined that a *Working Plan for Local Pilot ETS* would have to be developed by the local government as a guideline for the implementation of the ETS (approved by the NDRC). However, a working plan is not a law but a guideline for regional political strategies. Only in the next step, when the pilot ETS began operation, were regional governments expected to draft and create the local regulations. There are two different lawmaking approaches that the pilot regions can adopt to ensure that their efforts have a proper legal basis, the 'Provisions Mode' and the 'Interim Measures Mode'. While the 'Provisions Mode' is akin to a policy guideline that only provides broad legal references, the 'Interim Measures Mode' contains more specific and detailed legislative guidelines. An example of the latter is the Shanghai pilot ETS, which began operation at the end of 2013, while Shenzhen serves as an example of the former. In the next section, we explore the implications of these different approaches for the development of a legal basis in the various pilot schemes (Figure 8.2).

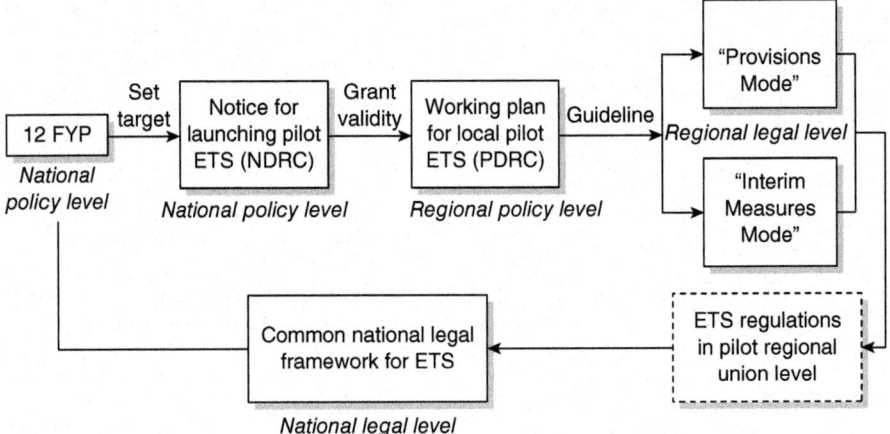

Figure 8.2 ETS roadmap based on the interaction between law and policy

Source: T. Qin's elaboration

With the inauguration of the pilot schemes, each of the seven pilot municipalities and provinces worked to design their own local ETS regulations and policies in their respective administrative areas. However, due to the differences in previous carbon trading experience, industrial structure, economic development and capacities to develop climate change policies, the actual progress varies across regions.[13] Despite these differences, the local lawmaking is currently the most dynamic and vigorous aspect of the pilot ETS phase.

Analysis

Policy goals and lawmaking in the creation of the pilot ETSs

As discussed above, five official documents were issued during this phase: the *12th FYP*, including its *Working Plan on Controlling GHG Emissions during the 12th FYP*, the *Notice for Launching Pilot ETSs*, the *Working Plan for Local Pilot ETS*, the *Several Provisions on Carbon Emissions Management*, and the *Interim Measures on the Management of Carbon Emissions*. At the national policy level, the task of the *12th FYP* is to establish mandatory targets for the pilot ETSs. The *Notice of Launching Pilot ETS* establishes the legal basis for the seven pilot areas; it cannot be regarded as a specific guideline but is rather an enabling policy document, meaning that the central government grants the local governments the right to launch a pilot ETS and develop the necessary legal framework. At the regional level, which is currently the core of the development of the ETSs, a technical support committee, the consultative and advisory body for the pilot ETS (selected by the Provincial Development and Reform Commission (PDRC)), is established to play a key role in drafting the ETS policies and laws in various administrative areas. In other words, the technical support committee, which includes experts and researchers in the fields of energy, climate change, economics and law, creates the draft version of the *Working Plan*, *Several Provisions* and *Interim Measures*.

Although the pilot ETSs are at a very early stage, certain problems are apparent when one examines the lawmaking process: they concern the transparency of lawmaking and the effectiveness of legal enforcement, incentive-inducing policy tools and, as mentioned above, the lack of a unified ETS framework at the inter-regional level. We can specify these issues more clearly by comparing Shenzhen with Shanghai: while Shenzhen employed the 'Provisions Mode' for developing a regional legal framework, Shanghai adopted the 'Interim Measures Mode'. We now briefly discuss the variations that result from this difference.

Specifically, regarding the *several provisions* document, this type of local regulation is not a specific and workable legal basis in the true sense but rather a macro-policy guideline, albeit one typically approved by a local legislative body. *Several Provisions on Carbon Emissions Management in the Shenzhen Special Economic Zone* is an obvious example, which simply defines a general legal framework for the ETS without providing concrete rules and regulations governing the creation of the market and trading activities. Statistics indicate that prior to September of this year, the total trading volume in the Shenzhen carbon market was only approximately 120,000 t.[14] There is a marked lack of transparency in the lawmaking process and in the effectiveness in legal enforcement. Neither the experiences with CDM (Schroeder 2009) nor the creation of the pilot ETS solicited the critical contributions of firms or civil society organisations. In contrast to the process in the EU, no firms were observed engaging in substantial lobbying efforts during the process of drafting the *Shenzhen Several Provisions* document. Although the technical support committee includes some experts, the members of this committee are selected by the ruling party to ensure that its intentions and plans will be effectively executed. Both the work of technical support committee and the lawmaking process are closed to the public and relevant companies. ·

However, the *Interim Measures on the Management of Carbon Emissions in Shanghai* contains specific provisions concerning legal liability, punishment mechanisms, a workable system for allowance allocation, the required management systems, etc.[15] It can be regarded as the first true local ETS regulation in China. The process in Shanghai also differed from that in Shenzhen in terms of transparency, in the sense that an exposure draft was circulated to seek opinions from the public. While this cannot be considered a thorough and sound public engagement mechanism (which would involve, for example, public hearings, participative decision-making and feedback systems), it at least represents a degree of public engagement. Moreover, the *Shanghai Interim Measures* document includes a special funding mechanism for managing carbon emissions to assist participating companies in improving their technical ability to reduce carbon emissions (Article 35). Another article (Article 36) suggests that participating companies should apply for investment projects subsidised by the central government's energy conservation fund. Although these financial incentives are mentioned, Shanghai nevertheless lacks a specific and workable mechanism to achieve and enforce articles 35 and 36.

Finally, the *Shanghai Interim Measures* document also concerns an inter-regional ETS. Article 30 of the Interim Measures suggests establishing an inter-regional carbon market and encouraging companies located in other regions to participate in the Shanghai ETS.[16]

ETS and the price-controlled energy sector

The energy sector and the way in which it is incorporated into the pilot ETSs are of particular importance in the context of the Chinese ETS experiments, as 75 per cent of the electricity supply comes from thermal power generation. However, the development of the electricity industry is considered a crucial precondition for China's economic growth. The development potential of the Chinese electricity industry will likely double over the next decade. Several problems derive from including the energy sector in the pilot ETSs. First, the electricity industry operates in a market with a clear growth orientation, in which electricity prices are controlled by the NDRC, whereas the ETS is a restricted market with an open price mechanism for carbon allowances. Crucial but unresolved issues include how the government will attempt to balance the two different types of markets and how the ETS will include rules to prevent adverse impacts on the development of the electricity industry. From a legal perspective, state-owned companies (SOCs) and private companies have equal rights and liabilities in an ETS. However, in a practical sense, two factors create important differences. The first is the administrative role of SOCs: because of the country's political system, officers of the SOCs receive administrative rankings. For example, Sinopec and Huineng Group are at the vice-ministerial level. Namely, the positions of the presidents of the two SOCs are comparable to that of a vice minister. Thus the SOCs will be able to exert substantial influence over the decision-making processes of central and local governments. The second is the monopoly positions enjoyed by certain SOCs: due to the political system and state capitalism policy, the SOCs occupy a dominant position in the energy and power sector. They can easily control the energy market and have the potential to manipulate electricity prices.

The participation of companies is a crucial factor for the success of pilot ETSs in China. In Shenzhen, the official plan called for more than 800 companies to participate in the ETS, most of which is on a compulsory basis, but some participate voluntarily because their CO_2 emission levels fell below a certain threshold. Only 635 companies are actually participating, representing 26 per cent of Shenzhen's GDP in 2010.[17] Many of the companies that could have participated on a voluntary basis chose not to due to a lack of financial and technical support.

According to informal comments by managers, companies in Shenzhen are concerned that the ETS will increase their production costs or reduce profits. At present, the ETS is perceived as a kind of fine imposed on companies, and the distinction between a market approach and command-and-control has not been obvious to the actors involved.[18] A shareholder of a company in Hubei involved in the pilot ETS claimed that the firm had no expert on ETSs, the company lacked a voice in the allowance allocation process, and it simply received orders to comply.[19] The Guangdong Development and Reform Commission reportedly informed the participating companies, including state-owned power companies, that they must bid in the ETS auctions and their bids should reflect a minimum price of 60 Yuan ($9.88) (Chen and Reklev 2013). Industry representatives also report that companies require additional external incentives for carbon-reducing innovations in the ETS pilots. Without additional financial support to introduce low-carbon technologies, participating in the ETS will only be a social responsibility and the result of mandatory regulations. For example, in the EU, a special programme connected with the ETS provides financial support for the development of low-carbon technology; however, China has yet to introduce an equivalent programme.[20] Overall, according to Chinese managers, the transparency of the allocation process and trading rules is limited.

The example of the EU ETS, which was established in 2005, demonstrates that even in the presence of detailed and specific trading rules and regulations, ETSs can fail as a policy instrument if there is insufficient political will to limit the number of available allowances. In recent years, significant declines in the price of EU carbon allowances resulted from an overabundance of allowances in the market. Comparable price volatility would create particular problems for sectors in China subject to price controls. This is a particular challenge in the electricity sector, which is currently subject to price controls. The Chinese government is seeking ways to establish effective price controls, for example, by establishing and changing the number of available allowances.[21] Other unsolved problems concern the establishment of independent monitoring systems and the question of what government authority is ultimately responsible for managing the ETS. The central government selected the NDRC bodies at the state level and the PDRC at regional levels as the lead authorities responsible for managing the ETSs. According to statutes, particularly *The Law on Atmospheric Pollution*, which was issued in 2000, CO_2 is not considered a pollutant. Thus the environmental administrative department responsible for preventing and controlling environmental pollution and ecological damage would lack a mandate to manage ETSs. However, the NDRC is the most powerful ministry of the State Council. It is responsible for state affairs concerning social and economic development and climate change. Specifically, the energy and power sectors are controlled by the National Energy Administration (NEA), which is a department affiliated with the NDRC. As the pilot ETSs in China are a complex innovation involving social and economic issues, the NDRC might be able to promote the ETSs and cope with the barriers they face in China.[22]

Summary and conclusions

Abstracting from whether the creation of carbon markets in China is desirable, we have attempted to describe and analyse how the politics of carbon markets in China produce a context-specific outcome. The analysis revealed that it will be challenging to make cap-and-trade a reality in China. Whatever specific form of carbon governance emerges in China in the next decade will have major implications on the future of global carbon trading. To analyse carbon governance through ETSs in China, we began with several assumptions: the Chinese system can be characterised as adaptive governance in the sense that new instruments and policy tools are often introduced as experiments in different provinces and regions. The intent is to generate practical experience through pilot projects to form the basis for decision-making at the national level. China is gradually transforming from a system of centralised planning to a market-based system, including some liberalisation of the financial sector, the opening up of economic sectors (including energy and logistics) to private and international investors and a strengthening of market mechanisms. Since 2007, Chinese energy policies have been increasingly reframed to address climate change and CO_2 intensity targets – the Chinese government seeks strategies to employ new approaches to energy based on improved efficiency and the increased use of non-carbon-based energy sources, including nuclear energy. The combination of these ongoing processes allowed for the introduction of ETSs in seven pilot cities and provinces, which will test the use of a carbon market as one instrument among many market-based approaches. The Chinese government has endorsed the adoption of market-based instruments as a complement to the hitherto dominant command-and-control-based instruments (see Lederer in this volume).

In our theoretical framework, we attempted to highlight the importance of institutional work to create carbon markets. We focused our analysis on the relationships among policy goals, lawmaking and market creation. We can now provide preliminary answers to our research questions:

- *To what extent have a transparent legislation and efficient legal enforcement for ETSs been achieved?* The legal framework for the ETSs is still under development. The specific rules for each pilot ETS and basic information on market participants remain opaque, both to market participants and external observers. One major challenge will be to create a coherent national ETS on the basis of divergent policy experiments in the various carbon markets and regional pilot ETSs. Our investigation demonstrated that transparent legislation and efficient legal enforcement have yet to be achieved, and there is no guarantee it will be in the near future. In the long run, the successful implementation of a national ETS will not only depend on explicit legal provisions at the level of the pilot regions, but also on a sound ETS legal framework at the national level.

- *What can we determine about the relationships among the policy goals, general legal framework and specific ETS regulations?* Specific ETS regulations are absent in most of the pilot ETSs. The roadmap for the creation of a legal foundation suggests that the national level only provides general rules. While the policy goals have been decided by the central government in a top-down manner, specific ETS lawmaking is organised in a decentralised, bottom-up manner. We identified two different approaches through which the pilot ETS regulations have been developed, and we demonstrated that the degree of specificity, the quality of the process and the resulting transparency vary substantially. It seems that potential market participants have difficulty understanding the new rights and obligations connected to emissions trading.

- *Which type of incentive structure is created for market participants and other interested actors? What types of lessons can be drawn from this for the introduction of a national scheme?* The ruling party provided strong intensity and efficiency targets and created the political space for experimenting with market-based policy tools. However, under the current governance framework, there is latitude for nearly everything – all types of different approaches are simultaneously tested in different locations. Which type of policy or combination thereof will become dominant following the test phase remains unclear. This is a difficult basis for the creation of effective markets and the institutional work required to do so. The uncertainty faced by both economic actors and local lawmakers is likely too high to encourage them to substantively invest in the necessary institutional framework. Therefore, we agree with Lederer (in this volume) in the sense that Chinese state actors provide substantial guidance for carbon markets, but we also demonstrate that the bottom-up implementation of the legal framework, including the lack of financial incentive mechanisms, is currently too weak to promote the frictionless adoption of carbon trading by companies. As the European experience with an ETS has demonstrated, a test phase predominantly serves to assist economic actors and administrations with establishing the technological and organisational infrastructures required for trading, monitoring and verification and establishing institutional capacities and expertise. Our impression is that the political and financial uncertainties remain too high to incentivise this type of institutional work.

- *What institutional imprinting can be observed in the creation of the pilot ETSs, and how does this influence the establishment of a price for carbon emissions?* At this early phase, the process of introducing pilot ETSs in China has resulted in strong mandatory targets and weak legal implementation. There is at least anecdotal evidence that relevant actors consider the ETS to be yet another form of a command-and-control system. The creation of carbon markets is occurring in the context of a more general transformation of the Chinese political and economic system. Some time is likely necessary before the institutional imprinting of the old system fades away and market schemes can be used to price carbon. It seems unlikely that the ruling party will not attempt to control the carbon prices. The central government played a decisive role in negotiating contracts and promoting the CDM and established a de facto price floor that has had global repercussions. The combination of an energy sector with price controls and a formally unregulated price for CO_2 allowances further strengthens this expectation. That the ETSs fall under the jurisdiction of the NDRC is a further sign of the strong role played by the central government in policy processes. If we attempt to compare the current state of the Chinese ETSs, accounting for the brevity of the observation period and the lack of more systematic data, with the experiences accumulated from the EU ETS, we observe complementary phenomena: the EU ETS created elaborate and highly specific regulations for emissions trading and achieved a participatory and relatively transparent policy process. This resulted from extensive institutional effort and the development of technological and organisational capacities to address the complexities of carbon markets. However, the EU ETS currently lacks the political means to establish strong absolute emission reduction targets, meaning that the market is oversaturated with emission allowances. The Chinese pilot ETSs, however, are designed to be tools to achieve highly demanding intensity targets that require important changes in the way that economic growth will be achieved in the future. However, the legal basis remains very weak and overly general, and the policy process is too uncertain to provide the necessary incentives for economic and administrative actors to substantially invest in the institutional effort required to create workable carbon markets. At present, it seems likely that trading activities in China's new pilot ETSs will be a façade – simply the submission of unsubstantial bids in response to the political pressure to actively participate in the trading while avoiding any institutional investments in the development of monitoring or management capacities, let alone investments in carbon-reducing innovations. This would more resemble a continuation of past experiences with SO_2 trading than the effective use of market-based policies to achieve difficult emissions targets. Thus, to respond to Lederer (this volume), while the observed carbon market practices correspond to the practice of state-led economic policy making in China and the Chinese government is better able to regulate the number of available emission allowances than the EU Commission, the lack of effective lawmaking and enforcement will likely prevent actual reductions in carbon intensity. If this proves to be the case, it will have consequences for the debate on alternative options for future carbon governance (Methmann *et al.* 2013).

Acknowledgements

We thank Richard Lane, Matthew Paterson and Benjamin Stephan for very helpful critical comments on earlier versions of this chapter, and Anika Hummel for helping with the manuscript.

Notes

1 E.g., *Several Provisions on Carbon Emissions Management of Shenzhen Special Economic Zone 2012*, (the standing committee of the People's Congress of Shenzhen), available at http://www.climaxmi.org/wp-content/uploads/2012/11/Comments-and-translation-

on-Shenzhen-ETS-Provision.pdf [accessed 25 July 2013]. *Interim Measures on the Management of Carbon Emission of Shanghai* (Shanghai government), available at http:// www.shanghai.gov.cn/shanghai/node2314/node2319/node12344/u26ai37414.html [accessed 30 November 2013].

2 In 1999, the US Energy Information Administration predicted that China would surpass the US as the largest emitter by approximately 2020 (Frankel 1999); however, this actually occurred in approximately 2007.

3 One of the key programmes to achieve this goal was the so-called 'Top 1000 Energy Consuming Enterprises Programme', which was modelled on international target-setting programmes and intended to reduce the energy consumption of major energy consuming industries by 100 million tons of coal equivalents by 2010 (Price *et al.* 2008).

4 According to Hofem and Heilmann (2013: 202), transnational actors introduced this term, especially the UK Department for International Development (DFID). In 2009, DFID defined low-carbon development as 'using less carbon for growth'.

5 The Second National Climate Change Report was released in 2011.

6 Non-fossil fuels include nuclear energy (Xinhua 2011).

7 For example, less developed regions such as Tibet and Qinghai are only expected to reduce their carbon intensity by 10 per cent, whereas booming regions in Eastern China such as Shanghai, Jiangsu, Zhejiang and Guangdong are required to achieve a carbon intensity reduction of 18 per cent (State Council 2011b).

8 The full passage on low-carbon development in chapter 6 'Green development' §21 reads: 'Explore the development of standards for low-carbon products, labels and a certification system, establish a complete monitoring system for GHG-emissions, gradually establish a carbon market. Establish low carbon demonstration sites' (PR China 12th Five Year Plan 2011).

9 Tianjin Climate Exchange Launch Carbon Contract (2009) China Tells, available at http://blog.chinatells.com/2009/09/1976 [accessed 25 July 2013].

10 Seven pilot cities and provinces: Shandong, Shanxi, Jiangsu, Henan, Shanghai, Tianjin and Liuzhou; see http://www.envir.gov.cn/info/2002/5/531670.htm [accessed 12 December 2013].

11 Source: internal report: *The key mechanism design for carbon emissions trading in Guangdong,* issued by Guangdong Academy of Social Sciences (GDASS), 2012.

12 Source: http://news.emca.cn/n/20130411110031.html.

13 Beijing, Shanghai and Guangdong issued their pilot programmes on 28 March, 16 August and 11 September 2012, respectively (State Council 2012). The *Several Provisions on Carbon Emissions Management of Shenzhen Special Economic Zone,* available at http:// www.climaxmi.org/wp-content/uploads/2012/11/Comments-and-translation-on-Shenzhen-ETS-Provision.pdf [accessed 25 July 2013] were approved and issued by the People's Congress of Shenzhen on 10 November 2012, and the local pilot ETS was officially launched by the Shenzhen government on 18 June 2013, becoming the first operational pilot scheme in China. Regarding Shanghai, its *Interim Measures on the Management of Carbon Emission,* available at http://www.shanghai.gov.cn/shanghai/node2314/node2319/node12344/u26ai37414.html [accessed 30 November 2013], which is regarded as the first true set of local ETS regulations was issued by the Shanghai government on 18 November 2013, and its local pilot ETS was officially launched on 26 November 2013. Other pilot regions are currently developing their programmes. Hubei province has completed an exposure draft of the *Interim Measures* document.

14 Luo Wenhui (2013): Current situation of pilot ETSs in Shanghai and Shenzhen. http:// finance.sina.com.cn/roll/20131128/022617458331.shtml [accessed 30 November 2013]. However, the Peopleer, the.sina.com.cn/roll/20131128/022617 'Interim Measures of ETS' to replace the 'Several Provisions'.

15 *Interim Measures on the Management of Carbon Emission of Shanghai* (Shanghai government), available at http://www.shanghai.gov.cn/shanghai/node2314/node2319/node12344/u26ai37414.html [accessed 30 November 2013].

16 *Interim Measures on the Management of Carbon Emission of Shanghai* (Shanghai government), available at http://www.shanghai.gov.cn/shanghai/node2314/node2319/node12344/u26ai37414.html [accessed 30 November 2013].
17 Source: http://www.cerx.cn/cn/trade_details.aspx?ArticleID=274
18 Sources: http://finance.sina.com.cn/money/future/fmnews/20130705/233916034038.shtml; http://www.sz.gov.cn/cn/xxgk/xwfyr/wqhg/20130521/.
19 Personal communication with Qin T.
20 See the NER300 programme. The aim of incentive mechanism is not only to avoid additional ETS costs and save energy, but also to provide the companies (especially private companies) with financial support to develop low-carbon technologies.
21 Yu Yuan, Rob Elsworth, *ETS in China, Sandbag Climate Campaign*, 2012, p. 19; http://www.sz.gov.cn/cn/xxgk/xwfyr/wqhg/20130521/.
22 As an interesting illustration of this, we note that the head of the Chinese delegation to the COP and international climate change negotiations is not from the Ministry of Environmental Protection but from the NDRC.

References

Alianca Redes de Cooperacao Comunitaria Sem Fronteiros Asociacion Ambiente y Sociedad *et al.* (2013) EU ETS myth busting: Why it can't be reformed and shouldn't be replicated. Available from: http://corporateeurope.org/sites/default/files/publications/eu_ets_myths.pdf (declaration signed by more than 40 NGOs) [accessed 26 February 2014].

Benecke, G. (2009) Varieties of Carbon Governance: Taking Stock of the Local Carbon Market in India. *The Journal of Environment & Development*, 18 (4), 346–370.

Callon, M. ed. (1998) *The Laws of the Markets*. Oxford: Blackwell Publishers.

Carbon Finance at the Worldbank/Ecofys (2013) *Mapping Carbon Pricing Initiatives: developments and prospects 2013*. Washington, DC: Worldbank.

Chen, K. and Reklev, S. (2013) *UPDATE 1: Guangdong may withhold free permits in China's 4th emissions scheme: sources.* Available from: Reuter.com [accessed 12 December 2013].

Ellickson, R. C. (1991) *Order Without Law: how neighbors settle disputes.* Cambridge: Harvard University Press.

Engels, A., Knoll, L. and Huth, M. (2008) Preparing for the 'real' market: national patterns of institutional learning and company behaviour in the European Emissions Trading Scheme (EU ETS). *European Environment*, 18 (5), 276–297.

Fligstein, N. (2001) *The Architecture of Markets: an economic sociology of twenty-first-century capitalist societies.* Princeton: Princeton University Press.

Fourcade, M. (2007) Theories of markets and theories of society. *Annual Review of Sociology*, 33, 6.1–6.24.

Frankel, J. A. (1999) Greenhouse gas emissions. *Brookings Policy Series.* Available from: http://www.brookings.edu/research/papers/1999/06/energy-frankel [accessed 22 July 2014].

Friberg, L. (2009) Varieties of carbon governance: the Clean Development Mechanism in Brazil – a success story challenged. *The Journal of Environment & Development*, 18 (4), 395–424.

Friedman, L. M. (1975) *The Legal System: a social science perspective.* New York: Russell Sage Foundation.

Fuhr, H. and Lederer, M. (2009) Varieties of carbon governance in newly industrializing countries. *The Journal of Environment & Development*, 18 (4), 327–345.

Han, G., Olsson, M., Hallding, K. and Lunsford, D. (2012) China's Carbon Trading. An Overview of Current Development. *FORES Study* 2012: 1. Available from: http://www.sei-international.org/publications?pid=2096 [accessed 29 July 2013].

Hatfield-Dodds, S., Nelson, R. and Cook, D. C. (2007) Adaptive Governance: An Introduction and Implications for Public Policy. Paper presented to the *51st Annual Conference of the Australian Agricultural and Resource Economics Society*. Available from: http://ageconsearch. umn.edu/bitstream/10440/1/cp07ha01.pdf [accessed 24 January 2014].

Heilmann, S. and Perry, E. (2011) *Mao's Invisible Hand: the political foundations of adaptive governance in China*. Cambridge: Harvard University Press.

Hofem, A. and Heilmann, S. (2013) Bringing the low-carbon agenda to China. *Journal of Current Chinese Affairs*, 1, 199–215.

Jordan, A., Wurzel, R. K. W. and Zito, A. R. (2003) 'New' instruments of environmental governance: patterns and pathways of change. *Environmental Politics*, 12 (1), 1–24.

Leibfried, S. and Zürn, M. (2005) *Transformations of the State?* Cambridge: Cambridge University Press.

Levin, P. and Espeland, W. N. (2002) Pollution futures: commensuration, commodification, and the market for air. *In:* A. J. Hoffman and M. J. Ventresca, eds., *Organizations, Policy, and the Natural Environment: institutional and strategic perspectives*. Stanford: Stanford University Press, 119–147.

Lewis, J. I. (2010) The evolving role of carbon finance in promoting renewable energy development in China. *Energy Policy*, 38, 2875–2886.

Methmann, C., Rothe, D. and Stephan, B. eds. (2013) *Interpretive Approaches to Global Climate Governance: (de-)constructing the greenhouse*. London/New York: Routledge.

MoST (Ministry of Science and Technology) (2007) Qihoubianhua Guojia Pingu Baogao 气候变化国家评估报告 (*National Climate Change Assessment Report*). Beijing: Kexue Chubanshe.

National Bureau of Statistics of China (2013). Available from: http://www.stats.gov.cn/tjsj/ndsj/2012/indexee.htm [accessed 9 August 2013].

NDRC (National Development and Reform Commission) (2007) *China's National Climate Change Programme*. Available from: http://www.china.org.cn/english/environment/213624.htm [accessed 18 July 2013].

NDRC (2010) Guojia Fagaiwei guanyukaizhan ditan shengquhe ditan chengshi shidian gongzuo de tongzhi 国家发改委关于开展低碳省区和低碳城市试点工作的通知 *NDRC Notification on the Launch of Low Carbon Provincial and Municipal Experimental Zones Document 2010* (No. 1587). Available from: http://www.gov.cn/zwgk/2010-08/10/content_1675733.htm [accessed 25 July 2013].

Oster, S. (2008) China expands markets for carbon trading. *The Wall Street Journal* (the US), 11 November 2008, A11.

PBL Netherland Environmental Assessment Agency (2012) *Trends in Global CO_2 Emissions*. Available from: http://www.pbl.nl/sites/default/files/cms/publicaties/PBL_2012_Trends_in_global_CO2_emissions_500114022.pdf [accessed 15 July 2013].

Price, L., Wang, X. and Yun, J. (2008) *China's Top-1000 Energy Consuming Enterprises Program: reducing energy consumption of the 1000 largest industrial enterprises in China*. Lawrence Berkely National Laboratory. Available from: http://china.lbl.gov/sites/china.lbl.gov/files/LBNL_519E._Top-1000_Energy_Consuming_Enterprises_Program._Jun2008.pdf [accessed 19 July 2013].

PR China 12th Five Year Plan (2011) 中华人民共和国国民经济和社会发展第十二个五年规划纲要 Zhonghua Renmin Gongheguo guomin jingji he shehui fazhan di shier ge wu nian guihua gangyao (Outline of the 12th Five Year Plan for the People's Republic of China's Economic and Social Development). Available from: http://news.xinhuanet.com/politics/2011-03/16/c_121193916_12.htm [accessed 25 July 2013].

Qin, T. (2012) *Climate Change and Emission Trading Systems (ETS): China's perspective and international experiences*. Konrad Adenauer Stiftung – Schriftenreihe CHINA 102 (en), Shanghai.

Schroeder, M. (2009) Varieties of Carbon Governance: utilizing the Clean Development Mechanism for Chinese priorities. *The Journal of Environment & Development*, 18 (4), 371–394.

State Council (2008) *China's Policies and Actions for Addressing Climate Change*. Available from: http://english.gov.cn/2008-10/29/content_1134544.htm [accessed 24 January 2014].

State Council (2009) *China's Policies and Actions for Addressing Climate Change*. Available from: http://www.ccchina.gov.cn/WebSite/CCChina/UpFile/File571.pdf [accessed 24 January 2014].

State Council (2011a) *China's Policies and Actions for Addressing Climate Change*. Available from: http://www.gov.cn/english/official/2011-11/22/content_2000272.htm [accessed 24 January 2014].

State Council (2011b) *Announcement of the State Council Concerning the Work Plan for the Control of Greenhouse Gas Emissions during the Phase of the 12th FYP*, (in Chinese) 控制温室气体排放工作方案的通知 Guowuyuan guan yu yinfa 'shierwu' kongzhi wenshi qiti paifang gongzuo fangan de tongzhi. Available from: http://www.gov.cn/zwgk/2012-01/13/content_2043645.htm [accessed 11 February 2014].

State Council (2012) *China's Policies and Actions for Addressing Climate Change*. Available from: http://www.ccchina.gov.cn/WebSite/CCChina/UpFile/File1324.pdf [accessed 24 January 2014].

State Council (2013) *China's Policies and Actions for Addressing Climate Change*. Available from: http://www.china.org.cn/government/whitepaper/node_7193982.htm [accessed 24 January 2014].

The Climate Group (2012) *Interim Measures for the Management of Voluntary Greenhouse Gases Emission Reduction Trading in China 2012 (People's Republic of China)*. Available from: http://thecleanrevolution.org/_assets/files/Interim-Regulation-of-Voluntary-Greenhouse-Gases-Emission-Trading-in-China(1).pdf [accessed 25 July 2013].

UNFCC (2013) *About CDM*. Available from: http://cdm.unfccc.int/Statistics/Public/CDMinsights/index.html [accessed 30 July 2013].

Wang, Q. (2013) China has the capacity to lead in carbon trading. *Nature*, 493 (7432), 273.

Xinhua (2011) Non-Fossil Fuels to Take up 11.4% of China's Energy Use. Available from: http://www.chinadaily.com.cn/bizchina/2011-03/04/content_12117490.htm [accessed 26 February 2014].

9 The currencies of carbon

Carbon money and its social meaning

Philippe Descheneau

Introduction

The rapid development of carbon markets has led to numerous studies on their environmental effects or economic consequences (e.g. Bachram 2004; Bumpus 2011). However, the representations and symbolic transformations of carbon[1] have been less thoroughly studied. The commodification of carbon is a complex social process that enables carbon to become a tradeable credit. I suggest that three particular moments in this process are crucial. The tonne of carbon is first *invented* and abstracted (Lovell and MacKenzie, this volume), second it is *monetised* into something sellable, and third it is *financialised*, transformed into a financial standardised product (see Paterson and Stripple, this volume). Through these processes, carbon takes a variety of forms, varying according to the social and power relations between actors involved. This chapter focuses on one of those moments: the monetisation process.

Some authors have gone further than focusing on carbon as a commodity, and have suggested that carbon can be considered as a form of currency (Button 2008; Victor and House 2004). This novel claim remains, however, to be elaborated fully. The legal character of carbon credits or allowances remains, in fact, rather weakly defined. There have been on-going debates in Europe and elsewhere about the accounting nature of the carbon credit[2] (MacKenzie 2009a, 2009b). As the many characteristics of the tonne of carbon allow it to take different forms (currency, financial instrument, potential liability), determining its status is far from straightforward and has important implications. For example, when Romania classified carbon allowances as financial instruments, it threatened the over-the-counter carbon trade in the country.[3] Studies invoking this claim about carbon as a form of money have focused either on the legal underpinnings of the credits (Button 2008) or on the need for an underlying political authority – States – to issue and guarantee the currency (Victor and House 2004). One of the principal aims of this chapter is to suggest that this conceptualisation is problematic since it implies a teleological vision of a global market and an abstract, asocial account of the production of money. As a consequence, the chapter offers an account of the process of social construction of money.

Here I attempt to problematise the currency dimension of carbon credits and its social underpinnings. Following actor-network theory (ANT) and literature in economic sociology, I analyse the political and social limits of fungibility and commensurability in carbon markets. This has important theoretical and practical implications. It can help us to better understand the development of carbon markets and the role of carbon currencies, as well as the possible links between carbon market systems. The carbon market is, in fact, characterised by rather fragmented systems where many standards and rules coexist. For example, just within the voluntary offset markets, there are at least 18 different standards applied to projects (Bayon *et al.* 2009: 18). Although there is some movement towards and desire for greater integration, there remain important hurdles that prevent a fully integrated global market.[4] The essay suggests that the largest hurdles are to be found less in the problem of international fungibility, and more in the way the carbon currency is socially constructed.

First, I will conceptualise the commodification of carbon in general, and the monetisation process in particular, using ANT's approach to the sociology of markets (Callon 1986, 1998). I then explore the characteristics of carbon as a currency and the limits of the way it is currently conceptualised as money. Third, I analyse the tools that agents use to make money from carbon as well as three important market devices that make carbon currencies possible. I conclude that the most important question regarding the fungibility of different carbon currencies is a social one, to do with the way they are assembled within actor-networks.

Commodification and the creation of carbon markets

The construction of carbon markets involves a complex agencement of many actors and devices. Agencement is similar to the more commonly used notion of 'assemblage' in the way that it refers to a collective of humans, technical devices, and discourses, that are relatively contingent in their 'assembling' – they cannot be reduced to an effect of social structure. But *agencement* also contains more of a sense of agency, both in the way the combination is put together, and in the effects it produces: the capacity for action. For a paradigmatic account, see the study of hedge funds made by Hardie and MacKenzie (2007). Agencement is a process that relies partly on discursive and sociotechnical foundations but is also and foremost a political choice: the choice of using large companies to try to internalise the externalities created by greenhouse gas (GHG) emissions by capping those emissions. The buying and selling of emissions permits can help reduce the costs associated with these caps, allowing flexibility for enterprises that have to reduce their emissions. They can exchange their allowances on a national or regional market such as the European Union Emissions Trading Scheme (EU ETS) or get credits from investing in international projects. The Kyoto Protocol's Clean Development Mechanism (CDM) allows credits for GHG-reducing projects in developing countries that can be used in international carbon markets.

The process of creating carbon allowances or credits involves different moments reflecting the positions of actors towards the symbolic status of the tonne of carbon. *Invention* refers to the inscription of the tonne that has the impact of commodifying carbon.[5] *Monetisation* represents the inscription of the value in the tonne via its sale, while *financialisation* refers to the complete disconnection or abstraction of the commodity from the tonne of carbon that gave rise to it. It is worth noting that throughout these moments, resistance appears from different agents along that process.

These moments (invention, monetisation and financialisation) are not mutually exclusive and not necessarily chronological. Following a tonne through that process can demonstrate how it is densely integrated into a network that allows it to function, how it becomes an abstract unit of market exchange. But that status is also fragile, as the actor-network mobilised for the transformation of carbon into commodity is never fully stabilised.

Unlike the classical political economy depiction of commodification from Marx or Polanyi (e.g. Lohmann 2010), this conceptualisation focuses more on the practices and micro-dynamics at play.[6] Indeed, by looking at how a market and a medium of exchange are constructed, we can better understand the overall process and politics of commodification. It also allows us to question some of the assumptions of neoclassical economics such as the motivations and behaviour of actors, the so-called rational economic agent.

ANT considers the 'calculative rationalities' of economic agents as being social rather than individual (Callon 1998). Instead of rational individuals taking decisions, calculation has to be understood as a process that involves indexes, statistical tools and metrological instruments as an *agencement* or assemblage (MacKenzie 2009b). The rationality of the calculation is embedded, among other sites, in the calculation of the additionality that occurs in several sites in the case of carbon markets.

Accounting tools are crucial to determine the additionality of a carbon offset project, and therefore to monetise carbon. They have, however, a performative value that is central: 'Not only do accounting tools contribute very largely to the performance of calculative agencies and modes of calculation, while allowing the constant reconfiguring of those agencies and modes of calculation, they also contribute directly to the shaping of a discourse through which these agencies account for their action' (Callon 1998: 26).

For ANT, the intervention of socio-technical devices is central to understanding markets. In the case of the CDM, the monetisation of the tonne is enabled by the transformation of carbon from a PDD (project design document) through to a registry that records the existence of a specific Certified Emissions Reduction unit (CER, the unit associated with CDM projects), with the overall process enabled by the formula of the Global Warming Potential (MacKenzie 2009a; see also Paterson and Stripple, this volume). Thus, the study of some of the devices in carbon markets can help us to understand how they have performative functions in the shaping of carbon commodification.

But ANT remains sometimes silent on the social relations and relations of power. For example, MacKenzie (2009a) suggests that the conversion rates between different sorts of carbon commodity are a social factor and not a political one. However, monetisation, far from being a mere technical process, has important political effects that can alter the evolution of carbon markets. Looking beyond the devices makes us examine how actors construct meanings for carbon. Some authors have stressed that importance. For example, for Descheneau and Paterson (2011), carbon markets are in tension between different objectives such as finance and environment. For Bumpus (2011), the tonnes of carbon have a materiality – referring both to the concrete character of specific projects that generate credits, and the way that carbon credits affect companies' balance sheets – that is normally ignored.

Carbon as money

So far, the legal status of the tonne of carbon $(tCO_2e)^7$ as a tradeable instrument remains contested and is changing according to different jurisdictions. It can be understood to have different characteristics such as commodity, currency or financial security (Button 2008). The case for regarding carbon markets as generating a sort of a carbon currency has been put forward by Button (2008) and Victor and House (2004). However, these analyses fail to explore the social implications of such a way of understanding carbon. Indeed, if we have to understand the social underpinnings of carbon markets in general, then this applies also to the understanding of carbon as money.

For Button, the main legal provisions for carbon credits have thus far defined 'the tradable unit not in terms of what the unit is, but what it entitles the holder to do' (Button 2008: 574). The legal status of the carbon credit itself remains ill-defined. Button (2008), for example, argues that a carbon credit is 'a *sui generis* right which, depending on the regulatory market under which it is created, exhibits characteristics of both a commodity and a currency' (Button 2008: 573).

Carbon credits can be considered as a form of money as they display many of its standard characteristics. It may not of course be as fungible as 'conventional' money and there are a number of characteristics of money that it lacks. But the point here is that it has the potential to evolve towards the money form – it contains the 'logic' of money immanently within it. Money is conventionally understood to 'act as both a medium of exchange and as a store of value. In performing these roles, money necessarily takes on two additional roles, as a unit of account and as a means of payment' (Leyshon and Thrift 1997: 5). Carbon credits can also act as a medium of exchange and a relative store of value that is guaranteed by the government.[8] It is a unit of account as shown by the multiple national registries and exchange platforms, such as the interaction transaction log and the Community Independent Transaction Log (CITL) for the EU. It is also used as a means of payment in some contracts where CER may be a form of payment for project developers or brokers.[9]

Even if, to date, carbon has been conceptualised principally as a commodity, considering carbon as a form of currency is fruitful for the analysis of carbon markets. In fact in a currency, the role of the state is central to guarantee the value of a specific currency, and of money in general. While standard commodities (wheat, oil, etc.) have a use value on their own (Button 2008), the carbon credit is essentially a fictitious one and only gains value via the cap, and thus market scarcity, established by governments. The role of the regulator in carbon money is thus central, and takes on similar forms to for conventional money – essentially that of establishing the overall amount of money to be available in the system.

The idea of carbon as a currency helps to make visible and comprehensible the transformations of carbon from an abstract reduced tonne in a project or as an allowance to a concrete unit in an account. However, what devices make the carbon currency possible? The next section focuses on this question.

Money from carbon

To the extent that carbon is taking on some of the qualities of money, this arises out of corporate strategies to make money, in the more conventional sense, out of carbon markets. As carbon markets develop, many devices are needed to make it function. Even though states create the caps in the first place, the evolution of carbon markets is marked by an important degree of implication of market actors and the (re)creation of specific market institutions. The market's devices allow carbon currencies to emerge, but also allow companies to make money from it. A major criticism of the EU ETS is, for example, directed at the windfall profits that the companies made from 'cashing in' their free allowances. Ellerman, Convery and de Perthuis (2010) mention that windfall profits could amount to some 19 billion euros.

The devices described below make the carbon market possible but have also a more particular effect: positioning large companies as new owners of 'carbon accounts'. The banking of allowances in the EU ETS illustrates well how companies have started to use carbon as a store of value, as they use the banking provisions in the EU ETS as part of their financial strategies. Allowances can be converted into dollars that, in turn, can finance investment projects. Credits can be easily converted into money and are seen as having this performative function.[10] But at the same time, this means that carbon emerges as a store of value, one of the key qualities of money.

Actors make sense of the carbon market by deploying the many market devices that they adapt or translate from other financial markets. They have to see it as an abstract product that they have encountered elsewhere. For market actors, the 'appearance' of carbon credits in market technology that is already familiar is sometimes presented as something magical:

> A purchase can be completed in seconds – a process that can take weeks or months OTC. The 'mystery' of price or quality of VERs is taken away. . . . For Brokers, finally the answer to their prayers – a web based user-friendly trading screen, and a continuous trading contract with easy access for them and their clients.
>
> (Carbon Trade Exchange 2010)

While these institutions create the possibility for some actors to profit from the market, they also inscribe more permanent institutions in carbon markets and in the economy. They also have the effect of rendering the accounting of carbon reductions more technical (Lovell and MacKenzie 2011). Three serve as examples here: registries, exchange platforms and the carbon currencies themselves.

Registries

A registry is a virtual or material place where reduction units are counted, in order to verify that each reduction is real. The CDM Executive Board rules require CERs to have a serial number in order to make it possible to track them. The international transaction log also tracks every carbon transaction.

Many companies have tried to standardise the registration of GHG-reducing projects. Some companies also offer a system that is close to a script for establishing carbon as money. For example, APX aims at:

> Providing a bank and mint for environmental certificates, APX solutions are trusted to create, track, manage, and retire renewable energy certificates (RECs), energy efficiency and conservation certificates, carbon offset credits such as voluntary emissions reductions (VERs), and greenhouse gas emission allowances.
>
> (APX 2010)

The centralisation and standardisation of carbon exchanges has been one of the avowed aims of many. While for some the role of voluntary markets is essential for innovation, there is nevertheless a danger of having too many registries that could, in turn, hinder the quality of credits by undermining the transparency of the overall process (for example, risking the possibility of double-counting of a particular project's emissions reductions).

Exchange platforms

Exchange platforms are central in assembling carbon markets as they provide a centralised means of trading allowances and credits. They also establish the rules and channels by which the credits can be exchanged. They help to abstract carbon and make it resemble other products facilitating the development of financial products (derivatives). One of those platforms, Bluenext, was founded by New York Stock Exchange (NYSE) Euronext and the Caisses des dépôts. It was the largest spot trade for carbon in the world and offered a number of services ranging from future contracts to spot trading before its closure. Several other exchange platforms have been created especially in developing countries, such as the China Beijing Environmental Exchange (CBEEX 2011) or MexiCO$_2$ (Alire Garcia 2013) which mainly facilitates the exchange of CDM credits.

The technologies used in the carbon market have the effect of reinforcing the idea of carbon as a currency. Registries and exchange platforms both follow specific rules (see UNFCCC 2009) that allow carbon to be counted and exchanged. There have, however, been cases of double-counting (Murray 2007) and fears that might happen again, possibly in the voluntary markets, where there are a significant number of different registries. In part in response to this problem, financial companies have also developed formal exchanges to trade carbon allowances and credits and systems of credit tracking, modelled on those in other financial markets.

In the regulated markets, there is some more centralisation and the numerous logs have implemented several checks, but the system is not immune to questionable transactions. For example, Hungary[11] has used surrendered CERs in an attempt to take advantage of better prices than those for AAUs (Kyoto's unit for direct exchange of emissions allowances between states). VAT fraud and other scandals recall the necessity of considering the security implications of such a massive amount of money transiting[12] even if, for the moment, most of the transactions remain for immediate 'consumption', either for compliance purposes or for direct offsetting by consumers with many 'over-the-counter' (OTC) transactions. A platform, the Carbon Trade Exchange aims to be the 'world's most inclusive trade platform' with members in 25 countries (Carbon Trade Exchange 2014).

The carbon currencies

The use of carbon credits in different systems led authors to describe them as a series of carbon currencies (Victor and House 2004), with direct analogies to the varied currencies produced in national money systems. Some organisations are pushing towards a fully integrated global market. For example, the International Emissions Trading Association (IETA), the main lobby group for those involved in emissions trading, states the development of a global trading regime as one of the elements of its goals.[13] But to understand how carbon currencies are constructed, we have to first look at the development of other currencies. The development of virtual currencies such as Bitcoin has sparked an interest in a better understanding of the underpinnings of currencies.

McNamara (1998), for example, has looked at the role of ideas in the construction of the euro. She notes that the importance of the changes in the international economic environment and the development of ideational factors among the elite were key in the adoption of the European monetary integration. The process of policy failure, paradigm innovation and policy emulation were also three conditions for such development (McNamara 1998: 5).

The evolution of the euro also gives us some clues to understand the evolution of carbon currencies and the dilemma between enlargement and integration between different carbon currencies. As with the euro, the need for integrated policies is important but not present at the moment. Arbitrage can thus take advantage of the discrepancies between different carbon currencies and the lack of management between the different systems. As with traditional currencies, there are many problems associated with the control and production of those currencies. For example, in the nineteenth century, the existence of many centres of production of money posed an important challenge in the US, as it enhanced problems of counterfeiting and with the control of the overall volume of money available. Thus the territorialising and centralising of its monetary functions has helped the state's control over the national economy (Helleiner 2003).

To understand how different carbon currencies can be traded with each other, we have to understand the functioning of the systems and what are the possibilities of linking them together. The linking between emissions trading systems has to take into account different features. For Ellis and Tirpak, six factors are crucial: 'How targets are expressed (e.g. fixed or indexed), the presence of price caps, non-compliance provisions, banking/borrowing provisions, commitment period lengths and starting points, eligibility of offsets and permit allocation methods in different countries' (2006: 6). Linking can be direct or indirect via the inclusion of offsets. It can also be unilateral or bilateral, where both systems accept the credits of each other (Ellis and Tirpak 2006: 8). The limited exchanges that have occurred so far concern indirect and unilateral linkages such as the inclusion of the CER in the EU ETS. The leakage issue and the fears of potential commercial disadvantage are pressures that favour better integration of different markets.

There is still considerable work to do on the subject of integration as the carbon trading systems are sometimes of a different nature: voluntary or mandatory, Kyoto or non-Kyoto, project-based or allowance-based. The exchange between mandatory and voluntary systems is possible but very limited at the moment.[14] There are also many possible benefits of having different systems; the voluntary markets might produce innovations that the mandatory markets can adopt later (Bayon *et al.* 2009).

To enable the linking of different systems, it is thus important to look at a number of considerations. To ensure the environmental integrity of credits, for example, there has to be harmonisation of various aspects of each market. The registries and transaction logs need to be standardised to avoid double-counting and leaks. Measurement, reporting and verification of emissions (MRV) also need to be consistent across systems.

But these conceptions of currencies rest on a territorialised conception of money. Some carbon currencies such as the CERs are somewhat deterritorialised given the CDM Executive Board's ability, in effect, to create credits that can be monetised. At the moment, project credits act as the principal linkage between markets (Buen 2008). But this carbon currency is not created *ex nihilo* and has to follow a lengthy bureaucratic process (Michaelowa and Jotzo 2005). The environmental integrity of the system is also vital for the value of credits and is clearly a political process (Michaelowa *et al.* 2008).

The possibility of exchanging different carbon units could grow in the future, making them appear more like currencies; forms of money. Examining the key currencies can also help us understand better the political economy of carbon trading and the quality of different units. If offset units can serve as a safety valve or a way to minimise the costs of compliance, it is not impossible to think of eventual exchange rates between currencies. The Waxman-Markey Bill (US Senate 2009), for example, proposed a discount rate of 5/4 for offsets and many studies have suggested other discounting approaches (Kollmuss *et al.* 2010). There are also several aspects that could influence the future evolution of carbon markets such as regional markets and international commitments.

Some proposed carbon units go somewhat further than cap-and-trade and project mechanism credits in the direction of carbon as money. For example, the International Institute of Monetary Transformation has proposed a carbon-based reserve currency, the Tierra (International Institute of Monetary Transformation 2010). The idea that we can fix the amount of money as a function of natural elements, such as carbon emissions, is reminiscent in some ways of the fixed reserve of money of the gold-standard era.

Other proposals for personal currencies, such as personal carbon allowances, show the potential and the interest for the treatment of carbon as a currency. Former UK politician David Miliband, acting as Environment Minister, suggested we 'imagine a country where carbon becomes the new currency. We carry bank cards that store both pounds and carbon points. When we buy electricity, gas and fuels we use our carbon points, as well as pounds' (Miliband 2006, quoted in Paterson and Stripple 2010: 355–358). But that possibility remains a long way from being applied, and its equity dimension is questionable.

The social meaning of carbon money

Carbon trading has become steadily more hegemonic as a political and economic approach to address climate change. A focus solely on the ideological desire for carbon markets, or on the socio-technical devices through which the markets are assembled, masks a different dimension to these markets. Beside the economic considerations of efficiency, the social meaning of carbon as money must also be considered. In fact, a variety of social limits to carbon commodification are apparent. In particular, several aspects need to be addressed, such as the control of money, the possibility of earmarking (see below), and the critique of carbon money, in order to better understand the social character of carbon markets.

The legal status of the carbon credit has important consequences and has been the subject of various controversies. The ownership of carbon credits may also entail a number of other rights. But beyond its legal status, carbon exhibits various monetary characteristics and roles by the way it is perceived and can be used by actors. It is also limited as a symbolic monetary value by the number of people that have access to it, mainly project developers, verifiers or large enterprises.

The question of the symbolic politics of money has been widely explored.[15] Money is an item that incarnates value and becomes an object of desire per se.

Carbon money may represent a step further in the commodification of carbon and can trigger gloomy visions such as Simmel's (2007) conception of money as a faceless item. Nevertheless, carbon currencies are political creations that can only be understood in a social process; the process of monetisation of carbon. Classical and contemporary studies on the political economy of money (e.g. Helleiner 2003; Leyshon and Thrift 1997; Zelizer 1994) have explored the social meaning of money and its political dimensions.[16]

Carbon currencies are usually understood as mainly determined by a state power dimension (House and Victor 2004) in a way similar to a neo-realist approach to money employed by Kirshner (1995). While this dimension is important, a look at the evolution of the markets tells us that non-state actors and social dynamics also play an important role.

Before asking the question of the international fungibility of credits, we must look at how carbon market institutions are made. By looking at some practices[17] that shape carbon money and the resistance towards it, we can understand how market actors frame carbon and how it is constructed inter-subjectively. Descheneau and Paterson (2011), for example, have shown how financial carbon markets are culturally constructed in advertisements and events such as Carbon Expo.[18] The cultural underpinnings of carbon money can be shown, for example, in the idea that trees can be transformed into money. Finite Carbon, a US company, insists on the monetisation in what they see as an objective chain of value: 'Finite Carbon is the forest carbon development company that offers landowners a single-source, end-to-end solution to create and monetise carbon offsets' (Finite Carbon 2010).

Considering carbon as a currency has also been invoked by IETA as a way to present carbon trading in a better light:

> Although intellectually most people can see that emissions reductions can be treated like a commodity, they are not entirely comfortable with the idea – there is an intrinsic lack of substance and convincing tangibility about emissions reductions units. This surfaces as suspicion about the baseline from which the reductions are computed and the reliability and durability of the reductions. Maybe the basic analogy is pushed too far: if emissions reductions were instead treated analogous to a currency, whose intrinsic worth can change from one jurisdiction to another, some of these problems would disappear, though others arise.
>
> (Henry Derwent, IETA 2010b: 12–13)

The then CEO of IETA acknowledges, here, the many problems associated with the commodity model. He suggests that seeing carbon credits as a currency would help to conceptualise the emissions reductions. But perhaps it is also because carbon money is still abstract and not experienced daily by people that it is hard to conceptualise. On the other hand, personal currencies, such as the one proposed by Miliband, risk an individualisation of the fight against climate change (Paterson and Stripple 2010) and a banalisation of the meaning of carbon. But the uses and the ways in which carbon money is created also remain unquestioned. Who gets the keys to the accounts and should we give them to some actors only? Does that framing as 'intrinsic worth' not risk also creating knock-on problems?

The representation of carbon as money is not merely a conceptual one. It poses the question of who should be responsible for carbon money. How can we understand value for nature's services? The monetisation of the environment is not only encountered in carbon markets. The question of value is also of great importance as it questions our representation of nature (Lovell and MacKenzie, this volume; McAffee, this volume). The logic is also the same for ecosystems services in general, as for carbon as a specific sort of 'ecological service'. For Sullivan (2009: 56):

> Behind this monetisation of environmental crisis is a logic and language that transforms the global environment – Nature – into a provider of services for humans. This conceptual capture, and the economic rationalisation of nature's value that it permits, is facilitating the creation of markets for the exchange of 'ecosystem services' in the form of Payments for Ecosystem Services (PES).

Carbon currencies can also be represented through different characteristics that involve tradeoffs between questions of quality, fungibility and traceability. The quality of the credits forms the basis value of the currency with, for example, the additionality criteria. Fungibility is a factor that allows for more efficiency and arbitrage will take place to take advantage of discrepancies between prices. Traceability is also important to avoid scandals of tax fraud that plagued the EU scheme in 2009–2010. While we do not find yet characteristics of advanced currencies such as a currency relying on a basket, or pegged currencies,[19] such possibilities are not excluded.

The possibility identified by Button (2008) of having the International Monetary Fund (IMF) supervising a currency based on carbon, akin to special drawing rights (SDR),[20] might be a little far-fetched. Even with this possibility, without proper regulation currencies could still be the target of speculators.[21] Button also suggests that the currency would have the effect of accelerating the process of commodification (Button 2008: 587) but could also change the character of resistance to it.

The framing of carbon as a universally fungible item has also led some people to forget that important differences exist between units used (project-based or allowances). The importance of this exchange between different systems also triggers more opposition to carbon offsetting. The commensurability of reductions between different gases provided through the Global Warming Potential (see MacKenzie 2009a) does much of the work of placing value on credits or allowances, a point sometimes overshadowed by the focus on the fungibility between systems.

The unit of value has to be interiorised and has to be linked to a system of value and to its environmental and social benefits. Zelizer (1994: 5) has shown that even if money is the most fungible item in exchange, people still create social usages that associate certain activities with certain revenues: 'In their everyday existence, people understand that money is not really fungible, that despite the anonymity of dollar bills, not all dollars are equal or interchangeable. We routinely assign different meanings and separate uses to particular monies'. For instance, money that comes from a lottery win would be used for fun, as opposed to work money which would be used for supporting children. So there will always be social strategies to earmark carbon and give it certain meanings, as there are for money.

We must also acknowledge, and maybe emphasise more fully, the different contexts in which emissions reductions take place and thus consider the international equity questions often raised about carbon markets. Carbon is a fictitious commodity and does not involve any concrete product. It rests largely on the plausibility of scenarios generated about real, verifiable and additional reductions, many of which are questionable.

There has been an important financialisation of carbon markets (Descheneau and Paterson 2011; FOE 2010) with various financial products that allow actors to hedge risks. As financial products become more and more complex, there seems to be a disconnection between the nature of emission reductions and the financial products involved in carbon markets. The market is characterised by uncertainties and the commensuration of tonnes across different types of projects and the credits they produce can wrongly translate many diverse dimensions of climate change into the same system. Furthermore, tying a financial product to an emission reduction would have just the effect of adding a layer of complexity to it and augmenting the risk of distracting attention from the main goal of reducing emissions.

The construction of carbon money as a progressive translation from a negative externality to a carbon currency is thus a long process that can be determined by many factors (on the development of externalities, see Lane, this volume). The social uses of carbon money matter as much as the policy experiment. In this construction, the initial translation from an abstract tonne to a monetary carbon unit should be reconsidered rather than solely focusing on the fungibility of carbon currencies.

The hardest limits to fungibility of credits or allowances from different systems are, then, not so much the formal integration between the systems but rather the social limits of translating the emissions reductions into money. We can see those limits in three important regards. First, there are different objectives involved in the creation of the credits. The development and environmental objectives of the CDM are not always taken into account to determine what an acceptable project is. Unlike regular money, a carbon credit holds the promise of an environmental benefit. Earmarking, that is making sure they are used only for environmental purposes, could potentially achieve a better environmental performance but potentially at the cost of economic profit and efficiency. Second, the fate of the currency is only determined by some actors that are well positioned structurally and that have the possibility to have financial leverage. The currency has also not been developed and interiorised as its future is uncertain and it is sometimes hard to know what hides behind it. Finally, the commodity is already contested by organisations such as Carbon Trade Watch and others interested in the ethics of carbon trading. The idea of carbon money may be resisted for similar reasons.

Conclusion

The development of carbon markets has entailed the construction of processes and practices that combine to allow carbon to be traded. Different systems, standards and units characterise the market and, despite the calls for a globally integrated market, the possibility of completely linking different systems is still far from sight. In the process of commodifying carbon, monetisation is a key part of that dynamic where an emissions reduction is transformed into a monetisable credit.

There is still, however, a debate to be had concerning the nature of a carbon unit. Is it best thought of as a currency, commodity or financial instrument? This chapter has attempted to show that carbon does show similar functions to money, however, what becomes clear is that it does so in specific ways and with particular limits. Whether carbon does take the form of money, and in what way it does so, is a political question that will depend on the conflict between groups with differing interests in a particular conception. For example, the financiers would like a model more closely associated with their conceptualisation of carbon as a financial product.

Carbon has enabled many companies and intermediaries to make money via market devices such as registries, exchange platforms and currencies. As much as the carbon market is fragile and subject to contestation, the burgeoning social meaning of carbon as money highlights an important change in the conception of carbon. It defines the limits to fungibility, the importance of embedded routines in actors and the possibility of earmarking.

As the carbon market experiment continues, there is certainly a need for flexibility and a true learning-by-doing process. Different objectives exist in the carbon markets and those must be acknowledged. Furthermore, there may be problems if only a few structural dominant actors decide at this time on the fate of the currency. In such a context, the carbon money will be difficult to stabilise and will always be contested (Buen 2008).

The conceptualisation of carbon as a currency has so far failed to address what are the social implications of the commodification of carbon. That conceptualisation should go beyond a state-centric construction where competition is the main focus (Victor and House 2004) of a new legal opportunity to foster the creation of markets (Button 2008). Carbon currencies are constructed through market devices but never fully questioned. Are carbon currencies able to produce meaningful behavioural change or are they merely another business incentive not earmarked for any environmental benefit?

Allowing only a relatively small number of economic actors to participate in carbon markets, also allows only them to make sense of carbon. The contestation of carbon markets is getting stronger, as many feel left out of environmental decisions. The political questions concerning carbon markets become more acute: How far should the commodification of carbon go? And who should decide this?

Acknowledgements

This chapter is a revised version of Philippe Descheneau (2012) The currencies of carbon: carbon money and its social meaning. *Environmental Politics*, 21 (4), 604–620. The research has benefited from a doctoral grant from the *Fonds Québecois de recherche sur la société et la culture*. I would also like to thank Matthew Paterson and the editors for useful comments.

Notes

1 The concept of carbon is used here as synonymous to tCO_2e (tonnes of carbon dioxide equivalent), a construction that is, itself, part of the symbolic transformation (see Paterson and Stripple, this volume).

2 The debates on whether carbon should be treated as an intangible asset or as a financial product have led to the directive IFRIC 3 (International Financial Reporting Interpretations Committee) that is not followed by everyone: 'It remains permissible to treat carbon in this way: as inside an economic frame, but in a sense invisibly so, since no accounting recognition is needed if the above conditions are met. Some market participants seem to do just that' (MacKenzie 2009a: 449).

3 'Romania's ruling on EUAs threatens OTC', *Carbon Market News* 24 February 2010.

4 This is even more acute with the financial crisis and the difficulties of the EU ETS.

5 Inscription is defined by Walters (2002: 84) as: 'the material practices of making distant events and processes visible, mobile and calculable in terms of documents, charts, forms, reports, signs and graphs'.

6 For another case study exploring these micro-dynamics, focused on the commodification of wetlands, see Robertson (2007).

7 Except for the Regional Greenhouse Gas Initiative (RGGI), which uses the short ton, the metric tonne of carbon is the main unit in circulation.

8 It is relative as a store of value, because it may expire after a certain time.

9 The distribution of CERs varies according to non-disclosed contracts. For an example see the template CERSPA Certified Emission Reductions Sale and Purchase Agreement (2nd version) (CERPSA 2011).

10 Following the performativity described in MacKenzie (2006), the carbon credits model follows some of the functions of money.

11 Hungary raised HUF 1 billion ($5.1 million) from selling CERs surrendered by ETS companies taking advantage of higher AAU prices. See 'Hungary sells "recycled" CERs', *Point Carbon*, 11 March 2011.

12 Europol estimate the money lost to be €5bn over the past 18 months (ENDS 2009: 8).

13 A critical aim for IETA remains the linking of trading regimes among Annex I countries, and its significance for the GHG market (IETA 2010a).

14 There has been an exchange of 100 $GtCO_2e$ between the EU ETS and CCX (2006), which looked more like a public relations deal than a commitment to fungibility between the systems, but the CCX has since been shut down.

15 It would take too long and is outside the scope of this article to try to present those views, but the reification of money is a common theme (see, for example, Simmel 2007; Zelizer 1994).

16 As Boyer (1986: 48, my translation) emphasises, 'money is not a particular commodity but a form of relation between centres of accumulation, workers and other market actors'.

17 See also Lederer, this volume.

18 'Carbon Expo is the "Global Carbon Market Fair & Conference", where around 3000 carbon traders, investors, lawyers, project developers, market infrastructure developers, consultants, and policy-makers meet annually' (see Descheneau and Paterson 2011).

19 Although there is a strong correlation between the secondary CER and the EUA, mainly because of the demand in the EU.

20 The Special drawing rights are an alternative unit of account designed by the International Monetary Fund, their value defined by a weighted basket of four currencies; the euro, the US dollar, the British pound and the Japenese yen (IMF 2011).

21 The Waxman-Markey Bill did contain such regulation for controlling financial speculation.

References

Alire Garcia, D. (2013) *Mexico Launches First Carbon Exchange to Cut CO₂ Emissions*. Available from: http://in.reuters.com/article/2013/11/27/mexico-carbon-idINL2N0JB1JL2013 1127 [accessed 4 March 2014].

APX (2010) *Who is APX?* Available from: http://www.reuters.com/article/2009/03/17/idUS10404+17-Mar-2009+BW20090317 [accessed 29 February 2012].

Bachram, H. (2004) Climate fraud and carbon colonialism: the new trade in greenhouse gases. *Capitalism, Nature, Socialism*, 15 (4), 5–20.

Bayon, H., Hawn, A. and Hamilton, K. (2009) *Voluntary Carbon Markets: an international business guide to what they are and how they work*. London: Earthscan.

Boyer, R. (1986) *La théorie de la régulation: une analyse critique*. [The regulation school: a critical analysis]. Paris: La Découverte.

Buen, J. (2008) *Linking: is CDM up to it?* Oslo: Point Carbon.

Bumpus, A. (2011) The matter of carbon: understanding the materiality of tCO_2e in carbon offsets. *Antipode*, 43, 612–638.

Button, J. (2008) Carbon: commodity or currency? The case for an international carbon market based on the currency model. *Harvard Environmental Law Review*, 32, 571–596.

Callon, M. (1986) Some elements of a sociology of translation: domestication of the scallops and the fishermen of St Brieuc Bay. *In:* J. Law, ed., *Power, Action and Belief: a new sociology of knowledge?* London: Routledge, 196–223.

Callon, M., ed. (1998) *The Laws of the Markets*. Oxford: Blackwell.

Carbon Trade Exchange (2010) *Press Release*. Available from: http://www.carbontradexchange.com/userfiles/file/CTX%20Launch%20Press%20Release.pdf [accessed 29 February 2012].

Carbon Trade Exchange (2014) *About Us*. Available from: http://carbontradexchange.com/about/about-us [accessed 5 March 2014].

CERSPA (2011) *Certified Emission Reductions Sale and Purchase Agreement*. 2nd version. Available from: http://www.cerspa.com [accessed 6 March 2014].

China Beijing Environment Exchange (CBEEX) (2011) *CDM Projects Operational Procedure*. Available from: http://en.cbeex.com.cn/article/CDMInformationCenter/OperationalProcedure/ [accessed 29 February 2012].

Descheneau, P. and Paterson, M. (2011) Between desire and routine: the assemblage of finance and the environment. *Antipode*, 43 (3), 633–668.

Ellerman, A. D., Convery, F. and de Perthuis, C. (2010) *Pricing Carbon: the European Union Emissions Trading Scheme*. Cambridge: Cambridge University Press.

Ellis, J. and Tirpak, D. (2006) *Linking GHG Emission Trading Systems and Markets*. Paris: OECD/IEA.

ENDS (2009) Carbon trading scam reaches epidemic levels. *ENDS Report 419*, December 2009.

Finite Carbon (2010) *Connecting Forestry and Carbon Finance*. Available from: http://www. finitecarbon.com [accessed 29 February 2012].

FOE (Friends of the Earth) (2009) *Subprime Carbon? Re-thinking the world's largest new derivatives market*. Washington, DC: Friends of the Earth.

Hardie, I. and MacKenzie, D. (2007) Assembling an economic actor: the agencement of a Hedge Fund. *The Sociological Review*, 55 (1), 57–80.

Helleiner, E. (2003) *The Making of National Money: territorial currencies in historical perspective*. Ithaca: Cornell University Press.

IETA (2010a) *About IETA*. Available from: http://www.ieta.org/emissions-trading [accessed 29 February 2012].

IETA (2010b) *What's Wrong with Emission Trading?* Available from: http://www.ieta.org/ index.php?option=com_content&view=article&id=231:whats-wrong-with-emissions-trading&catid=26:reports&Itemid=93 [accessed 6 March 2014].

IMF (International Monetary Fund) (2011) *Factsheet. Special Drawing Rights*. Available from: http://www.imf.org/external/np/exr/facts/sdr.htm [accessed 29 February 2012].

International Institute of Monetary Transformation (2010) *The Tierra Carbon-Based Monetary Standard with Tierra as Unit of Account*. Available from: http://timun.net/content. php?nID=7&cID=23 [accessed 29 February 2012].

Kirshner, J. (1995) *Currency and Coercion: the political economy of international monetary power*. Princeton: Princeton University Press.

Kollmuss, A., Michael, L. and Smith, G. (2010) Discounting offsets: issues and options. *Working Paper* WP-US-1005. Stockholm: Stockholm Environment Institute.

Leyshon, A. and Thrift, N. (1997) *Money/Space*. London: Routledge.

Lohmann, L. (2010) Uncertainty markets and carbon markets: variations on Polanyian themes. *New Political Economy*, 15 (2), 225–254.

Lovell, H. and MacKenzie, D. (2011) Accounting for carbon: the role of accounting professional organisations in governing climate change. *Antipode*, 43 (3), 704–730.

MacKenzie, D. (2006) *An Engine, Not a Camera*. Cambridge: MIT Press.

MacKenzie, D. (2009a) Making things the same: gases, emission rights and the politics of carbon markets. *Accounting, Organizations and Society*, 42 (3–4), 440–455.

MacKenzie, D. (2009b) *Material Markets: how economic agents are constructed*. Oxford: Oxford University Press.

McNamara, K. R. (1998) *The Currency of Ideas: monetary politics in the European Union*. Ithaca: Cornell University Press.

Michaelowa, A. and Jotzo, F. (2005) Transaction costs, institutional rigidities and the size of the clean development mechanism. *Energy Policy*, 33 (4), 511–523.

Michaelowa, A., Flues, F. and Michaelowa, K. (2008) UN approval of greenhouse gas emission reduction projects in developing countries: the political economy of the CDM Executive Board. *CIS Working Paper*, 35.

Miliband, D. (2006) Government to look at personal carbon allowances to combat rising domestic emissions. London: Department for Environment, Food and Rural Affairs.

Murray, J. (2007) *EU Trading Scheme Slammed for 'Double Counting' Carbon Credits*. Available from: http://www.businessgreen.com/bg/news/1807026/eu-trading-scheme-slammed-double-counting-carbon-credits [accessed 27 February 2012].

Paterson, M. and Stripple, J. (2010) My Space: governing individuals' carbon emissions. *Environment and Planning D: Society and Space*, 28 (2), 341–362.

Robertson, M. (2007) Discovering price in all the wrong places: commodity definition and price under neoliberal environmental policy. *Antipode*, 39 (3), 500–526.

Simmel, G. (2007) *Philosophie de l'argent* [Philosophy of money]. Paris: PUF.

Sullivan, S. (2009) Constructing Nature as Service Provider. *In:* S. Böhm and S. Dabhi, eds., *Upsetting the Offset: the political economy of carbon markets*. London: Mayfly Books, 255–274.

UNFCCC (2009) *Data Exchange Standards for Registry Systems under the Kyoto Protocol*. Available from: http://unfccc.int/kyoto_protocol/registry_systems/items/2723.php [accessed 7 April 2009].

US Senate (2009) US Senate. HR 2454. *American Clean Energy and Security Act of 2009*. Available from: http://frwebgate.access.gpo.gov/cgi-bin/getdoc.cgi?dbname=111_cong_bills&docid=f:h2454pcs.txt.pdf [accessed 29 February 2012].

Victor, D. F. and House, J. C. (2004) A new currency: climate change and carbon credits. *Harvard International Review*, 26, 56–59.

Walters, W. (2002) The power of inscription: beyond social construction and deconstruction in European Integration Studies. *Millenium*, 31 (1), 83–108.

Zelizer, V. (1994) *The Social Meaning of Money: pin money, paychecks, poor relief, and other currencies*. New York: Basic Books.

Part III

The politics of carbon after carbon

10 The politics of researching carbon trading in Australia

Clive L. Spash

Introduction

In recent times, conducting environmental research has become an increasingly dangerous activity for the researcher. Revealing the biophysical and social reality of our economic systems threatens a range of vested interest groups who rely upon resource exploitation, pollution and environmental destruction as necessary business practices. The capital-accumulating growth imperative dominates modern society – East, West, South and North – and anything that is deemed a threat to the beneficiaries of this system is subject to their attack, by whatever means necessary.

Carbon markets create a new set of beneficiaries who include fossil fuel corporations, large polluters, financial intermediaries, banks and speculators. They are a neoliberal triumph supporting the idea that markets can solve environmental problems, and show the way forward for addressing other problems such as biodiversity loss (Spash 2011). Research on carbon markets inevitably either supports, makes apologia for, or criticises a set of institutions and organisations in society. Criticism, no matter how legitimate or revealing of the inability of carbon markets to address climate change, means becoming a target for political attack.

Undermining the credibility of the natural and social scientific basis for environmental concern can be seen as aiding the maintenance of a weak market-based regulatory approach. This has been most prominent in the human-induced climate change policy arena due to the rise of climate denialism and its sponsorship by right-wing think-tanks from the USA (Jacques *et al.* 2008). The campaign of denial waged against the science that supports action to prevent human-induced climate change is merely the latest form of corporate-sponsored attempts to avoid radical greenhouse gas mitigation. Direct political lobbying by corporations in the USA to fight the Kyoto Protocol in the late 1990s amounted to at least US$ 100 million (Grubb *et al.* 1999: 112). Supposedly scientific economic studies emphasising control costs and downplaying the benefits of mitigation (i.e. avoided damages) were funded by electric power generators from the USA and then cited as evidence against taking action (Chapman and Khanna 2000; Spash 2002: 160). Exxon corporation, copying the tactics of the tobacco industry, has been involved in documented campaigning to spread misinformation on human-induced climate change (Union of Concerned Scientists 2007). Yet all this has not seemed enough. When corporate lobbying and media campaigns proved inadequate, the tactics moved to discrediting scientists as untrustworthy. For example, the equivalent of industrial espionage was employed in the 'Climategate' case of the University of East Anglia email theft and publication, and this was followed up with an internet campaign where numerous right-wing bloggers claimed climate science is nothing but a religion (Nerlich 2010). Attempts to silence climate researchers have also involved direct harassment, threats of violence and death threats (Hamilton 2011). In Australia, public debate has been subject to use of the full arsenal of weaponry supplied to the climate deniers by their corporate backers who control or are connected to powerful media outlets (as detailed in a series of articles for the Australian Broadcasting Corporation by Clive Hamilton (2010a, b, c, d, e)).

While the level and open vehemence of the orchestrated international attack on human-induced climate change research appears new, the general phenomenon is far from uncommon. Questions over the scientific credibility of environmental arguments pointing out the harms posed by new technologies have been repeatedly raised with respect to diverse subjects, from nuclear power (Carter 1987) to genetic modification (Burgess 1999; Robins 2012; Sarewitz 2004). For years industry has backed the organised denial of health impacts from DDT, smoking, asbestos, lead and so on (Markowitz and Rosner 2002; Oreskes and Conway 2010). Research exploiting scientific uncertainty has been funded by corporations to delay government regulation of, and action against, harmful company products and practices (Michaels 2005).

Researching human-induced climate change and addressing how to control greenhouse gases is inherently about revealing the structure of power relationships in society. In this context, a highly relevant concern is that of Galbraith (2007 [1967]) for the undemocratic power wielded by professional corporate managers (whom he termed the technostructure) and their ability to capture those agencies meant to be regulating the corporations within which the managers operate. The institutions – conventions, norms and formally sanctioned rules – of scientific research seem to provide poor protection against vested interest groups for environmental researchers entering the minefield of public policy. Indeed researchers often seem unaware that the science–policy interface is a battlefield where the contest is for control over the role and status of knowledge in modern society.

Rather than abstractly speculating on these matters, I will detail the treatment of my own critical research on carbon emissions trading in order to reveal how political sensitivities can arise in research. More specifically I will report on my personal experience working as a Science Leader, and Senior Civil Servant, in Australia for the Commonwealth Scientific Industrial Research Organisation (CSIRO) between 2006 and 2009. I had been head hunted for this role within their Sustainable Ecosystems Division. As an economist with an interdisciplinary research record, I was promised an open remit to build up, develop and lead 'blue skies research' on environmental values and policy. The CSIRO, a leading national and international scientific research organisation (employing 6,500 people), was and remains an organisation dominated by those trained in the natural sciences and engineering, supplemented by a handful of mainstream economists and a small sprinkling of other social scientists. The dominant belief within the organisation is in a modernist tradition that regards science as creating an objective, value-free knowledge on the basis of empiricism. In 2007, a corporate trained manager, Dr Megan Clark, became the Chief Executive Officer (CEO) of the CSIRO. She had previously worked in the Australian mining industry with its heavy involvement in the high greenhouse gas emitting aluminium and coal sectors.

In detailing the treatment of my work critical of carbon trading, by Dr Clark and others, I will show how, in practice, norms supposed to achieve quality control can be perverted to achieve justifications for censorship and suppression of politically sensitive work. In this case, the CSIRO was brought into disrepute in public media and the Australian Senate for allegedly engaging in censorship. However, there was no impact on the structures used to control and manipulate information, and indeed they were reinforced by the *a posteriori* defence of the need for even greater direct managerial 'quality' control. This can, then, be seen as providing a carte blanche for future potential censorship.

The case study raises general questions about the foundation of scientific 'evidence-based' policy, how Western democracies can suffer from the suppression of information of substantive public interest, and how the spreading role of managerialism from the corporate world is being used to control research and researchers. The case study shows how the mythical fact–value dichotomy, presumed by the naïve objectivism of much contemporary natural scientific and economic research,[1] collapses and why the sociology and politics of science cannot be ignored. The attempt to totally separate climate science (as fact) from policy (as value laden) is then revealed as deeply flawed and also a poor defence for researchers, especially social scientists. This raises questions as to how regulatory design, such as carbon markets and alternatives to them, can be debated in an open and explicitly value-laden research context, and how the necessary societal debate should be conducted concerning the future of our social, ecological and economic systems.

Carbon trading and the science–policy interface in Australia

Among OECD countries, Australia is the highest per capita emitter of all greenhouse gases combined, and in 2009 overtook the USA as the highest per capita source of CO_2 emissions (Lo and Spash 2012). The Kyoto Protocol allowed the country to increase emissions by 8 per cent over 1990 levels during the first commitment period (2008–2012). Australia only ratified the treaty in 2008 after a change of national government from John Howard's Liberal Party to Kevin Rudd's Labor Party. The Rudd government then staked its reputation on getting an emissions trading scheme in place. This was an idea originally floated by the Howard government. Entitled a 'Carbon Pollution Reduction Scheme', it was contentious and, in 2009, struggling to get passed into legislation by the Senate. In Australia coal is king and the mining sector is powerful and politically influential. The political support for the scheme was shifting and corporations exploited this to the full by negotiating massive multi-billion dollar permit bonanzas for the worst polluters. Despite this effort, by Rudd, to buy off the big fossil fuel interests and their political allies, the scheme was voted down by the Senate in the middle of 2009. At the end of that year, when my censorship case was at its height, a second vote was to be held and Senate debates were a hot topic for Australian media attention. The proposed free permit transfers to the corporate greenhouse gas emitters were being increased, but media attention seemed more preoccupied with climate change denial. Tensions were high when the Senate narrowly voted the scheme down for a second time and dealt a major blow to the credibility of the Rudd government.

Julia Gillard was then able to ascend to the prime ministerial role in Australia after an internal Labor Party coup. That coup was inspired by the failure to get the carbon emissions trading scheme into legislation, and the disaffection of the mining sector due to a proposed tax on their profits. Gillard substantially revised the mining tax in consultation with corporate leaders. She then called a general election in July 2010, that resulted in a hung parliament. The Labor Party managed to form a minority government supported by the Australian Greens and three Independents who favoured carbon pricing. This has been described as Australia's climate change election, because the expectation at the time was that Labor would be forced by the Greens to adopt a stronger environmental position (Rootes 2011).

The resulting compromise scheme was a carbon tax for three years leading to an emissions trading scheme in 2015. Like its predecessor it offered large financial transfers to the worst polluters. Australia had done nothing to limit emissions to the Kyoto mandated increase of 8 per cent on 1990 levels, and the Gillard scheme limited itself to reducing the rate of increase (i.e. choosing a base of 5 per cent reductions now on 2000 levels rather than 1990 levels). While touted as a major environmental success the reality was not impressive at all and seemed unlikely to do much to control Australia's runaway greenhouse gas emissions (Spash and Lo 2012).

In 2013 Gillard lost a leadership election within the Labor Party and Rudd returned to become prime minister once again. A national election soon followed that Rudd and the Labor Party lost to Tony Abbot and a right-wing coalition government. Abbot vowed, as one of his first acts, to remove the carbon tax/trading scheme established by the Labor coalition.

The brave new world of carbon trading

In February 2009 I completed drafting an article on carbon emission trading schemes for a special issue of *New Political Economy*, a leading academic journal in the field of economics and public policy. The article argued that carbon emissions trading schemes are fundamentally flawed, and it critically analysed support for their recommendation. I used a descriptive, institutional and logical analysis to deconstruct claims made by economic theorists and carbon market designers. The argument was not directed at any specific country or scheme design, but written as a general critique using a variety of examples to illustrate key points. A common position among economists was, and remains, that emissions trading schemes are the most efficient regulatory control for pollution, and any implementation problems they encounter can be solved by simple redesign (see also Lane and Stephan, this volume). My paper criticised both the claim of economic efficiency for carbon emissions trading and argued redesign was not feasible given the structure of the real economy (e.g. the power of corporations, the banking and finance sectors), which such economists fail to take into account. The paper also raised a variety of concerns over the operation of carbon markets including greenhouse gas accounting, permit allocation, the Clean Development Mechanism (CDM), and the role of voluntary offset markets.

As a Science Leader within the CSIRO my work was subject to an internal review process prior to sending an article off to a journal. This was often treated as no more than a formality, especially for a senior researcher, and the process was not anonymous. However, on this occasion, prior to completion of the internal review process, my divisional manager, Dr Daniel Walker, intervened citing concerns over the political sensitivity of the work.

At the time, the Australian Senate was soon to conduct its first debate over whether to pass the Carbon Pollution Reduction Scheme, which aimed to establish greenhouse gas emissions trading. The divisional manager noted a requirement within the CSIRO that the Minister for Innovation, Industry, Science and Research (then Senator Kim Carr) be informed when politically sensitive research was due to appear. That was, apparently, meant to be merely an information process not a control or influence on the research findings. However, the divisional manager also expressed his concern that the contents be judged of high quality and thus be defensible against the minister. To that end he requested a higher standard than internal review, cited international peer review as such a standard and agreed, with myself and my co-author, that this would be most easily achieved by immediate submission to the international journal for which the paper had been written.

The paper was, therefore, formally submitted to the journal *New Political Economy*. This meant it was sent out for international peer review by the journal while the internal CSIRO review was still on-going. Shortly after, the internal review (now superseded by the international review process) was completed and, while noting the political sensitivities, approved publication with some minor suggestions. A working paper would normally have then been published and made available online, but in this case was delayed awaiting the international review. The journal's international anonymous refereeing process resulted in my receiving a report that noted the positive contributions of the paper, made some critical remarks and suggested several revisions and amendments. All the points made by the international referees' report were then addressed and responses to each and every point noted in a cover letter to the journal editor, who accepted the revised paper. My divisional manager was informed of the successful forthcoming publication and, as a courtesy, sent the referees' report, response and revised version.

At this point something unusual happened. A few days later I received an email from Dr Walker stating that the publication should be withdrawn from the journal to undergo further internal review and discussion. A totally ad hoc procedure was suggested involving a further three unnamed managers, as well as input from his superior Dr Andrew Johnson and from those driving the CSIRO's Carbon Strategy. Research at the CSIRO involved investigating various schemes that might have been expected to benefit directly from making carbon into a valuable commodity through emissions trading (e.g. carbon sequestration in forestry and agriculture, carbon capture and storage for coal-fired power stations, clean coal technologies and more). The reasons stated for stopping publication were that: 'This paper deals with an extremely important public policy issue for Australia (and globally) and raises significant considerations. The proposed ETS is extremely politically sensitive at present' (email, 24 June 2009).

I refused to withdraw the paper from the journal and stated my intention to publish without CSIRO affiliation. This stalemate continued through various meetings until Dr Walker himself wrote directly to the journal stating the paper must be withdrawn for not having completed 'internal CSIRO approval and review processes'.

Over the course of several meetings and correspondence this same manager repeatedly made clear that the reason for stopping publication was purely concern over unspecified political sensitivities. Three options were at one point tabled: publication with CSIRO affiliation, publication without affiliation, no publication. However, after conferring with his superiors, only the last option was deemed suitable. This led to the CSIRO discussing their claim over anything I might write (including in a personal capacity) which would have had the effect of allowing them to prevent me from publishing the paper even as a private citizen. My co-author, also a CSIRO employee, had several visits from management, feared for his job and withdrew from the paper.[2]

In light of previous concerns over gagging of climate scientists within the CSIRO, freedom to 'speak' about research of public interest was still deemed permissible. An understood option was to claim no affiliation and present ideas as an individual. The CSIRO Charter states 'Researchers who speak as individuals should not claim to represent an organisation'. So, about five months after the publication ban was applied, I submitted the paper to an international conference for anonymous review as an independent researcher (with no CSIRO affiliation) so it could be accepted for presentation. I was later criticised by management for not seeking approval to present. Many CSIRO staff presented at the same conference and I know of only one person having sought official approval for doing so,[3] despite (unlike me) presenting their work under the CSIRO banner.

A few days after the conference presentation, the matter became national news, with *The Australian* (2009) running a front page headline 'CSIRO Carbon Trade Dissenter Silenced'. The story was quickly picked up and spread. The media concern was the attempted suppression of information of public interest by CSIRO management. That led to the CEO, Dr Clark, calling me to a meeting at CSIRO headquarters. During this meeting she expressed little concern over the content of the paper and stated only 'tiny changes' would be required to meet her requirements, namely conforming to 'the Charter' (the need for tiny changes was also reported in *Nature*, see Pincock 2009). The CSIRO then made a press release stating the paper would be published after all.

The Charter is an agreement that had been signed with Science Minister Senator Carr. This was meant to be a document that affirmed the freedom of speech of CSIRO researchers and, as Senator Carr had stated in a press release, defended 'frank and fearless' debate. This implied debate without ministerial or government intervention. However, the text ends with the following sentence: 'As CSIRO employees, they should not advocate, defend or publicly debate the merits of government or opposition policies (including policies of previous Commonwealth government, or State or local or foreign governments).'

This sentence leaves much open to interpretation. It could be seen as merely stating 'avoid being partisan', or it could be taken as a total ban on any research deemed as touching on any government policy anywhere in the world at any time. Taking the latter position would mean banning most, if not all, CSIRO research. For example, among other things, the organisation advocates clean coal technology, carbon sequestration, Green jobs, extracting wealth from the oceans, adaptation to climate change, biotechnology and publishes food diet books. All these are quite clearly areas of government policy. In addition, the CSIRO takes specific positions on water and land use, farming practices, management of pests, wildlife and biodiversity, and much else.

This one sentence of the Charter provides a key instrument for control of information flowing from the organisation. At the time of my meeting with the CEO, I was led to believe a few minor word changes would address the Charter. Within a week of that meeting I was given an ultimatum to accept an anonymously changed paper which (among other things) had cut crucial sentences and text (11 per cent in total) and halved the concluding section.[4] I was called to a meeting of senior managers, including Andrew Johnson, at which the CEO informed me to either accept all the changes as they were, with no input on my part, or have the paper ban imposed once again and also be personally banned from speaking to the press. I chose the bans.

The essence of the censored version of the paper was to substantively change the argument concerning carbon emissions trading. My argument involved a series of points to establish that emissions trading is fundamentally flawed as an approach. This involves failures by mainstream economists to take into account economic and social reality such as corporate power, strong uncertainty and the exploitation of the poor to establish carbon offsets. The censorship removed key sentences and paragraphs so that the revised argument took the familiar mainstream environmental economist's line that, while emission trading has its problems, these things require more research and, through redesign, over enough time, such schemes can be improved and all issues resolved. Strangely enough, putting the cut text together created a coherent critique of emissions trading.[5]

The issue did not stop with the ban by Dr Clark, because the Green party lobbied the Australian Senate to hold a debate about the paper and the issue of government censorship at the CSIRO. That debate led to the passing of a motion which instructed Senator Carr to table the paper in the Senate as a matter of public interest, something which he initially refused to do.[6] When he acquiesced, after further Senate pressure, the paper appeared in a version which had been submitted to the aforementioned international conference for peer review. Senator Carr (using parliamentary privilege) also simultaneously tabled a personal attack on my character in the form of a letter written by Dr Clark. In that letter she threatened me with unspecified punishment for having presented the paper at the conference and included copies from a webpage showing it had already been published online.

Indeed, the paper had miraculously appeared as an online publication, without my knowledge, just a few hours before Senator Carr presented it in the Senate. A conference proceedings website had suddenly released the paper without seeking or being given any permission, and despite the ban status of the paper being common public knowledge.[7] A mystery surrounds how the CEO of the CSIRO knew about this release and, within a few hours of its posting, was able to write a letter about it and supply a copy of both to the minister in time for his Senate appearance. Personally, I only heard of the online posting during the public broadcast of the Senate tabling of the paper by Senator Carr, and I had it removed within hours after contacting the website manager.

I resigned within a week of these events, having concluded the organisation to which I belonged lacked all credibility and its scientific integrity was being compromised by senior management. This ensured the publication of the internationally peer-reviewed version of the paper in an unadulterated form.[8] I was also free to speak about the contents, which I did in a series of public lectures around Australia.[9] The publication was officially released as an independent discussion paper (Spash 2009) and later as a journal article (Spash 2010a). However, the story did not stop there.

Quality, peer review and political control

On 10 February 2010, a few months after I resigned, the CSIRO, represented by its CEO, Dr Clark, and Head of the Environment Division, Dr Andrew Johnson, appeared before the Senate Estimates Committee. During discussion of my case, Senator Carr went on record making direct derogatory comments about the paper. In addition, he supported the latest position of the CSIRO management as to why the paper could not be published.

Now, apparently, the paper had been too low in quality. In the Official Hansard transcript of that meeting Dr Clark, stated:

> This was always an issue of quality; it was always an issue of maintaining the standards of the organisation. We always encouraged Dr Spash to publish the paper. I personally encouraged him to do so.
>
> (Australian Senate 2010: E48)

Later she stated that:

> In this case, the scientist was not prepared to make those changes to meet the quality. But there was always the encouragement to publish this work and to get it out there into the arena with quality changes that we required.
>
> (Australian Senate 2010: E50)

Senator Carr intervened in the questioning of Dr Clark and made two claims. First, he tried to claim the paper had been published as a conference proceeding so there was no issue of a ban. Second, he tried to deride the paper itself by quoting a confidential referees' report from the journal in which it was accepted. This was the report I sent my divisional manager, Dr Walker, along with my comprehensive response to referees' comments, prior to the first banning of the paper. Clearly CSIRO managers passed the confidential peer review to Senator Carr. In his attempt to deride my work the Senator chose half sentences and bits and pieces of the confidential referees' report. After quoting these he stated: 'In my judgement, this is a clear case of CSIRO defending the brand name of this organisation and has absolutely nothing to do with the personal political opinions of the author of this paper' (Australian Senate 2010 E49).

This is rather strange as prior to my resignation the CSIRO management claimed the right to stop me publishing the paper even without their brand name on it. In addition, he chose to cite the external referees' report because there was no document or statement from the CSIRO showing any internal concern over the publication's quality prior to my resignation.

Senator Carr did not stop there in his attempt to claim his support for quality assurance at the CSIRO. He stated: 'As a former schoolteacher, I really wondered whether or not this was the sort of thing we would be employing people to write on behalf of the CSIRO. The quality just was not there' (Australian Senate 2010: E50).

This from a Science Minister who, when launching Charters to protect the integrity and independence of public research agencies, had gone on record supporting the scientific peer-review process, the importance of avoiding government intervention in that process and the need for 'frank and fearless debate'. He had then stated (press release 16 January 2008): 'The value of scientific endeavour and importance of vigorous and transparent public debate, unfettered by political interference but subject to peer review, is something I have advocated for my entire public life.'

When questioned by Mark Colvin of the ABC (broadcast 24 February 2010) about the treatment of my paper, the office of Science Minister Carr stated that: 'Clive Spash may have made some revisions but the revisions did not address all the concerns of the reviewer and still failed to meet the standards of quality required of a CSIRO paper.'

This baseless claim is flatly contradicted both by the support for publication of the paper from CSIRO's internal reviewers, the point-by-point revisions address-ing all referees' comments and subsequent acceptance for publication by the journal editors at *New Political Economy*. The fact that the journal is internationally peer reviewed had actually been regarded as a higher standard than that of the CSIRO internal review. Indeed, *New Political Economy* was ranked 'A' class by the Australian Research Council in their 2010 Excellence in Research Assessment report.

This direct political intervention in the case raises serious concerns about the independence of scientific review in Australia, especially by its government agencies and, specifically, the CSIRO. Senator Carr had intervened in a pro-cess to deride a paper, prior to its actual publication, on the basis of comments which had led to its revision and subsequent acceptance. In a letter (dated 24 November 2009) from Professor John O'Neill to Senator Carr the editors of *New Political Economy* had made clear their position.[10]

> The CSIRO is asking not for minor but for major changes in the central arguments of the paper. This is clearly unacceptable to the author. I should add that is also unacceptable to me as the editor of the special issue. It involves interference in our own peer-reviewing procedures that would be incompatible with academic integrity of the journal.

The letter from the journal also addressed the role of Dr Clark in demanding changes to the paper.

> What is clearly improper is for her to use her position to insist on changes to the paper which alter its conclusions prior to publication. No international journal would accept a paper under those circumstances. Neither would or should any academic scientist be expected to agree to such alterations to his or her work.

Both Senator Carr and the CSIRO management took a stand that placed their own (non-specialist) opinion above that of the expert peer-review pro-cess the Science Minister had himself claimed was the ultimate test of quality for engaging in public debate. Their claims of poor quality and no political motivation for the banning of the paper must confront the facts. None were prepared to publicly debate the contents of the paper or even specify any issues of concern.

The case reappeared in later Senate Estimates Committee meetings and the CSIRO was forced to explain itself further. This involved a new elaboration of why the paper was banned compared to other similar work, and specifically the public advocacy of the Gillard government's climate policy by CSIRO's Dr Wonhas (see also endnote 4). In response to a question (SI-71, 19 October 2011) by Senator Colbeck, the CSIRO reiterated claims of the need for quality control and meeting the Charter, but also stated:

> The issues related to Dr Spash were not about the content of his paper, nor were they related to any public comments regarding his paper. . . . CSIRO's internal review concluded that the original paper did not report new research or present empirical evidence to support all of the authors' conclusions. The paper was also viewed as offering opinion on matters of government policy by applying a critique of neoclassical economic theory to the ETS. Therefore it was not approved for publication.

So now, besides falsely claiming the CSIRO only publishes evidence based on empiricism, criticism of neoclassical economic theory had also become a reason for banning research work on emissions trading! What is this if not a matter of content?

During the Senate Estimates discussion, of February 2010, an interesting remark came from Dr Clark in terms of the control management have over the science coming out of the CSIRO:

> Our processes are very consistent across all our scientists in terms of working with them to make sure they are published in the most appropriate journals, making sure that the science is robust, making sure that the conclusions can be drawn upon that science.

My former line manager wrote to Dr Clark about her Senate Estimates statements. His letter was entitled 'Misinformation given to Senate Estimate committee' (dated 1 March 2010). In this letter he made clear that, in his opinion, Dr Clark and Senator Carr had misled the parliamentary committee and Dr Clark had abused the peer-review system in releasing confidential referees' reports to the Senator. In addition, he also made clear his belief that the quality of the paper was actually diminished after CSIRO editing.

Even more seriously, this raises questions regarding what goes on behind the scenes at the CSIRO. Prior to becoming CEO of CSIRO, Dr Clark built her career in the mining industry and was a vice-president at BHP Billiton, one of the largest resource extracting multinationals in the world (which publicly proclaims its active engagement in extracting coal, petroleum and aluminium, among other natural resources). This has inevitably led to expressions of concern and speculation over the influence on the CSIRO of corporate interests, the powerful Australian mining sector and specifically 'big coal' (Manning 2010).

In all this some things are very clear. Massive financial transfers to polluters have been an essential part of the design being employed by emissions trading schemes to get buy-in from powerful vested interest groups. The Australian Carbon Pollution Reduction Scheme was no different, and built in generous 'allowances' for the coal, aluminium and petrol-driven transport sectors (Spash 2010a). This was repeated under the scheme passed into legislation by the Gillard government (Spash and Lo 2012). Similarly, the gains to be made by the finance and banking sectors in running a new multi-billion dollar commodity market have been touted as a great economic opportunity (Stern 2006: 270), rather than a massive transaction cost to society. My work indicated how these financial flows are more appropriately characterised as side payments, bribes and deadweight loss. As this work brought the development of carbon markets into question, it became an act threatening the potential for those massive transfers, speculative gains and rent capture to take place or continue. Rather than enter into 'frank and fearless debate', the minister and CSIRO management sought to discredit the work. The strategies employed escalated from changing the content, to suppressing the whole work, to attacking the author and his reputation, to finally questioning the quality of the work and its scientific credibility.

The sociology and politics of knowledge

Knowledge (whether regarded as objective, subjective or something else) is necessarily embedded within social institutions. Scientific knowledge is produced by epistemic communities of scholars that have developed elaborate intellectual and social organisations to demarcate what is to be regarded as communal activity and how it should be conducted. Essentially, such scholarly communities, as social groups, control the way work is carried out, the goals of that work and who is employed to conduct such work. Interdependency creates community-based research standards as to competent use of research techniques to enable knowledge transfer. Researchers who fail to fit within this structure of dependency are regarded as not producing the right kind of knowledge and can be marginalised within or excluded from the community.

The epistemic community of scholars, investigators, analysts and others in any given research field can be regarded as being united by a set of norms of social behaviour, i.e. institutions of group and individual control over, and validation of, information. The range of practices for establishing hard-core 'scientific' credibility are then often regarded as self-evident, including such things as: peer review, mathematical formalism, experimental method, employing data and statistics, writing in the third person, avoiding explicit engagement in politics and trying to exclude value-laden statements. For many, perhaps most, within a community the conventions, norms and rules are not even perceived as such, they are just part of what defines their occupational practice.

Lee (2009: 12) explains this social organisation of science, and the interdependency of its members, as involving several factors including: the nature of the audience for which the scientific output is intended; control over (i) the means of production of scientific knowledge, including the equipment, techniques and labouring skills, (ii) the format in which it is reported, and (iii) the communication outlets, such as journals; the role of individual and institutional reputations in affecting both the production (e.g. what is accepted) and the goals of scientific knowledge; and the role of State and other organisational power outside the science community in legitimising, supporting or otherwise affecting scientific knowledge, its goals and the reputations of specific individuals. Thus, a range of quite different social systems (e.g., hierarchical managerially controlled vs. diffuse locally coordinated) can exist for the production of scientific knowledge, with implications both for the type of work produced and how it is employed. Indeed, the intellectual and social organisation of such work, and who it is deemed to involve, is historically and consciously determined by its participants and the recipients of scientific knowledge (Lee 2009: 13).

Both the social and natural sciences share this communal framing of knowledge but have developed different practices of work. An important institution they share in common is independent and generally (but not always) anonymous peer review for quality evaluation. In democratic states, the intellectual and social organisation of work for both has been a matter for the scientific communities involved. Yet, as the case study indicates, this process is also susceptible to control or manipulation by those in powerful positions and gate keeper roles.

Which journals are 'appropriate', who decides and on what basis? Who selects referees and how? When is managerial consistency in publication practice merely code for uniform control and manipulation of what gets published? Why should empirical data and quantification be regarded as the test of 'truth'? On what substantive grounds is qualitative data or descriptive analysis regarded as less robust than quantitative data? Why is work of social scientists typically treated as less 'hard' or 'objective' than that of natural scientists and on what basis? Who should be the judge and who the jury in these matters? These questions relate to the philosophy and conduct of science as well as its relationship to public policy and political process in a democracy.

All social and natural scientists produce their work within a specific social and institutional context. At its best this can provide an empowering collegiate atmosphere where challenging common goals are set and tackled through collaboration among senior and junior researchers alike. At its worst this can be a disempowering hierarchical system of control where research managers force colleagues into serving narrow politically motivated ends. Derision of politically unpalatable research findings is then carried out via ad hominem remarks, harassment, bullying and questioning the scientific quality and credibility of the results on totally unscientific grounds.

The problem facing us today is the extent to which the latter approach is becoming dominant and there is creeping censorship of science in research by official government process in supposedly democratic states (Spash 2010b). In 2009, while I was fighting my case in Australia, Professor David Nutt was getting sacked by the UK government for speaking openly on the relative merits of drugs versus alcohol (Tran 2009). Then in the US there was the case, closer to my own, of two EPA lawyers told to remove and then edit parts out of a video critical of emissions trading (Broeder and Kaufman 2009). More recently, in Canada public muzzling of government scientists seems to have become a legitimate political policy (Ghosh 2012). These high-profile cases involving government employees are likely the tip of the iceberg.

As neoliberal governments are elected around the world, the rhetoric of austerity is employed to shut down research which asks too many unwanted and difficult questions. Indeed, my own former social science division at the CSIRO, Sustainable Ecosystems, was quickly closed after I left, and the social scientists merged with entomologists within a natural science dominated Ecosystem Science division. This is now led by my former divisional manager, Dr Walker, who advocates an evidence-based science approach.

Government agencies and institutions, as well as individual civil servants, that try to protect the environment and society against damaging development, harmful technologies and pollution are subject to being removed under the guise of reducing red tape and increasing free market competition, boosting growth and providing jobs. Employing those with corporate experience to manage research (as in the CSIRO) further muddies the waters between the independence of science and the objectives of vested interest groups. In such a climate of fear civil servants are most vulnerable and self-censorship prevails. Yet the myth remains that if research is factual, empirically based, and avoids comment on policy, then researchers have nothing to fear.

The artificial division of science from policy and facts from values

In 2010, Australia's chief scientist, advising the government on all scientific issues, was Penny Sackett. She took a very traditional view of the science–policy interface. That is, one which claims the trust of people in science as progress is evident in their adoption and use of technology. The only real problem for science then is one of communication, i.e. the public disagree with scientists because they are ill-informed and misunderstand scientific facts. Her belief in scientists directly observing facts as truth reflects a naïve objectivist position (see Sayer 2010). In a TV interview on ABC's Lateline (broadcast 18 March 2010) she stated that: 'a CSIRO scientist can and should speak out clearly on matters of scientific evidence'. Yet when asked directly about my case she stated:

> The question is, when that crosses lines of policy there are matter [sic] – employment practices in the CSIRO that draw a distinction between matters of policy and matters of science and those matters are handled internally by the CSIRO. But I think one thing we can feel quite confident of is that the quality of research that is done by those researchers is recognised the world over as of the highest international standard.

She went on to draw a distinction, or division, between 'the science of climate change as opposed to a particular political solution to address climate change'.

Suppression and manipulation of research of the former was a concern, but not the latter. So social scientists and policy analysts within the CSIRO are fair game, but leave the natural scientists alone!

Simply calling for 'evidence-based science' as if this were an answer to the way in which science should interact with policy is to totally misunderstand the problem. Drawing such simplistic divides as policy vs. science comes across as a purely rhetorical argument, meant to convince without any substantive meaning. Take, for example, the introduction of genetically modified crops, which has been hotly debated in many countries including Australia (e.g., see Robins 2012). Where is the divide between evidence-based science and policy advocacy? That senior scientists within the CSIRO have heavily promoted research on such crops leaves many questions unanswered. We might ask who funds the research, who benefits from the results and what norms and beliefs drive that research? Why do these researchers not promote organic farming, permaculture or other alternatives?

The presumption of evidence-based policy is that there is no community of scientists or institutional context within which science is conducted, and that facts always speak for themselves. Compare this with an article in *Nature* on nanotechnology which explains how science embodies the perspectives of the researchers involved. That article concludes:

> Different positions on fundamental questions, such as the relationship between humanity and nature, permeate technology development and social debate. . . . [They] are much more than simple factual descriptions . . . each of them rests on different assumptions, supports different beliefs and leads in different directions
>
> (Wickson 2008: 315)

Once we move from the natural to the social sciences the claims of a dichotomy between facts and values becomes even more starkly ridiculous. Mainstream economists who push for carbon trading as 'the answer' to human-induced climate change are implicitly criticising a range of social institutions which offer different regulatory options (e.g., taxes, direct regulation), other pathways in society (e.g., degrowth) and alternative social, ecological and economic futures outside of their abstract conceptions of human society as a money-obsessed, competitive trading enterprise. That the CSIRO chooses to support some avenues of research rather than others is a direct reflection of its values, or today those of its managers. The CSIRO Charter and the positions of those like Clark, Carr and Sacket, if taken seriously, would ban all social science research from the CSIRO, which would also need to remove all connections with industry, including the use of the word 'industrial' in the organisation's title. Clearly none of these people would seriously consider closing down all CSIRO research advocating new industrial technologies, despite these clearly creating societal path dependency and crossing the imagined science–policy divide. Yet all seem to advocate placing social science within the confines of evidence-based fact finding, naïve objectivism and value-free knowledge.

In conducting social science research, such as that on carbon markets, there is no value-free information. All such research in promoting one set of institutional arrangements over another must involve a complex of facts and values. This does not mean there is no factual knowledge or social reality, but rather that knowledge requires conceptualisation and leads in specific directions which are far from value free.

In the end public policy requires ethical judgement and this cannot be avoided. We need institutions that allow for open debate and democratic decision processes. As Alan Holland (2002: 33) has stated:

> The penalty for not developing institutions in which ethical and other deeply felt concerns can be properly voiced will be residues of grievance, mistrust, injustice and guilt which are as corrosive of the civic body as are pollutants in the natural environment.

Conclusions

The development of public policy on climate change has been heavily influenced by a scientific credo from the outset of the Intergovernmental Panel on Climate Change. This provides a specific instance in which science becomes involved in answering political and ethical questions using a set of practices and methodology ill-suited to the task, and employing people formally trained to believe their work, as scientific research, can and should be isolated from societal context and implications. This creates a major weakness that has left climate science open to attack, because scientists cannot hope to live-up to their own claims of being providers of value-free knowledge. Rather than scientists identifying the failings of their own approach, admitting their own ignorance and limits to knowledge, they too often persist in an extreme and untenable claim to having access to the truth about reality based upon empiricism. So a second major weakness has been the inability to claim total certainty. This has left the door open for those who want to delay action on greenhouse gas control by allowing them to claim the need for more research to provide the ultimate proof.

Scientists who present evidence of fossil fuels being a danger to humanity and Earth's climate are directly criticising a set of institutions and social organisation (facts are connected to values). Thus, the resource extraction and energy industries, so closely associated with major emission sources of greenhouse gases, and other pollutants, are inevitably the enemies of a science which exposes their complicity in creating social and environmental harm. Similarly, all those citizens who like gas guzzling cars, Formula 1 motor racing, monster trucks and so on will feel themselves the subject of criticism. Indeed, anyone living a typical Australian, American, European or rich Chinese, Brazilian or Indian lifestyle is being held up to scrutiny for their high energy and material throughput. That researchers might believe their factual data and findings are free from association with politics and social values is clearly naïve.

In effect, climate science tells us that society must remove the vast majority of fossil fuel combustion activities from the economy and do so quickly to avoid scenarios involving substantive and extensive damage and loss of life. This science is a major criticism of modern industrial society, fossil fuel-based economies, the competitive capitalist growth society and the institutions of progress it has constructed. Ideas of freedom and progress have become integrated with massive energy and material consumption per capita. Policy action based upon this science is then necessarily engaged in challenging some of the most powerful organisations in the capital-accumulating modern world, both within the structures of Western-style democracies and the totalitarian regimes of Asia and elsewhere. Scientifically pointing out the causal mechanisms is then akin to a political act. Claiming policy and politics are things 'scientists' can ignore, while they get on with the real work, is quite simply sticking your head in the sand.

Discovering knowledge about the world has implications for the way humanity organises its activities and interacts with the non-human world. The advocacy of carbon markets is based upon the naïve objectivity encapsulated in neoclassical economics and its pursuit of efficiency. The lie of evidence-based science, like that of positive economics, is the claim that all information can be neutral and separated from the implications it has for the society within which that information is produced. If facts were so easily identifiable and separable from values there would be little need for epistemic institutions (e.g., peer review), ethics in scientific conduct or critical academic debate.

Research on biophysical reality now interacts very directly and immediately with the social and economic world. Attempts to close-down and/or control critical science–policy debates are all too evident. Scientists of both the social and natural type must take responsibility for the institutions within which they work, the uses employers make of their work and the implications of their work for society. That includes exposing abuse of peer-review processes, speaking out against inappropriate managerial and political control of research organisations and universities, and avoiding work with clearly unethical consequences. This also means being open about the processes used by the community within which they work to create knowledge by bounding it, and the associated problems of partial ignorance that this creates.

The susceptibility of traditional institutional processes of quality control to manipulation and the spread of misinformation should not be underestimated. Choosing the 'right' referees, selecting the 'right' journal for publication, requiring use of the 'right' method of research can all be open to abuse aimed at controlling and suppressing information. The natural sciences are just as vulnerable as the social sciences.

More crudely, climate deniers try to harass and intimidate scientists and threaten their families in the hope that they will stop their work and others will be driven away from the field to less contentious research. Many corporations and national governments have vested interests in continuing emissions of, and destroying sinks for, greenhouse gases. Due to the prevalence of naïve objectivism they can exploit scientific uncertainty and fund research to increase doubt. The game is one of disempowering those who oppose their operations and threaten their sources of wealth and power.

The concern then switches to how we, the researchers, can conduct investigation of a complex problem such as controlling greenhouse gas emissions in a way that builds understanding. Scientists arrogantly claiming the truth about facts in the face of strong uncertainty do not help. Neither do economists arrogantly claiming the truth about the most efficient solutions. None of this means reverting to an extreme postmodern doctrine of belief in strong social constructivism or radical relativism that denies the existence of reality or treats knowledge as a form of storytelling.

Instead we can accept our fallibility and require a far more nuanced comprehension of how knowledge is created and used in modern society. We can recognise there is much consensus on biophysical reality and that this has proven easier to achieve than consensus over economic and social reality. The social world changes faster than the natural world and therefore needs to be reconceptualised more often to be understood. Some facts, relationships and patterns are more uncertain and contestable than others. Some uncertainty is also not susceptible to reduction to a scientific consensus via more research. We then need institutional processes that help address strong uncertainty openly and allow for critical reflection and meaningful public engagement and debate.

Knowledge is powerful and the creation of knowledge can be seen as a political act most clearly where it directly empowers some and disempowers others. Control over information and its sources then becomes central to holding power in society. The implication is that knowledge is suppressed when it has unpalatable social and economic implications for those in powerful positions.

What this means in the context of carbon markets is that the more established they become the harder they will be to criticise. However, the inherent contradictions of promoting markets to solve the problems of the market economy can only be tolerated as long as the system remains functional. While truths can be obscured, messages rewritten and messengers made to disappear, the resulting loss of knowledge and denial of biophysical and social reality will, in the end, lead to failure. Over time this failure becomes increasingly evident socially, ecologically and economically.

Notes

1 For further explanation of the meaning and use of the term naïve objectivism and its relationship to the claim that values can be separated from values see chapter 2 of Sayer (2010).

2 Since this case, an increasingly vocal group of CSIRO employees have been making complaints about bullying and harassment within the organisation (e.g., Victims of CSIRO http://victimsofcsiro.com/). The CSIRO's managers were forced to respond after being held in breach of regulations by Comcare (the official government health and safely at work watchdog), and in 2013 initiated their own inquiry, which reports back to themselves. The consistent suppression and dismissal of whistleblowers and their reports left little doubt among victims that this would be a management whitewashing exercise.

3 The one person was my former co-author, who was then asked by his line manager why he was bothering to seek permission? He was told this was unnecessary, but as he insisted he got official approval.

4 Dr Clark indicated, at our meeting, that the referees would include Dr Alex Wonhas, a physicist, appointed in August 2009, to lead the CSIRO Energy Transform Flagship, from a job at the management consultants McKinsey & Co. Wonhas had previously co-authored a report on greenhouse gas control costs that claimed to be presenting the 'factual basis' for policy while advocating carbon trading and the benefits of engaging in the CDM (Lewis *et al.* 2008). My paper, besides being critical of emission trading and specifically the CDM, criticised this type of economic cost calculation and efficiency analysis.

5 The edited text the CSIRO wanted cut from the document can be found online at http://www.clivespash.org/OrwellianGuidetoCarbonETS.pdf.

6 This was happening at exactly the same time as the Carbon Pollution Reduction Scheme was being debated in the Senate.

7 The organisation, the Australian branch of the society for ecological economics, responsible for releasing the paper online was at the time headed by a senior CSIRO employee. The vice-president, and former president of the society, had also been a senior CSIRO employee, only recently having moved to the Department of Climate Change to directly advise the Labor government on emission trading. In a televised broadcast of the Senate debates on emissions trading in November 2009 he could be seen supplying Senator Penny Wong with answers to various questions. Both the society's president and vice-president were fully aware of the ban on the paper and had previously spoken directly with me about it.

8 The journal's publishers, Routledge Journal Division of Taylor & Francis, fearing litigation, wrote to the CSIRO seeking their approval before they would proceed with publication. My previous divisional manager sent them approval on the basis that I had resigned and there would be no CSIRO affiliation, although he simultaneously claimed the paper had already been published; presumably hoping this would affect publication.

9 Links to a full-length lecture presentation of the paper, radio interviews, news articles and related documentation, as well as the paper itself, can be found online at www. clivespash.com.

10 In response to a question by Senator Eggleston in Senate Estimates, 31 May 2010, Senator Carr's Office denied any knowledge of the letter sent to him by Professor O'Neill on behalf of the journal, *New Political Economy*, or that he had read it, and so denied being able to table the letter as requested.

References

Australian Senate (2010) *Senate Economics Legislation Committee Estimates*. Canberra: Official Hansard, 10 February 2010.

Broeder, J. M. and Kaufman, L. (2009) Environment agency warns 2 staff lawyers over video criticizing climate policy. *The New York Times*. New York.

Burgess, J. (1999) Frankenstein's footsteps: Science, genetics and popular culture. *Public Understanding of Science*, 8 (2), 142–144.

Carter, L. J. (1987) *Nuclear Imperatives and Public Trust: Dealing with Radioactive Waste*. Washington, DC: Resources for the Future.

Chapman, D. and Khanna, N. (2000) Crying no wolf: Why economists don't worry about climate change and should. *Climatic Change*, 47 (3), 225–232.

Galbraith, J. K. (2007 [1967]) *The New Industrial State*. Princeton/Oxford: Princeton University Press.

Ghosh, P. (2012) *Canadian government is 'muzzling its scientists'*. Available from: http://www.bbc.co.uk/news/science-environment-16861468 [accessed 22 February 2012].

Grubb, M., Vrolijk, C. and Brack, D. (1999) *The Kyoto Protocol: A Guide and Assessment.* London: Earthscan and Royal Institute of International Affairs.

Hamilton, C. (2010a) Bullying, lies and the rise of right-wing climate denial. *The Drum Opinion,* Australian Broadcasting Corporation.

Hamilton, C. (2010b) Think tanks, oil money and black ops. *The Drum Opinion,* Australian Broadcasting Corporation.

Hamilton, C. (2010c) Who is orchestrating the cyber-bullying? *The Drum Opinion,* Australian Broadcasting Corporation.

Hamilton, C. (2010d) Manufacturing a scientific scandal. *The Drum Opinion,* Australian Broadcasting Corporation.

Hamilton, C. (2010e) Who's defending science? *The Drum Opinion,* Australian Broadcasting Corporation.

Hamilton, C. (2011) Windsor receives death threats as climate of hate ramps up. *Crikey.*

Holland, A. (2002) Are Choices Tradeoffs? In: D. W. Bromley and J. Paavola, eds., *Economics, Ethics, and Environmental Policy: Contested Choices.* Oxford: Blackwell Publishing, 17–34.

Jacques, P. J., Dunlap, R. E. and Freeman, M. (2008) The organisation of denial: Conservative think tanks and environmental scepticism. *Environmental Politics,* 17 (3), 349–385.

Lee, F. (2009) *A History of Heterodox Economics: Challenging the Mainstream in the Twentieth Century.* London: Routledge.

Lewis, A., Görner, S., Downey, L., Slezak, J., Michael, J. and Wonhas, A. (2008) *Carbon Abatement Cost Curve for Australia.* Melbourne/Sydney/Perth: McKinsey & Company.

Lo, A. Y. H. and Spash, C. L. (2012) How Green is your scheme? Greenhouse gas control the Australian way. *Energy Policy,* 50 (November), 150–153.

Manning, P. (2010) CSIRO in bed with big coal. *The Sydney Morning Herald,* Sydney.

Markowitz, G. E. and Rosner, D. (2002) *Deceit and Denial: The Deadly Politics of Industrial Pollution.* Berkeley: University of California Press.

Michaels, D. (2005) Doubt is their product: Industry groups are fighting government regulation by fomenting scientific uncertainty. *Scientific America,* (June), 96–101.

Nerlich, B. (2010) 'Climategate': Paradoxical metaphors and political paralysis. *Environmental Values,* 19 (4), 419–442.

Oreskes, N. and Conway, E. M. (2010) *Merchants of Doubt: How a Handful of Scientists Obscured the Truth on Issues from Tobacco Smoke to Global Warming.* New York: Bloomsbury Press.

Pincock, S. (2009) Australian agency moves to calm climate row. Researcher will be allowed to publish his paper after making 'tiny' changes. *Nature,* 13 November.

Robins, R. (2012) The controversy over GM canola in Australia as an ontological politics. *Environmental Values,* 21 (2), 185–208.

Rootes, C. (2011) Denied, deferred, triumphant? Climate change, carbon trading and the Greens in the Australian federal election of 21 August 2010. *Environmental Politics,* 20, 410–417.

Sarewitz, D. (2004). How science makes environmental controversies worse. *Environmental Science & Policy,* 7 (5), 385–403.

Sayer, A. (2010) *Method in Social Science: A Realist Approach.* London: Routledge.

Spash, C. L. (2002) *Greenhouse Economics: Value and Ethics.* London: Routledge.

Spash, C. L. (2009) The brave new world of carbon trading. Munich: *Munich Personal Research Papers in Economics Archive* (MPRA).

Spash, C. L. (2010a) The brave new world of carbon trading. *New Political Economy,* 15 (2), 169–195.

Spash, C. L. (2010b) Censoring science in research officially. *Environmental Values,* 19 (2), 141–146.

Spash, C. L. (2011) Terrible economics, ecosystems and banking. *Environmental Values*, 20 (2), 141–145.

Spash, C. L. and Lo, A. Y. H. (2012) Australia's carbon tax: A sheep in wolf's clothing. *Economic and Labour Relations Review*, 23 (1), 67–86.

Stern, N. (2006) *Stern Review on the Economics of Climate Change*. London: UK Government Economic Service.

The Australian (2009) CSIRO carbon trade dissenter silenced. *The Australian*, 2 November 2009, front page.

Tran, M. (2009) Government drug adviser David Nutt sacked. *The Guardian*. London.

Union of Concerned Scientists (2007) *The Heat is On: Global Warming, the Greenhouse Effect & Energy Solutions*. Cambridge, MA: Union of Concerned Scientists.

Wickson, F. (2008) Narratives of nature and nanotechnology. *Nature*, 3 June, 313–315.

11 Dialogue of the deaf?

The CDM's legitimation crisis

Peter Newell

Introduction: Carbon markets are dead.
Long live carbon markets!

We are currently at a key juncture in the debate about the role of carbon markets in responses to climate change. On the one hand, the low price of carbon and the declining interest on the part of financial capital, thought to be one of the main beneficiaries of carbon trading (Spash 2010) and repeated scandals which have engulfed the Clean Development mechanism (CDM) and the EU's Emissions Trading Scheme (ETS), have led to claims of their imminent, inevitable and (for some) welcome demise (Clark 2012). On the other hand, carbon markets, judged by their ongoing proliferation, are flourishing. At the UN level the negotiations are focused on the construction of new market mechanisms. Governments in Annex I countries continue to construct mechanisms and streams of finance to support the growth and inter-linkage of carbon markets (such as the UK government's Carbon Market Finance Initiative). And at regional and national levels a swathe of new carbon markets and ETSs are coming online in countries such as China, South Korea, Vietnam, Thailand, Australia and Mexico (see Lederer and Engels *et al.*, this volume). It is in the context of this curious anomaly of crisis and expansion that this chapter assesses the role of the CDM Policy Dialogue.

Established as an effort to take stock of experience to date with a view to reforming the CDM, the dialogue presented itself as an opportunity to engage critics and articulate a response to the multiple crises facing the CDM at the time of the dialogue's launch at the end of 2011. It provides an interesting case study on the one hand of how the tension between accumulation and legitimation plays out in climate politics (Paterson 2010), and on the other of civil society's engagement with carbon markets and the merits or otherwise of advocacy aimed at exposing the shortcomings of carbon markets by engaging in debates about their governance and reform, as opposed to rejection of them in their entirety (Lohmann 2012). To explore these themes the chapter draws on a diverse literature on legitimacy and legitimation crises to explore the dynamics that unfolded in the case of the CDM Policy Dialogue, and combines this with neo-Gramscian understandings of hegemony and accommodation to comprehend the politics of granting concessions in order to preserve the legitimacy of an historic bloc (see also Matt and Okereke, this volume), or as it is applied more concretely here, the practice of 'disciplinary neo-liberal participation' (Newell 2010). Critically, then, the chapter draws attention to the ways in which civil society organisations have both managed to create a legitimacy crisis for the CDM and then also played an important role in shaping the nature of its (attempted) resolution.

Background to the CDM Dialogue

The CDM was brought into life by the 1997 Kyoto Protocol embodying a compromise between richer countries, particularly the USA, wanting flexibility in how they met their emission reduction targets and poorer Non-Annex I countries insisting that they got something in return in the form of a commitment that projects should also deliver sustainable development benefits to the communities hosting them. The resulting Article 12 of the treaty thus states the purposes of the CDM as being: 'to assist Parties not included in Annex I in *achieving sustainable development* and . . . to assist Parties included in Annex I in *achieving compliance with their quantified emission limitation and reduction commitments*' (UNFCCC 1998). The speed with which the CDM was included and the extent to which it departed from original proposals from Brazil for a Clean Development Fund led to it being labelled the 'Kyoto surprise' (Werksman 1998). Though the USA then walked away from the Kyoto deal, the CDM market took off rapidly. The global value of primary offset transactions grew to US$ 7.2 billion in 2008, more than a ten-fold growth from 2004, largely due to the CDM. Under the CDM, Certified Emissions Reductions (CERs) amounting to more than 1.8 billion tonnes of carbon dioxide (CO_2) equivalent were produced in the first commitment period of the Kyoto Protocol (2008–2012) (UNFCCC 2012a). The revenues of the CDM constitute the largest source of mitigation finance to developing countries to date (World Bank 2010) and over the period 2001 to 2012, the CDM is claimed to have spurred US$ 215 bn in investment (UNFCCC 2012a).

Yet the CDM has come in for repeated criticism on a number of grounds that conspired to produce a legitimacy crisis which the dialogue was meant to alleviate. First, doubts have been raised about the additionality of many CDM projects (Schneider 2007) compounded by revelations in 2011 from WikiLeak cables in which government officials claim that none of the CDM projects from countries such as India (the world's second largest host of CDM projects), can be considered genuinely additional (Yan 2011).

Second, critics have pointed to the failure of the CDM to deliver sustainable development benefits to host communities. This was the other half of the deal negotiated in Kyoto: lower cost emissions reductions in return for development benefits in the form of jobs, sourcing of local materials, technology transfer or other health and environmental benefits. The evidence is overwhelming that this has not occurred on anything like the scale anticipated (Olsen 2007; Schneider 2007; Sutter and Parreño 2007). A flurry of exposés about acts of dispossession and allegations of human rights violations, such as those surrounding the Barro Blanco dam in Panama and the Bajo Aguan biogas project in Honduras, for example, intensified pressure on the CDM Executive Board (EB) to respond with a review of mechanisms and procedures.

Third, a series of scandals about instances of fraud and gaming by market actors and evidence of collusion and corruption, detailed at length in the 'Global Corruption Report' on climate change (Transparency International 2011) and picked up by media such as *The Guardian* (2011) in the UK and *Times of South Africa* (2011) forced the CDM to demonstrate it could prevent such abuses in future. The critique emanating from opponents of carbon markets and critical researchers was gaining widespread coverage in international media such that the chair of the EB, Martin Hession, was being continually dragged before TV cameras to account for the CDM's implication in human rights violations, land dispossessions and the like.[1]

Because of the spiralling sense of crisis surrounding the CDM's credibility and ability to simultaneously secure additional emissions reductions and sustainable development benefits, the CDM was forced onto the offensive to make and re-state the case for the centrality of the CDM to global efforts to tackle climate change. The call for the dialogue lists these criticisms explicitly, even referring directly to the analogy with the 'new indulgences' used by carbon market critics (Smith 2007) and noting that 'it has been alleged that CDM projects have been the scene of environmental damages or human rights abuses' (CDM Policy Dialogue 2011). This public outreach to regain legitimacy has continued unabated after the completion of the dialogue process. The CDM passed the 6,000 project milestone in February 2013. To mark this event the UNFCCC communications office issued the following statement: 'In the past 10 years, CDM projects have delivered 110,000 MW of renewable energy capacity. That is roughly equivalent to the total power generation capacity of Africa' (UNFCCC 2013a). The release continued with a statement from Mr Stiansen of the UNFCCC secretariat that

> The effects of climate change are already being seen, so increased action on climate change is inevitable. Tools like the CDM will become indispensable. . . . The CDM is a functioning and effective tool that has proved its worth. Given the tangible threat that we face from climate change, it makes sense that we use the tools that we have at our disposal.
>
> (UNFCCC 2013a)

It is this mixture of optimism, defiance and appeal for renewed interest that formed the background to the CDM Policy Dialogue of 2011. This was a year-long initiative that grew out of the 64th meeting of the CDM EB whereby, in the face of uncertainty regarding its future post-2012, the EB sought to initiate a debate about the CDM's current and future role. The idea for it was conceived during a private retreat of the Board to which select specialists were invited to submit comments and recommendations about how to maximise the CDM's contribution to reducing emissions and delivering sustainable development benefits, and how some of the barriers to promoting these could be overcome. The key outcome of the retreat was a decision to launch the dialogue. The EB decided that a CDM Policy Dialogue should be conducted in 2012 and that the dialogue should be organised separately from the EB. The intention was to generate 'recommendations regarding how to best position the CDM to respond to future challenges and opportunities and ensure the effectiveness of the mechanism in contributing to future global climate action' in an 'independent', 'transparent' and 'balanced' manner (CDM Policy Dialogue 2011: 1). The practice was somewhat different, as is explored in the following sections. But first, some attention to the concepts and theories that can help to account for this legitimacy crisis and the politics which produced it and shape its management.

Crisis? What crisis? Thinking about legitimation, participation and hegemony

It is suggested here that we can account for the formation and conduct of the CDM Policy Dialogue as one expression of a broader legitimacy crisis as perceived by a global managerial elite charged with the responsibility for the day-to-day governance of the CDM, but acting on behalf of 'climate capitalists': the community of traders, auditors, financiers and project developers with a material interest in securing ongoing support for carbon markets as a central pillar of the international community's response to the threat of climate change. Approaching the experience of the dialogue this way reveals the vulnerabilities of carbon markets and the actors that promote them, even where support for market-based mechanisms appears to be almost hegemonic and persists in spite of their many failings. Hence, one reaction to the anomaly outlined above between an unwavering commitment to new carbon markets in the face of a track-record, to date at least, of under-achievement, is to fuel claims of their zombie-like status, staggering onwards, damaged but undeterred (Reyes 2011), driven by an unfailing ideological commitment to market-based solutions.

Another is to read from this the need, even of powerful actors, to secure legitimacy and public approval – not from a notional general public, but from key concerned stakeholders. Interpreted this way, such fragilities and the processes they give rise to might create potentially productive openings for critics to refine or re-position the debate about *who* and *what* carbon markets are for. It invites further analysis of the relationship between legitimation, on the one hand, often sought in response to a crisis of legitimacy and pursued through enhanced efforts to engage publics in participation in debate and dialogue, and hegemony on the other, which provides an understanding of *how* and *by whom* the terms of engagement and the scope for manoeuvre are determined, or what in other contexts I have described as 'disciplinary neo-liberal participation' (Newell 2010). This refers to the ways in which participation in invited spaces takes the form of discussion around a pre-proscribed set of themes where the range of possible outcomes is delimited to ones acceptable to those organising the process. It draws attention to the issues of power at play in these processes, despite attempts to de-politicize them, and sheds light on the strategies employed by policy elites to protect established accumulation strategies . . . in the face of potential public challenges to claims made . . . regarding an array of social and environmental benefits.

(Newell 2010: 472)

This is a far cry from the Habermasian notion of 'public reasoning', 'an attitude toward social cooperation, that of openness to persuasion by reasons referring to the claims of others as well as one's own' (Habermas 1998: 244). Such engagements take place on a terrain of (unequal) power where conflicting discursive frames, in the case of the CDM about 'efficiency', 'equity' and 'justice', vie for attention underpinned by sharply unequal material power and levels of institutional access to key decision-makers. Just as with other controversies (around GMOs or nuclear power, for example) where legitimation exercises are often triggered by a perceived loss of legitimacy and the need to contain crises aimed at protecting a particular strategy of accumulation, so it is the case with the CDM Policy Dialogue. Engagement has to be managed on terms that de-limit the range of challenges that can be posed to prior commitments.

In relation to the nature of the responses to the legitimacy crisis and the precise form that the dialogue took, these can be understood at the level of the specific strategies invoked to manage and restrict participation in 'invited spaces'. This is a distinction employed by Gaventa (2006) to denote the fundamental difference between processes that allow people to participate in making and enforcing policy themselves, as opposed to selecting from a pre-selected menu of future policy paths which they have played no part in designing. This resonates with the emphasis on 'disciplinary neo-liberal partici- pation' sketched above, the 'posing all the questions around which the struggle rages' (Gramsci 1971: 181–182) by key social groups. It is also captured in the dynamic Habermas describes whereby administrative systems, in an attempt to avoid legitimation problems, adopt 'the symbolic use of inquiries, expert opin- ions [and] advertising' which '[b]y resorting to emotional appeals . . . arouse unconscious motives, occupy certain contents positively and devalue oth- ers . . . pushing uncomfortable themes, problems and arguments below the threshold of attention' (1973a: 657). The CDM Policy Dialogue process might be thought of as one such inquiry and the push to produce videos, host CDM photo competitions and other means of using media to communicate the ben- efits of the CDM to wider audiences beyond those with the technical capacity or inclination to pore over project design documents (PDDs) was explicitly aimed at countering the negative publicity surrounding the CDM documented above, as it lurched from one crisis to another in the run up to the announce- ment of the Policy Dialogue.

But they can also be understood as attempts at co-optation and accommoda- tion, either engaging in the rituals of reflection and engagement before proceed- ing as before, having provided some concessions of an 'economic-corporate kind' to adversaries, and with some minor reforms in place, or as part of a broader war of position in Gramscian terms which aims at reinforcing hegemonic social forces and re-taking control over the terrain of political conflict (Morton 2007: 94) which, in this instance, was perceived to have been ceded to critics of the CDM. The weakness of those charged with the public authority to govern carbon markets is perhaps more characteristic of the process of 'passive revolution' which Gramsci uses to describe a process of reformist change from above, which entails extensive concessions by relatively weak hegemonic groups, in an effort to preserve the essential aspects of social structure. In other words:

> Actors part of this bloc have an interest in ensuring that carbon markets are the preferred and most profitable way of responding to climate change. This means they have to engage in accommodation of critics, working with civil society organisations in standard-setting processes and demonstrating to publics and consumers of their services, that emissions reductions are genuine and can come with social benefits. Attempts to secure the legitimacy and salience of market based responses by carbon brokers, traders and other beneficiaries of the carbon economy can be seen as attempts to preserve the hegemony of the material base, institutional apparatus and ideational claims of neo-liberalism.
>
> (Newell and Paterson 2011: 39)

What accounts which emphasise the zombie-like quality of carbon markets (Reyes 2011) often miss, is the way they have to reinvent themselves and undertake political work to maintain the political support of the balance of social forces that sustain them, and, as we saw above, on whose legitimacy they depend. Hegemony is then contingent and unstable. As Gramsci reflects: 'What is this effective reality? Is it something static and immobile, or is it not rather a relation of forces in continuous motion and shift of equilibrium?' (1971: 172).

In his work on legitimation problems in late capitalism Jürgen Habermas showed how legitimation crises stem from the fact 'the economic system keeps creating new and more problems as it solves others' (1973a: 644). Creating carbon markets appears as a solution to climate change that is preferable for many businesses to tax and regulation, but then is dogged by inevitable dynamics of gaming the system, fraud and acts of dispossession that are also persistent (some would suggest endemic) features of capitalism (Harvey 2010) which give rise to a crisis of legitimacy about the value of carbon markets and the ways in which they are currently organised. For Habermas such crises are prone in 'legitimation systems' that require loyalty (or acts of faith) without direct participation (as is the case with buyers of carbon credits), where 'The private autonomous decision about investments is complemented by the civil privatism of the population' (1973a: 648), a situation that might be thought to describe the CDM which oversees investments on behalf of some imagined public for whom it is meant to act to ensure fair play and the delivery of the public goods of climate protection and sustainable development.

The sense of overstretch on the part of administrative structures trying to reproduce the conditions for capital accumulation in Habermas's work also usefully characterises the CDM's attempts to govern locally from afar or globally. For Habermas a 'legitimation deficit means that it is not possible by administrative means to maintain or establish effective normative structures to the extent required' (1973b: 47). Legitimation crisis can occur then when a governing structure still retains the legal authority by which to govern, but is not able to demonstrate that its practical functioning fulfils the end for which it was instituted. In the case of the CDM this gap underpins calls from market participants for a full-time board with the relevant expertise and capacity to process project approvals in a more timely and efficient manner. The CDM is also vulnerable to such crisis, since it is necessarily distant from the sites where the projects it approves are located and is dependent upon the auditing and verification bodies to whom it delegates authority, to serve as its street-level bureaucrats (Lipsky 1980) ensuring that the rules and norms that ostensibly govern the system operate in practice. This leaves the CDM significantly exposed to allegations of wrong-doing and false accounting for which it is forced to accept responsibility, but over which it exercises very little control.

Legitimacy crises also stem from a loss of public confidence when institutions make decisions seen as mistaken, unpopular or controversial, whereupon publics begin to lose faith in the governance arrangement to act efficiently and effectively. The scandals engulfing projects accused of being non-additional or complicit in violations of social and human rights gave rise to such a situation in the case of the CDM, as did unpopular and controversial decisions to reject a series of Chinese wind projects at the time of the Copenhagen summit in December 2009 that so irked the 'market makers'. The crisis was then produced by a diffuse but prevailing sense of dissatisfaction with its performance, and for very different reasons, from carbon traders and critics of carbon markets alike, articulated at a critical and vulnerable moment for the CDM where a window of opportunity existed to reflect on its workings to date and articulate an alternative vision for its future. Reus-Smit (2007: 157) claims:

> An actor or institution experiences a crisis of legitimacy . . . when the level of social recognition that its identity, interests, practices, norms, or procedures are rightful declines to the point where it must either adapt . . . or face disempowerment. International crises of legitimacy can be resolved only through recalibration, which necessarily involves the communicative reconciliation of the actor's or institution's social identity, interests, practices, norms, or procedures, with the normative expectations of other actors within its realm of political action.

The dialogue might be understood precisely in terms of an attempted 'recalibration' of identity, practices and procedures. For 'Institutions without developed organizational structures . . . it is not their right to act that is in question, but the rightfulness of the norms, rules, and principles that constitute them' (Reus-Smit 2007: 159). This explains the heavy and continued emphasis on the governance dimensions of the CDM, those rules and procedures which might be revised, improved or built upon, rather than whether the mechanism itself was fit for purpose. To reconstitute its legitimacy, an actor or institution must first recalibrate the relationship between its social identity, purposes, and practices, and the prevailing social norms that define the parameters of rightful agency and action; and, second, realign its realm of political action with its social constituency of legitimation (Reus-Smit 2007: 166–167).

Paterson (2010: 345) also identifies a 'recurrent tension within capitalism between accumulation and legitimation' which he explores in relation to the role of carbon markets, including the CDM, as prevailing responses to the threat of climate change. 'The principal tension is that while the pursuit in general of climate governance is brought about by the search for legitimacy, the specific mechanisms developed as private climate governance strategies are informed primarily by the search for accumulation' (Paterson 2010: 359). Put differently:

> The key tension for policy makers lies between the need to create cycles of growth for particular sectors of the economy in order to sustain climate policy, and the pursuit of the environmental integrity on which the stability of the political coalition that carbon markets have enabled depends.
>
> (Paterson 2012: 83–84)

What results, Paterson suggests, is that:

> [I]f rules governing offsets become more stringent . . . it will be precisely because the critics of carbon markets are in fact participants in the process by which the actually existing markets are constructed – their attempts to delegitimise such marketisation leads to governance practices to shore up their legitimacy which may (repeat, may) lead to emissions reductions. But it is the critical projects of de-legitimisation that produce this result.
>
> (Paterson 2010: 363)

The CDM Policy Dialogue process in practice: the politics of inclusion and exclusion

Who and what is the CDM for?

In concrete terms, then, how was the CDM's Policy Dialogue conducted? The notion of a dialogue implies exchange. Dictionary definitions of the term refer to 'an exchange of ideas or opinions on a particular issue, especially a political or religious issue, with a view to reaching an amicable agreement or settlement' (Online Dictionary 2013). This is helpful in understanding the nature of the CDM process. Even if an amicable agreement might be an unrealistic expectation, an exchange of ideas and opinions about political issues did take place, notwithstanding the 'post-political' terrain on which much of it took place (Swyngedouw 2010). Definitions of dialogue also note an intent 'to discuss areas of disagreement frankly in order to resolve them' (Online Dictionary 2013). This again points to dialogue being a means to an end: to resolve, agree or settle something. Indeed it may have been the aspiration of the EB that in setting up a policy dialogue some of the heat of controversy would be channelled into this process and away from board members as the public face of the world's most important carbon market mechanism, reflected in the call for a 'positive' dialogue. Perhaps they believed it would even bring closure to an ongoing and increasingly heated set of exchanges and would do so through a tried and tested tactic of inquiry plus resort to selectively commissioned expert reports as a means of reasserting control over the unruly spectacle which carbon markets had become in the public domain.

Instead of facilitating an exchange of views, what happened in practice was that a series of parallel conversations took place simultaneously in different places, notionally about the same subject, though considering the content of them you would be entitled to wonder if that was the case. In Bonn, earnest talk of redress mechanism and righting wrongs was the order of the day at the side-event at the climate negotiations organised by CDM Watch at which I presented and which some members of the High-Level Panel attended. Meanwhile at precisely the same time on the same day in May 2011 in central London the talk was all about securing returns and reducing transaction costs, in which discussion of sustainable development was seen as an irrelevance: at best a distraction and a side-show in a meeting dominated by investors and project developers. This basic divide originates in the tension that results from trying to deliver both of article 12's mandates of efficiency and equity simultaneously: of squaring the global North's demand for flexible and cheaper emissions reductions options with the global South's insistence that they get paid a development dividend in return. It is reflected in the EB's 'assessing impact' report which notes that 'while greater scrutiny of sustainable development is important, a more rigorous system might be counterproductive and have a negative impact on the market' such that 'incorporating sustainable development criteria into the verification process would increase transaction costs, which are already one of the most important barriers to the CDM' (CDM Policy Dialogue 2012b: 35).

The recommendations summarised for the EB regarding the modalities of decision-making included that 'the Board should become more strategic, and less enforcement/regulatory-focused' (UNFCCC 2013b). This forms part of a wider chorus of opinion that the answer to the problem of a backlog of projects and limited capacity for direct oversight is to delegate authority downwards to designated national authorities (DNAs) to approve projects. Knowing full well the problems of weak and limited capacity that beset DNAs in many countries, since it forms precisely the rationale for World Bank and other donor capacity-building programmes such as Carbon-Finance Assist, and the nature of political networks characterised by close links and documented instances of collusion between project proponents and regulators (Newell 2013), such proposals inevitably amount to a further weakening of oversight and quality control mechanisms.

In a further move to place constraints around the EB's freedom of manoeuvre it is suggested: 'The Board should retain only supervisory functions and its executive functions should be limited to final approval of recommendations from the secretariat, with the authority to intervene when the Board considers it necessary' (UNFCCC 2013c). This could sound like a means to strengthen the accountability between the CDM Executive Board and the UNFCCC secretariat from whom it derives its mandate. But its intention to circumscribe meddling in markets becomes clear in the call to expand one area of authority for the EB where the demand is to: 'Add a new responsibility to the Board relating to the protection of the credibility of the CDM in international markets and ensuring that certified emission reductions (CERs) are in demand and not subject to further Party-driven qualitative exclusions' (UNFCCC 2013c). Underscoring a classic imperative of institutions in managing neo-liberalism, the role of international bodies is to trump attempts by governments to intervene in markets by reference to a greater good that is assumed will be achieved (in this instance) by letting the market identify the cheapest and most efficient sources of emissions reductions. The call came on the back of EU restrictions on the purchase of CERs from controversial but highly profitable industrial gas projects and, from 2013, to only source credits from countries that are 'least developed' that caused considerable disquiet among carbon traders.

Lest there should be concern about the balance of actors and interests represented on the EB, equity of representation could be preserved in the following way, the report suggested: 'The Board could be expanded to include six extra members: three seats for buyers of certified emission reductions (CERs), and three seats for sellers of CERs!' And with regard to evidence of Board members unduly disproportionately supporting projects in their own countries, came the concession that: 'Board members should not be staff members of any DNA, thus avoiding conflicts with activities like attracting CDM projects or approving CDM projects' (UNFCCC 2013b). Acknowledgement of the effect of the critiques around the governance of the CDM came in the recommendations proposed for the SBI (Subsidiary Body on Implementation) to consider:

> The Board recommends strengthening the requirement . . . to require the DOE to validate that the local stakeholder consultation was carried out in accordance with host Party laws and regulations, and to evaluate how the content of the comments was duly taken into account.
>
> (UNFCCC 2013c)

This goes to the heart of the schizophrenic personality of the CDM and whether two sides of the 'gemini' CDM, of equity and efficiency, can live together. The dialogue process did not manage to bridge this divide. Rather, it had to offer up simultaneous concessions for the traders and to certain elements of civil society in order to maintain their respective interest in engaging with the CDM. For while project developers and financiers enact the CDM on an everyday basis, it also relies on activists to bring attention to wrong-doing and malpractice given the range of projects it is meant to oversee, but for which it does not have the resources. This is the fragile coalition that had to be upheld, a point to which I will return later.

Legitimation brokers

The invitation to engage in the dialogue was an open one, e-mailed out to thousands of members in the community of climate policy watchers that subscribe to climate-l, for example, and placed prominently on the UNFCCC web site.

The process of selecting who would sit on the High-Level Panel was both politicised and revealing of which actors are seen as legitimate, authoritative and with a genuine commitment to the CDM, recalling again the call for a 'positive dialogue'. The panel was made up of 11 members drawn from the public, private and civil society sectors (see Table 11.1). The panel members were nominated by Martin Hession, then chair of the EB, and Christiana Figueres, head of the UNFCCC, in consultation with the EB. Nominations were meant to represent a broad geographic area and gender balance. Further, the nomination process was meant to avoid members who are currently engaged in the carbon markets. The panel members nevertheless ended up being either architects of the CDM (UN bureaucrats), more conservative civil society organisations in favour of market-based instruments, or individuals with links to firms that benefit from the CDM (see Table 11.1 for notes on their background and networks). Strong connections exist among many of the panel members to business beneficiaries of carbon markets, NGOs supporting them or global institutions (such as the World Bank) financing and enabling them. It hardly represents a genuine cross-section of opinion about the performance of the CDM. In other words, this was always going to be a dialogue on a limited range of issues about a proscribed set of themes, the outcome of which would only ever be incremental institutional innovation rather than fuller critical reflection about modalities and effectiveness.

The lack of representation on the panel of those voices that took a more critical view of the CDM's performance to date and, therefore, future role, quickly attracted disquiet among civil society organisations, as was apparent from conversations in the negotiating halls of the Durban summit in 2011. The distinction implicitly at work was one which other international institutions facing legitimacy crises routinely employ: to demarcate a difference between 'insider' legitimate/constructive/engaged civil society and 'outsider' critical, disengaged, and by implication illegitimate, civil society. In the case of the dialogue, the justification for the exclusion of many NGOs, communicated

Table 11.1 Members of the High Level Panel on the CDM Policy Dialogue

Panel Members	Current Roles	Notes
Valli Moosa	Chair of the panel Chairman of the World Wide Fund for Nature South Africa, Former Environment Minister of South Africa	Principal of the Lereko Metier Capital Growth Fund Lereko Metier Sustainable Capital Serves on the boards of Lereko Investments, Sun International, Anglo Platinum, Sanlam and Imperial Holdings Previously served as Chairman of Eskom the fossil fuel para-statal and host of CDM projects
Joan MacNaughton	Vice-Chair President of the Energy Institute, Executive Chair of Energy and Climate Policy Assessment, World Energy Council Global Advisor Sustainable Policies, Alstom, United Kingdom	Serves on the boards of IETA the carbon trading lobby group and CCSA (Carbon Capture and Storage Association) Alstom is a key beneficiary of CDM projects
Luciano Coutinho	President of the Brazilian Development Bank, Brazil	
Maggie L. Fox	President and CEO of The Climate Reality Project, formerly known as the Alliance for Climate Protection, United States of America	Currently serves on the board of the Green Fund Spent 20 years working at the pro-market Sierra Club, including five years as its Deputy Executive Director Worked with Hewlett Foundation, the Western Conservation Foundation, the Energy Future Coalition, and Western Resource Advocates
Ross Garnaut	Distinguished professor of economics at Australian National University, Climate Change Advisor to the Australian Government until June 2011, Australia	Director on the Board of Ok Tedi Mining Limited (Papua New Guinea)

(continued)

Table 11.1 (continued)

Panel Members	Current Roles	Notes
Prodipto Ghosh	Distinguished Fellow at The Energy & Resources Institute, Former Secretary to Ministry of Environment and Forests, India	Co-Executive Secretary of the Indo–US Economic Dialogue; and Member Secretary of the Prime-Minister's Economic Advisory Council and Trade & Industry Advisory Council Consultant to World Bank, and Asian Development Bank
Yolanda Kakabadse	President of the World Wide Fund for Nature, Former Minister of Environment, Ecuador	President of the World Conservation Union (IUCN) from 1996 to 2004, and Member of the Board of the World Resources Institute (WRI) Chaired the Scientific and Technology Advisory Panel of the Global Environment Facility (STAP / GEF) from 2005 to 2008 Member of the Presidential Commission of the Yasuni Initiative promoting the concept of payment for non-emission of CO_2
Margaret Mukahanana-Sangarwe	Permanent Secretary, Ministry of Tourism and Hospitality Industry of Zimbabwe, Former AWG–LCA chair, Zimbabwe	Lead negotiator under the United Nations Framework Convention on Climate Change and the Kyoto Protocol since the preparations for the Rio Summit
Paul Simpson	CEO of the Carbon Disclosure Project, United Kingdom	Previously worked with Chesham Amalgamations & Investments Ltd Former Director of the Social Venture Network
Nobuo Tanaka	The Institute of Energy Economics, Former President of International Energy Agency, Japan	Executive Director of the International Energy Agency (IEA) from 2007 to 2011 Former Ministry of Economy, Trade and Industry (METI) Led many trade negotiations at the World Trade Organisation (WTO) and for bilateral Free Trade Agreements
Changhua Wu	Greater China Director of The Climate Group, China	Executive Director of China Operations of ENSR, working closely with multi-national corporations to support their business development in China Programme director at World Resources Institute Consultant for World Bank

to them privately (though not publicly stated as such), was the requirement for a high level of 'technical' knowledge of the CDM and carbon markets. Those that were welcomed, as is apparent in Table 11.1, were the NGO elite, or what Leslie Sklair would call the 'global environmental elite' (Sklair 2002: 276). In this instance it was made up of representatives from WWF, former WRI (World Resources Institute) staff, and those with the links to the 'big 10' Washington-based NGOs such as the Sierra Club, or business-oriented NGOs such as The Climate Group whose CEO Mark Kenber was a leading architect of many of the governance instruments developed for carbon markets (such as the Voluntary Carbon Standard and Gold Standard) and those with close links to global neo-liberal institutions such as the World Bank and its regional counterparts or the GEF. The alleged requirement for technical knowledge of carbon markets as a prerequisite for representation on the panel had to be squared, nevertheless, with the need to find individuals that were not seen to have a direct interest in the CDM, for reasons of 'independence'.

This, as the NGO CDM Watch notes, brought with it other problems: 'limited opportunities for input and stakeholder meetings and inadequate support for travel made it almost impossible for civil society representatives to participate in stakeholder meetings that were heavily dominated by business lobbyists' (CDM Watch 2012). One participant, Dr Leena Gupta, Senior Scientist (Senior Programme Coordinator), Society for Promotion of Wastelands Development, New Delhi, India, who participated at the stakeholder meeting in Delhi on 16 July 2012, noted:

> It was really sad that there was such a low civil society presence in a crowd of companies, consultants and FICCI employees. An official from DNA-India was also present. Social and ecological aspects were neglected in the discussion and it felt as if the policy dialogue was focusing on the commercial aspects of the CDM only.
>
> (CDM Watch 2012)

Another participant at the same meeting noted:

> The Policy Dialogue meeting was more like a commercial meeting with most of the industry asking for quick money and much smoother process and less time. There was hardly any discussion on sustainable development . . . . Only 3 civil society representatives were present, making it a one sided, biased industry meeting. Also, looking at the panel there was no civil society representation. There was government and industry. It is almost as if civil society had to forcefully enter the dialogue. Still, the Policy Dialogue was constructive in the sense that the issues we raised were given due respect and we were also given time to present our issues in writing in 3 days which we did. This would have been more constructive had we had longer notice and had more civil society organisations from other states be present in the meeting.[2]
>
> (CDM Watch 2012)

The time frame for written submissions was indeed short. It was launched on 3 December 2011. All submissions had to be made by 16 January 2012, coinciding with the festive holidays in many parts of the world. This inhibited the collation of evidence, the formation and articulation of common positions and the widespread diffusion of the call for inputs to relevant groups and affected stakeholders. I received both individual and open-ended requests through listserves such as 'no-carbon trade l' to contribute to civil society-based critiques of the CDM organised by groups such CDM Watch and others. Time for reflection, discussion, let alone deliberation was severely restricted. A similar time frame was put in place for consultation on modalities and procedures (M&P): the Conference and Meeting of the Parties (CMP) launched a call for public inputs on possible changes to the CDM M&P based on the experience of stakeholders in implementing the CDM. This call for inputs was open from 17 December 2012 until 23 January 2013 and, perhaps unsurprisingly, received just 11 responses.

Beyond the organisation of the consultations, calls for written submissions and the appointment of 'independent' experts, a further strategy deployed for managing the terms of debate was to pull on the authority and intellectual resources of an expert or epistemic community (Haas 1990) of carbon market knowledge brokers. Commissioning a series of reports that were meant to evaluate the performance of the CDM around key sites of controversy, including contributions to sustainable development and technology transfer, the secretariat trumpeted a 'wide-ranging research programme' (UNFCCC 2012b) that assessed the impact of the CDM. Less was said about the independence of some of the report authors who were largely CDM consultants and former architects of the system such as Erik Haites, carbon market consultant, employed by the UNFCCC secretariat in the design of the Kyoto mechanisms and Axel Michaelowa, academic, project developer and member of the CDM's own methods panel. The epistemic community was constituted by an incredibly tight knit community of architects turned defenders of CDM who could be relied upon to rally around in its hour of need. The authority of these authors as the right ones to provide the knowledge base for the panel's decision-making was reiterated and reinforced through heavy self-citations throughout the reports. To take one example: 'The majority of the studies agree that the CDM does have a positive impact on the various facets of sustainable development in the host countries!' (CDM Policy Dialogue 2012b: 5).

Discursive strategies of containment and legitimation

This section probes further some of the devices use to shore up the credibility of the CDM in the eyes of its key stakeholders and to attempt to reconcile some of the tensions between accumulation and legitimation which derive from the competing mandates described above around incentivising 'efficient' emissions reductions while delivering tangible development benefits. In particular it draws attention to some of the discursive strategies employed by the EB and UNFCCC secretariat to acknowledge problems, discredit dissenting views but then enrol them in calls for better 'governance' of the CDM. It focuses on the release and content of the report of the High-Level Panel, which was released at the 69th CDM Executive Board meeting held in Bangkok on 11 September 2012. It constituted the main and final outcome of the year-long process described above whose aim was to recommend how to reform the design and decision-making process of the CDM, and to envision a way forward to prevent the global carbon market from further disintegrating.

One of the first things you can observe is the alarmist language in the report and the strength of the call to arms. This was about raising the stakes. It was no longer about salvaging the CDM or even carbon markets, but the fate of the climate itself. The problems facing the CDM should be experienced, therefore, as a general crisis of concern to all, posing the questions on 'a universal plane' (Gramsci 1971: 94) which makes opposition harder to maintain. The executive summary report of the High-Level Panel sets the tone in this regard: 'Global carbon markets – an important policy instrument that the international community has developed over the past decade to facilitate real-world emissions mitigation – are collapsing with potentially devastating consequences' (CDM Policy Dialogue 2012a: 2). Then come the central ideological and normative claims which underpin the CDM's relevance and legitimacy: 'While not sufficient in themselves, well regulated *carbon markets have proved essential to addressing climate change*, and *nations must as a high priority restore faith in global carbon markets generally and in the CDM specifically* (CDM Policy Dialogue 2012a: 2; emphasis added) since the 'CDM is imperilled'. Recognition of shortcomings is followed by the familiar theme of 'learning by doing', that the CDM is a work in progress improving all the time: 'Although the CDM has been criticized for approving some projects with questionable environmental and sustainable development benefits, the CDM has *improved markedly in recent years* and its positive impact extends well beyond specific projects. The CDM has helped combat climate change by creating a *global culture for action* and by *mobilizing the private sector through markets*'. It continues: 'Some might not mourn the potential death of the CDM. But time is not on our side and the CDM is the only game in town' it is suggested. 'Yet, new solutions will take years to design and make operational. For the balance of this decade the CDM is likely to remain the world's foremost – and possibly sole – means of gaining the benefits of a truly global carbon market. This means that a strong CDM is necessary to support the political consensus essential for future progress' (CDM Policy Dialogue 2012a: 2). Like it or not, it's here to stay. TINA.[3]

What then follows is a series of demands to provide a life-line to the CDM. 'To avoid this self-inflicted wound, the international community must take four essential and mutually reinforcing actions as a matter of great urgency' (ibid.: 3). These included raising the level of mitigation ambition; expanding the role of the CDM through sectoral approaches and links to larger funding mechanisms (such as the Green Climate Fund); improving operating procedures in order to deal with the 'negative perceptions' of the CDM's contribution to emissions reductions and sustainable development which 'threaten the credibility of the CDM and the long-term viability of global carbon markets'; and strengthening and restructuring its governance to become 'a more accountable and efficient organization'. This is in order to respond to the '*perception* that it is slow, opaque, unresponsive and politicized' (ibid.: 3; emphasis added). Such measures were thought to be imperative because '[w]ell regulated global carbon markets can help to avoid the unacceptable risks of climate catastrophe' (ibid.: 3).

Having acknowledged that some projects with 'questionable environmental and sustainable development benefits' (CDM Policy Dialogue 2012a: 17) slipped through the net, an attempt is then made to discredit the basis of stories of negative projects that feature in critical activist and academic literature, and which were reported in the media, such as the landfill gas project on Bisasar Rd in South Africa (Bond *et al.* 2009), or the biogas project in Honduras where human rights violations allegedly took place, all the while accepting entirely the positive claims contained in PDDs about sustainable development benefits reported by project developers which go unverified. It seeks to convey a sense in which objective and value-free research is being brought to bear on some of the wild and loose claims made of CDM projects by their opponents. It is suggested:

> The objective was to identify literature focusing on the negative impacts of CDM projects on sustainable development. Reviewing the claims, and the nature of such claims, in turn led to the identification of specific registered CDM projects where concerns had been raised. Each identified project was studied in greater detail on the basis of both its PDD and associated stakeholder comments (local and global). The authors of the negative claims were contacted, as were the project owners. The responses received were then screened and further reviewed as part of the assessment.
>
> (CDM Policy Dialogue 2012b: 47)

The literature available on this issue is predominantly 'grey literature' (i.e. not peer-reviewed scientific literature) (ibid.: 48). Mention is not made of who 'screens' the responses or on what basis, though presumably the fact they do not count as 'peer-reviewed scientific literature' significantly reduces their chances of being taken seriously. It then goes on to mention by name GAIA (Global Anti-Incinerator Alliance), International Rivers Network, and the Centre for Civil Society at the University KwaZulu Natal, headed by carbon markets critic Patrick Bond. In some cases 'the reply to enquiries was insufficient to make an evaluation'. Elsewhere 'the issue raised regarding the CDM project related to a problem that existed before the CDM project was developed' (CDM Policy Dialogue 2012b: 49).

> For example, in the case of the Bisasar Road landfill in South Africa and Aguan biogas recovery from POME ponds and biogas utilization in Honduras, the problems cited by stakeholders existed prior to the CDM project activity. This means that the CDM projects themselves were not

the causes of the problems, nor was there evidence presented that the CDM projects worsened these pre-existing situations.

(ibid.: 49)

Interestingly, while absence of replies is read as a problem for supporting criticisms made of CDM projects, the absence of replies is not taken as being a problem when reporting positive cases. 'Only two cases indicated field visits and filed testimonies as the evidence for their claims (Xiaoxi CDM Hydropower Projects in China and Improving Rural Livelihoods through Carbon Sequestration in India)' (CDM Policy Dialogue 2012b: 49). We are not told how many field visits took place before arriving at positive stories of CDM cases that adorn the CDM web site and feature in video and photo contests aimed at enrolling publics in a communications outreach strategy that reinforces the multiple benefits the CDM can bring. The different treatment afforded to claim-making by project developers and carbon market advocates on the one hand, and critical NGOs on the other, and the very different standards of rigour applied to interrogating their claims is apparent, for example, in the reading of silences as evidence of a potentially even greater impact of CDM projects: 'because projects are not required to report technology transfer, a substantial portion of projects that do not explicitly claim this benefit may nevertheless involve some form of technology transfer' (ibid.: 7). In other words, though no evidence is provided, nor none required of projects which nevertheless claim to deliver sustainable development benefits, it is assumed that benefits are nevertheless forthcoming. Compare this with the dismissal of activist research on the basis that insufficient site visits were made!

This willingness to believe claims made in documents comes despite acknowledgement that 'on-the-ground examination' of the actual state of affairs with regard to benefits generated by CDM projects is 'indispensable' (CDM Policy Dialogue 2012b: 41) and noting the limits of relying on PDDs that many were at pains to highlight, including the fact that: 'most DNAs' sustainable development criteria lack transparency and clarity, which makes it easy for project developers to comply with the requirements. The process of stakeholder consultation is often "only rudimentary, completely unregulated and badly documented" (Sterk *et al.* 2009)' (CDM Policy Dialogue 2012b: 42). Again, the strategy of assuming the best even in the absence of evidence prevails. The minor 'limitations' mentioned are acknowledged where it is stated:

The source material for the analysis is the PDDs and therefore only positive contributions to sustainable development can be measured, since project developers are unlikely to write anything negative about their project. Furthermore, the descriptions of sustainable development contributions in the PDDs are only potential benefits and not the sustainable development benefits actually delivered.

Nevertheless, it is assumed that: 'All sustainable development benefits claimed are considered to be caused by the project, even though it is possible that some

of those benefits could have been realized without the CDM project activity'
(ibid.: 44).

Conclusions

According to Reus-Smit, an actor's legitimacy is only established and maintained
when its self-representations and institutional interpretations resonate with the
normative expectations of other actors. Crises of legitimacy are only resolved,
therefore, when these are reconciled, when the discordance that has eroded social
recognition is overcome through a similar process of communicative clarification
or reconstitution (2007: 172). The reports analysed above testify to the level of
effort to communicatively clarify and reaffirm the purpose and legitimacy of the
CDM, but there is less evidence of reconstitution. How far, then, did the CDM
Policy Dialogue resolve the crisis of legitimacy facing the CDM?

It is perhaps too early to answer definitively, but the dialogue did result
in concrete recommendations aimed at reasserting its right (and ability) to
govern, in particular by referring to improvements in governance. For example, the report on governance recommended that the CDM EB 'assist DNAs
(Designated National Authorities) in coordinating more effective local stakeholder consultations and improve the accountability of project proponents
to local communities throughout CDM project design and implementation'
(CDM Policy Dialogue 2012c: 8). The Final Report of the High-Level Panel
called on the CDM to 'strengthen and restructure its governance' by 'enhancing its openness and transparency, and opportunities for stakeholder participation, creating avenues to hear appeals and address grievances' (CDM Policy
Dialogue 2012a:19). This included increased monitoring of sustainable development benefits, efforts to strengthen national capacity in this regard and sanc-
tions in cases of harm caused by projects, plus guidelines on adequate local
consultation procedures. It suggested 'There is a clear need to improve the
reporting, monitoring, and verification of the sustainable development impacts
of CDM projects, and to implement safeguards against projects with negative
impacts' (ibid.: 43).

Indeed, in the wake of the report resulting from the Policy Dialogue and
NGO advocacy around it, state Parties to the UN climate negotiations called
on the CDM EB to act. It did so, conceding to its critics, by approving a
voluntary tool for describing sustainable development co-benefits of CDM
projects at its 70th Board meeting in November 2012, moving away from
original opposition to the need for any additional tool for recording such benefits, but falling short of the sort of binding mechanism sought by CDM Watch
and others. The CDM EB has also sought to encourage governments to share
experiences of local consultation processes given the sorts of shortcomings
highlighted by the dialogue as part of an ongoing review of the M&P of the
CDM aimed at strengthening its procedural legitimacy. For some this might
be taken as evidence of an ongoing dynamic within the CDM of adaptation to
new policy challenges and institutional reflexivity. As Adam Bumpus puts it:

Rolling with the punches of critics, and constantly being improved through policy decisions of its Executive Board, the CDM has evolved in a fast-shifting political and environmental landscape. It has been under constant amendment, change and evolution to achieve its two major goals: cost effective emissions reductions and contributing to developing countries' sustainable development.

(Bumpus 2012: 2)

However, if one of the aims of the CDM Policy Dialogue was to engage critics and show them that genuine reform of the CDM would result from the report it produced on the conclusion of the dialogue, it failed. Sunita Narain, Director General of the influential Indian NGO Centre for Science and Environment said: 'The report is singularly geared towards saving the ailing CDM and carbon markets, with not enough focus on how the social and environmental integrity of the mechanism should be addressed' (Mukerjee 2012). The CDM Policy Dialogue Report has a list of 51 recommendations which focus on saving the carbon market by increasing the demand and price of carbon credits. Others such as CDM Watch (now Carbon Market Watch) levelled their criticism both at the weak recommendations that were issued in relation to the reform of the CDM, (despite acknowledgement of many of the grounds for more far-reaching changes), as well as at failings in the process around appointments to the High-Level Panel and the ways in which the organisation of the consultations served to minimise engagement from civil society actors at the expense of business and state actors. For critics, therefore, there were perhaps few surprises.

Noting the relative weakness of the concrete reforms that the Policy Dialogue process has generated to date, it would be tempting to suggest that civil society organisations that engaged with the dialogue process were taken for a ride. But lack of ability to directly secure the degree of change sought, particularly when those changes have to be approved by all parties to the Kyoto Protocol, should not be confused with inability to produce the legitimation crisis in the first place and then shape its response to some degree. The research, watchdog activism and exposure that groups such as CDM Watch engaged in, played an important part in informing the creation of the dialogue and the parameters of its subsequent conduct. It is perhaps testimony to the success and effectiveness of their campaigning that there was a dialogue process at all and there is clearly a critical role for watchdog activism aimed at exposing, monitoring and holding to account as well as exploring alternatives. Yet producing and then participating in efforts to help resolve a legitimacy crisis, though not on terms of their choosing, is not free of trade-offs and dilemmas. Perhaps the key dividing line is engagement to what end: exposure to de-legitimate and discredit as a basis for advocating for alternatives, or in order to check excesses and improve accountability and performance.

There is not space here to enter into age-old debates about insider/outsider dynamics and whether the effect of more critical and oppositional advocacy of groups like Carbon Trade Watch is to make engagement with 'insider', or at least engaged groups such as CDM Watch more attractive and appealing. But it would, perhaps, be harsh and unfair to dismiss such groups as mere 'patzers' (Lohmann 2012) outmanoeuvred by more powerful groups whose institutions and processes they have lent legitimacy to without securing much in return. This is not just a way of saying things might have been even worse without their involvement, but that strategies of exposure, research and lobbying do have a role to play in shaping the organisation of (carbon) markets and the rules by which they are governed and the weight they give to equity as well as efficiency criteria. In other words *who* and *what* they are for.

It is the case though that the example of the CDM Policy Dialogue highlights, once again, the limitations of public participation 'from above'. The critiques of the process and the invited nature of participation lend themselves as an additional case study of the type of 'disciplinary neo-liberal participation' often overseen by global institutions and the constraints this imposes on more meaningful forms of citizen participation and deliberation (Newell 2010). While national and regional consultations were organised, their terms of reference, time frames and format were set from above in such a way that wider forms of participation and deeper forms of deliberation were effectively precluded. By default a limited sub-section of an already thin global civil society that has the resources and (political and financial) credentials to effectively participate and shape global processes such as the dialogue, become the interlocutors or intermediaries between a notional public and the public face of the UN global carbon market, the CDM EB. At minimum this invites some scepticism with regard to more cosmopolitan claims of the democratisation of global governance (Scholte 2011; Held and Koenig-Archibugi 2005).

The debate noted above about the tension between accumulation and legitimation, as well as the ways in which 'resistance makes markets' (Paterson 2009), might be one way of understanding the cue for the dialogue and some of the reforms it proposed. The lobbying and claim-making from different actors during the dialogue also lends weight to arguments about 'climate capitalism' (Newell and Paterson 2010); that carbon markets provide one means by which powerful fractions of capital are incentivised to put their immense economic and political power behind a collective project of decarbonisation. The process showed clearly that a constituency of powerful actors has a material interest in securing the CDM's future. Sunk costs of time and investment and political credibility invested in the CDM mean that many business associations rallied to its cause, even if their more fickle counterparts in finance were losing interest in carbon markets amid low prices. An International Emissions Trading Association (IETA) press release at the launch of the dialogue pleaded that:

> Any new calls for new market mechanisms by the UN or other bodies should recognize the huge investment of time and knowledge that has been put into the CDM and leverage on this when designing new market instruments. The CDM is in its 'teenage years' as an institution. It has built

its capacity around the world and now is the best time to explore how the CDM can 'grow up' in terms of enhancing its benefits on a global scale.

(IETA 2011)

Cries not to 'throw the baby out with the bathwater' (Buen 2013) came from all quarters. Echoing this, Peer Stiansen, the chair of the EB said: 'The CDM is the product of many years of hard work and improvement by countries, the Board, and the mechanism's many stakeholders, not least the thousands of project participants in more than 80 countries' (UNFCCC 2013d).

It seems clear that ongoing and future attempts to control the terrain of the debate about the role of carbon markets in responses to climate change will continue to reflect many of the dynamics highlighted in this chapter: acknowledging problems and claiming learning by doing; opening dialogue but delimiting its terms such that only incremental reforms can result; defining as authority brokers and legitimate participants in debate only those that have a prior commitment to carbon markets and 'positive dialogue' about constructive reform of the CDM; discrediting the value of 'lived' knowledge produced by 'non-experts' in 'non-objective' forms while privileging desk-based and quantitative assessments of performance; providing narrow windows of opportunity for effective engagement and meaningful deliberation; invoking the urgency of responding to climate change as simultaneously a rationale for saving the CDM; and excluding non-market-based policy and political alternatives. The ongoing tension between accumulation and legitimation will surely mean this is not the last crisis of legitimacy carbon markets will face.

Notes

1 See, for example, the films 'Carbon Markets: Trading with our future' http://vimeo. com/32995647 or BBC World, at: http://www.bbc.co.uk/mundo/noticias/2011/ 11/111117_carbono_mercado_am.shtml; *The Guardian* http://www.theguardian.com/ business/feedarticle/9947090; as well as stories in the *Daily Mail, Times of South Africa* and on several radio stations.
2 Tushar Pancholi, Paryavaraniya Vikas Kendra participated at the stakeholder meeting in Delhi on 16 July 2012.
3 'There is No Alternative' is a reference to a phrase often used by advocates and defenders of neo-liberal globalisation, that no viable alternatives exist to an economy organised along the lines of global capitalism, as a way of foreclosing debate about the merits of alternative economic systems.

References

Bond, P., Dada, R. and Erion, G. (2009) *Climate Change, Carbon Trading and Civil Society: Negative Returns of South African Investments.* Scottsville: University of Kwazulu Natal Press.
Buen, J. (2013) CDM criticisms: Don't throw the baby out with the bathwater. *FNI Climate Policy Perspectives No. 8.* Norway: Fridtjof Nansen Institute.
Bumpus, A. (2012) 'Fruitful Design: the CDM'. University of Melbourne. Available from: http://cdm.unfccc.int/about/dev_ben/CDM-Benefits-2012.pdf [accessed 13 February 2014].

CDM Policy Dialogue (2011) 'Terms of Reference for the Policy Dialogue on the CDM', 26 October, EB 64 Report Annex 1. Available from: http://www.cdmpolicydialogue. org/background [accessed 13 February 2014].

CDM Policy Dialogue (2012a) Climate Change, Carbon Markets and the CDM: A Call to Action. *Report of the High-Level Panel on the CDM Policy Dialogue.* Available from: http:// www.cdmpolicydialogue.org/report [accessed 13 February 2014].

CDM Policy Dialogue (2012b) *Assessing the Impact of the CDM.* Report commissioned by the High-Level Panel of the CDM Policy Dialogue, 15 July. Available at: http://www. cdmpolicydialogue.org/research [accessed 13 February 2014].

CDM Policy Dialogue (2012c) *Research Area: Governance,* 1 October. Available from: http:// www.cdmpolicydialogue.org/research [accessed 13 February 2014].

CDM Watch (2012) Watch this! NGO voices on the CDM. *CDM Watch newsletter,* August edition.

Clark, P. (2012) UN-led carbon market 'close to collapse'. *Financial Times,* 2 October. Available from: www.ft.com/cms/s/0/ee81799c-0c84-11e2-a776-00144feabdc0.html [accessed 13 February 2014].

Gaventa, J. (2006) Finding the spaces for change: a power analysis. *IDS Bulletin,* 37 (6), 23–33.

Gramsci, A. (1971) *Selections from the Prison Notebooks.* New York: International Publishers.

Haas, P. (1990) Obtaining international environmental protection through epistemic consensus. *Millennium: Journal of International Studies,* 19 (3), 347–363.

Habermas, J. (1973a) What does a crisis mean today? Legitimation problems in late capitalism. *Social Research,* 40 (4), 643–667.

Habermas, J. (1973b) *Legitimation Crisis.* Boston: Beacon Press.

Habermas, J. (1998) *The Inclusion of the Other: Studies in Political Theory.* Cambridge: MIT Press.

Harvey, D. (2010) *Enigma of Capital and the Crises of Capitalism.* London: Profile Books.

Held, D. and Koenig-Archibugi, M., ed. (2005) *Global Governance and Public Accountability.* Oxford: Blackwell.

IETA (2011) *The Launch of the CDM Policy Dialogue.* Durban, South Africa, 3 December 2011. Available from: http://www.ieta.org/assets/EventDocs/COP17-2011/cdm%20 policy%20dialogue%20launch_031211.pdf [accessed 13 February 2014].

Lipsky, M. (1980) *Street-Level Bureaucracy: Dilemmas of the Individual in Public Services.* New York: Russell Sage Foundation.

Lohmann, L. (2012) *Beyond Patzers and Clients: Strategic Reflections on Climate Change and the 'Green Economy'.* Dorset: The CornerHouse.

Morton, A. (2007) *Unravelling Gramsci: Hegemony and the Passive Revolution in the Global Political Economy.* London: Pluto Press.

Mukherjee, S. (2012) CDM policy dialogue report completely inadequate: CSE. *Times of India,* 14 September. Available from: http://articles.timesofindia.indiatimes.com/2012-09-14/bangalore/33843158_1_sustainable-development-clean-development-mechanism-climate-change [accessed 13 February 2014].

Newell, P. (2010) Democratising biotechnology? Deliberation, participation and social regulation in a neo-liberal world. *Review of International Studies,* 36, 471–491.

Newell, P. (2013) The politics and political economy of the Clean Development Mechanism in Argentina. *Environmental Politics,* 23 (2), 321–338.

Newell, P. and Paterson, M. (2010) *Climate Capitalism.* Cambridge: Cambridge University Press.

Newell, P. and Paterson, M. (2011) Climate capitalism. *In:* E. Altvater and A. Brunnengräber, eds., *After Cancún: Climate Governance or Climate Conflicts.* Berlin: VS Verlag, 23–45.

Olsen, K. H. (2007) The clean development mechanism's contribution to sustainable development: A review of the literature. *Climatic Change,* 84, 59–73.

Online Dictionary (2013) *Dialogue.* Available from: http://dictionary.reference.com/browse/dialogue?s=t [accessed 28 February 2013].

Paterson, M. (2009) Resistance makes carbon markets. In: S. Böhm and S. Dabhi, eds., *Upsetting the Offset: The Political Economy of Carbon Markets.* Colchester: MayFly Books, 244–254.

Paterson, M. (2010) Legitimation and accumulation in climate change governance. *New Political Economy,* 15 (3), 345–368.

Paterson, M. (2012) Who and what are carbon markets for? Politics and the development of climate policy. *Climate Policy,* 12 (1), 82–97.

Reus-Smit, C. (2007) International crises of legitimacy. *International Politics,* 44, 157–174.

Reyes, O. (2011) Zombie carbon and sectoral market mechanisms. *Capitalism Nature Socialism,* 22 (4), 117–135.

Schneider, L. (2007) Is the CDM fulfilling its environmental and sustainable development objectives? An evaluation of the CDM and options for improvement. *Öko-Institut e.V.: Report prepared for WWF.* Available from: http://www.wwf.org.uk/filelibrary/pdf/cdm_fill_objectives.pdf [accessed 13 February 2014].

Scholte, J. A. (2011) *Building Global Democracy? Civil Society and Accountable Global Governance.* Cambridge: Cambridge University Press.

Sklair, L. (2002) *Globalization: Capitalism and its Alternatives.* Oxford: Oxford University Press.

Smith, K. (2007) *The Carbon Neutral Myth: Offset Indulgences for your Climate Sins.* Amsterdam: Carbon Trade Watch.

Spash, C. (2010) The brave new world of carbon trading. *New Political Economy,* 15 (2), 169–195.

Sterk et al. (2009) *Further Development of the Project-Based Mechanisms in a Post-2012 Regime.* Wuppertal: Wuppertal Institute for Climate, Environment and Energy. Available from: http://www.jiko-bmu.de/files/inc/application/x-download/cdm_post_2012_study_wi.pdf [accessed 5 October 2013].

Sutter, C. and Parreño, J. (2007) Does the current Clean Development Mechanism (CDM) deliver its sustainable development claim? An analysis of officially registered CDM projects. *Climatic Change,* 84 (1), 75–90.

Swyngedouw, E. (2010) Apocalypse forever? Post-political populism and the spectre of climate change. *Theory, Culture and Society,* 27 (2–3), 213–232.

The Guardian (2011) 'Bribery, collusion hinder UN carbon scheme'. 15 November. Available from: http://www.theguardian.com/business/feedarticle/9947090 [accessed 20 September 2013].

The Times of South Africa (2011) 'Bribery, collusion hinder UN carbon scheme'. 15 November. Available from: http://www.timeslive.co.za/scitech/2011/11/15/bribery-collusion-hinder-un-carbon-scheme [accessed 17 February 2014].

Transparency International (2011) *Global Corruption Report: Climate Change.* London: Earthscan.

UNFCCC (1998) *The Kyoto Protocol to the United Nations Framework Convention on Climate Change.* Bonn: UNFCCC.

UNFCCC (2012a) *UNFCCC releases report on the benefits of the Kyoto Protocol's clean development mechanism.* Bonn: UNFCCC press office, 20 November. Available from: http://cdm.unfccc.int/CDMNews/issues/issues/I_KYD6PO19YS9DE7YGH894BJ9WRBCQ4Z/viewnewsitem.html [accessed 13 February 2014].

UNFCCC (2012b) *High-level panel releases large body of research on CDM*. Bonn: UNFCCC press office, 31 October. Available from: https://cdm.unfccc.int/CDMNews/issues/issues/I_R1JMKBLU98GHWLZ8I5C3CEQRV9BEVX/viewnewsitem.html [accessed 13 February 2014].

UNFCCC (2013a) 'Kyoto Protocol's Clean Development Mechanism surpasses 6,000 projects', 30 January. Available at: https://cdm.unfccc.int/press/newsroom/latestnews/releases/2013/02_index.html [accessed 7 October 2013].

UNFCCC (2013b) CDM-EB72-A01-INFO: Compilation of inputs considered by the Board in its review of the CDM modalities and procedures. Available from: https://cdm.unfccc.int/filestorage/r/a/extfile-20130315160425775-info_note41.pdf/info_note41.pdf?t=bXR8bjVpbGJzfDDw7hztAZEg0REvK9IdjLYe [accessed 13 February 2014].

UNFCCC (2013c) CDM-EB72-A02-RECO: Recommendation to the SBI: Possible changes to the CDM modalities and procedures. Available from: http://cdm.unfccc.int/filestorage/e/0/extfile-20130325161519814-Info_note42.pdf/Info_note42.pdf?t=dEh8bjVpbGd6fDC2Up40FjlYlqUq05U1KmgW [accessed 13 February 2014].

UNFCCC (2013d) Executive Board agrees plan to improve, position the CDM. Bonn: CDM Executive Board, 71st Meeting, 1 February. Available from: https://cdm.unfccc.int/press/newsroom/latestnews/releases/2013/03_index.html [accessed 13 February 2014].

Werksman, J. (1998) The Clean Development Mechanism: unwrapping the Kyoto surprise. *Review of European Community and International Environmental Law*, 7 (2), 147–158.

World Bank (2010) *Development and Climate Change*. Human Development Report, Washington: World Bank.

Yan, K. (2011) WikiLeak cable highlights high level CDM scam in India. *International Rivers*, 20 September. Available from: http://www.internationalrivers.org/node/6855 [accessed 1 November 2011].

12 The post- and future politics of green economy and REDD+

Kathleen McAfee

Green economy and the persistence of politics

During the past three decades, market-based management of nature, emphasizing monetary valuation and trading of so-called natural capital, has become the prevailing framework in international environmental governance and treaty negotiations (McAfee 2012a). Recently this approach has been framed in terms of 'green economy'. To their advocates, green-economy principles offer the best hope for overcoming a global 'triple crisis' of subsistence (hunger and extreme poverty), environmental degradation (global warming and species loss), and economic stagnation (Gills 2010; IMF 2012).

As elaborated by United Nations agencies, international financial institutions, global-North governments, and private conservationist organizations, green economy is intended to incorporate environmental factors into national and international strategies for economic growth and sustainable development (GEC 2013; Le Blanc 2011; OECD 2011; TNC 2013; UNDESA 2012). In short, green economy is a strategy to save globalized capitalism from its most ecologically and socially destructive consequences by constructing markets in environmental assets and deficits. This, it is said, will benefit investors and society as well as nature, thus helping to circumvent the financing short-falls, conflicts of interest, and political disputes that have blocked progress on climate change mitigation and biodiversity conservation. Green-economy strategies are said to promise 'new tools and a fresh approach for overcoming the gaps and challenges experienced over the past 20 years' (UNDESA 2013: 23).

Carbon-offset markets are one of the most vigorously promoted green-economy policies (World Bank 2012). When traded across borders, carbon offsets are essentially permits that allow continued emissions of certain amounts of carbon dioxide and other greenhouse gases (GHGs), mainly in the industrialized economies, in exchange for putative climate change mitigating activities elsewhere: biofuel production, introduction of greener manufacturing methods, energy production by less polluting means, for example, hydropower instead of coal, or preservation and expansion of forests and peat lands that sequester carbon. Because most of these activities can be carried out more cheaply, in monetary terms, in countries where incomes and prices are relatively low, investment in climate-saving actions in the global South is considered one of the least-cost means of addressing global warming (Stern 2009; Storm 2009). This representation of transnational carbon markets as a rational, efficient, scientifically based strategy, however, obscures the politics inherent in offset trading and market-based management of nature more broadly.

Among other things, carbon markets are expected to generate revenue for Reducing Emissions from Deforestation and Degradation (REDD+), a global program aimed at maintaining and enlarging forests as carbon sinks, primarily in the tropics.[1] Market-based REDD+ builds upon the model of Payments for Ecosystem Services (PES) projects in Latin America, Asia, and Africa since the late 1990s (Corbera 2012; McAfee 2012b).[2] The main institutional sponsors of REDD+, the World Bank and the UN Environment Program (UNEP), see carbon-offset markets as a major source of private financing for conservation (UNEP 2011). REDD+, along with the Kyoto Protocol's Clean Development Mechanism (CDM), are the most fully realized applications of green-economy policy at an international scale thus far.

By promising benefits to all parties from trade in carbon credits, both the CDM and market-based REDD+ attempt to skirt the tensions between two goals: slowing global warming and fostering economic growth and development as it is conventionally understood. Advocates hope that this approach can surmount the political discord that has thwarted renewal of the Kyoto Protocol and other international action on climate change. In theory, REDD+ payments could compensate developing countries for reducing their GHG emissions and for conserving carbon sinks in their territories. It would thus help to resolve disputes about the relative responsibilities of long-industrialized, global-North countries, countries now undergoing rapid industrialization, and would-be industrializing countries. Some also believe that REDD+ payments, if distributed strategically to land owners and the poor, can ease the conflicts over land and territorial rights that now seethe in many rural regions.

As a framing device in policymaking, green economy exemplifies the present, 'post-political' conjuncture (Mouffe 2005; Žižek 1999). It constructs an idealized, static, unitary, and authoritative Nature, in defense of which political conflicts must be set aside in favor of technocratic, environmental-economic management justified in terms of the purported common interest of all humanity (Swyngedouw 2011). For example, the policy documents of the UN Intergovernmental Panel on Climate Change, the main multilateral body that informs climate negotiations, reflect such a perspective, generally avoiding reference to the social and political dimensions of climate change, much less to the power-laden processes through which its own research agenda has been constructed. Thus environmental post-politics attempts to position environmental governance above the fray, in a realm of purportedly objective scientific rigor and technical expertise.

Despite such efforts, international climate policy is more contested than ever, as struggles between different classes, interests, and institutions over wealth, territory, resources, and legitimacy reemerge repeatedly. This was strongly evident at the 2012 Rio+20 Earth Summit, where green economy became a focal point of dissent among official delegations. While some international NGOs are campaigning for carbon markets and investing in REDD+, others oppose them just as forcefully. Academics and coalitions of indigenous and peasant groups have condemned REDD+ as cover for a new phase of land enclosure and dispossession of the poor (Fairhead *et al.* 2012; Rosset 2013). For the first time in decades, some governments are questioning whether neoliberal models of competitive, export-driven, private-sector-led growth can foster ecologically sustainable development. At the same time, non-market-centric alternatives envisioned by social-movement opponents of green economy are gaining attention.

The first section of this chapter describes how market-financed REDD+ and carbon trading are constructed in terms of the concepts, methods, and assumptions of green economy: market prices, the fungibility of natural assets, the representation of transcontinental trade in ecosystem services as an apolitical, 'win-win' strategy for both conservation and development, and the crucial role of the idea of scarcity in this discourse. The next section describes the evolution of international green economy thinking and its application to climate policy through carbon-offset trading and REDD+. I then summarize the critiques of carbon markets as a conservation-and-development strategy put forward by academic and other critics, myself included, with emphasis on the problem of inequality. The section that follows discusses the ongoing international political dissent about green economy, illustrated by controversies at Rio+20, and the counter-politics that have crystallized in response to PES and REDD+ projects and carbon-offset trade at local and national levels. Finally, I consider the alternatives to green-economy and market-led development being articulated by social-movement critics, noting how they challenge the representation of environmental crisis as a crisis of scarcity that can best be addressed by means of market exchange.

Market-based environmental governance in theory

A controversial set of presumptions links the logic of green economy to carbon markets and market-oriented versions of REDD+. As it is being advanced by environmental and development agencies, green economy is based on conventional environmental economics, which itself is built largely upon a foundation of neoclassical economics. Many environmental economists also apply the tools of institutional economics to analyze market 'imperfections', complexities, and transaction costs (Ferraro 2008; Swallow *et al.* 2007; Vatn 2005). Some who distinguish themselves as 'ecological' economists envision the economy as situated within and limited by the ecosphere, as put forward by Boulding (1966; also see Daly and Farley 2003). In either case, 'environment' and 'economy' are understood as distinguishable, at least conceptually (see also Lane, this volume).

Most versions of environmental and ecological economics depict nature as part of an immanent market-world. For instance, UNEP's Green Economy manifesto asserts that causes of crises such as looming food and water shortages, as well as 'persistent social problems', share a common feature: 'the gross misallocation of capital', exacerbated by '[e]xisting policies and market incentives' that allow 'unchecked social and environmental externalities' (UNEP 2011: 14–15). In this view, everything, including environmental assets and deficits, ought to be subject to monetary pricing, ownership, and exchange (Turner *et al.* 1994). Market-based conservation requires that the exchange values of so-called natural capital and the costs of environmental damages be determined by means of expert estimates or, preferably for some, that such values be 'discovered' through the workings of supply and demand. Once their values become known, then environmental externalities – the unintentional costs or benefits of economic activities – can supposedly be internalized into the accounting of individuals, enterprises, and states.

In line with the commonly asserted self-depiction of neoliberal doctrine, most advocates of environmental markets see competition among private actors as more conducive than state-centered 'command-and-control' to the efficient allocation of nature's resources and functions (Chichilnisky and Heal 2000). In keeping with neoclassical economic premises, they generally view land users, businesses, and consumers as rational individuals who respond mainly to material incentives. In theory, once property rights are established and transaction costs are minimized, voluntary trade in environmental goods and bads will produce optimal, least-cost outcomes with minimal state involvement (Coase 1960). In this way, the magic of the market is expected to achieve a maximally efficient allocation of land and other resources for manufacturing, energy production, and agriculture. Ecological degradation, particularly the damages caused by global warming, can thus be minimized by reducing conflict between different industries and interests.

In other words, markets can be used to supersede politics. That said, most green-economy theorists do not expect that the pricing, ownership, and trading of environmental goods and bads will arise spontaneously or will be sufficient to ensure that conservation and climate-mitigation goals will be achieved. Environmental markets must be supported and constrained by state interventions such as taxes on fossil-fuel extraction or use, fines or legal limits – 'caps' – on GHG emissions and other pollution, and tax reductions or other subsidies that increase the values of ecosystem services and resources and create incentives for their sustainable use.

Central to this world model is the idea of resource scarcity. In orthodox economics, markets are the means to manage unlimited human wants and desires for finite goods and resources, whether the goal is promoting economic growth, constraining it, or redistributing its products. Scarcity discourse is, likewise, fundamental to green-economy thinking, and this assumption invites the use of tools of quantification and market exchange. As economist Fred Luks writes, ecologically sustainable development 'is about ends (in other words, meeting the needs of people living today and future generations) and limited means (namely, the environment, technology, and institutions)' (2010: 93). Luks provides a cogent comparison of how '[in] both mainstream and ecological economics, scarcity is naturalized' (2010: 93).

International environmentalism has been long framed by beliefs about resource shortages and population excesses, especially since the 1960s. Concepts common in environmental policy conversation in the Anglophone global North, such as 'tragedy of the commons', 'lifeboat ethics', 'carrying capacity', and the I=PAT formula, hold a notion of absolute scarcity at their core.[3] As Lane explains, during the past 30 years the concept of scarcity has risen, died, and been revived in environmental policy. The idea that humanity faces tradeoffs dictated by biophysical realities has been 'translated' and deployed in recent years to justify the creation of markets in GHGs and other pollutants (Lane, this volume).

Given their shared presumption of scarcity, green-economy thinkers are variously optimistic or pessimistic about the extent to which technological innovations can enable expanded development without devastating resource wars or ecological collapse. Some are not dissuaded by the prospect of ecological limits because they assume either that natural resources are largely substitutable, so that a combination of technological advances and well-functioning markets can circumvent fraught choices, or that such choices are inevitable, but are the business of politics, not economics. In contrast, ecological economics takes scarcity as its starting point: living within the planet's limits is its analytical and programmatic priority. Building on analysis of thermodynamics and entropy and the idea that economy is enclosed within the ecosphere, ecological economists posit absolute limits to human production and consumption, or total 'material throughput' (Daly and Farley 2003; Georgescu-Roegen 1971; Martínez-Alier 2003).

But, as Mehta argues, scarcity is political: a socially constructed concept 'that emerges as a political strategy and is used to justify certain interventions over others' (2010: 4; also see Xenos 1989). From the green-economy point of view, humanity as a whole is running out of carbon sinks, farmland, water, minerals, and energy, making planetary catastrophe imminent. If this is so, then surely the concerns of particular groups and even democratic process should be set aside in favor of ecological and economic expertise. 'Everything happens as if Green politics had *frozen* politics solid' (Latour 2008: 2). Swyngedouw (2010, 2011) and others have pointed out the depoliticizing effects of the specter of social and economic collapse caused by climate change and resource limits. From this perspective, there is only one Nature, pitted against a generic Society. There are, therefore, no choices to be made about what sorts of nature – that is, which particular socio-ecological arrangements – are to be preserved.

'Ideological' – that is, political – discussion of how scarcities have been created, whose interests are thus served, and whether market-based management of nature is even possible, become mere distractions from the imperative of averting climate apocalypse (Swyngedouw 2010). The treatment of 'humanity' or 'society' as an undifferentiated whole obscures the facts that some people and places suffer directly from ecological degradation while others benefit immensely from the unsustainable use of natural resources and services. Meanwhile, most green-economy advocates take it for granted that radical redistribution of wealth and resource access is politically impossible and that Western-style 'development' is desirable and universally desired. Nearly all assume that economic growth is vital to the management of ecological crisis and that economic development must continue to be led by the private sector.

Green economy in practice: markets in ecosystem services and carbon offsets

Efforts to apply market logic to multilateral environment policy date back at least to the early 1980s, when the World Bank, pressed by US and European environmentalists, formed its own Environment Division (Wade 1997). Market-centered thinking was implicit in the influential report, *Our Common Future*, which held that conservation, economic growth, and poverty reduction are not only compatible but mutually supportive (Bruntland 1987).

Aspects of green-economy reasoning were present in the discourse that framed the first Earth Summit in Rio de Janeiro in 1992. To win the support of global-South countries, the conservationists and policymakers who promoted that event had added 'Development' to the title of what would have been the 'UN Conference on Environment'. The presumed lesser capacity of global-South states was recognized in the form of pledges by richer states to help them pay for conservation actions. But in the context of stagnant economic growth, strengthening neoliberal ideology, and tightening austerity, expectations soon faded for substantial North-to-South environmental aid flows and for compliance with Earth-Summit obligations by governments in either category. By 2002, market-centric environmentalism was in full bloom. It shaped the Global Compact for public-private partnerships, the main notable outcome of the Rio+10 Earth Summit in Johannesburg.

The original text of the Convention on Biological Diversity (CBD), launched at the first Earth Summit, reflected the idea that both environmental and development benefits can be captured from the profitable, private-sector exploitation of so-called genetic resources such as those found in medicines from rainforest plants (McAfee 1999). What I have called 'selling nature to save it' through biodiversity prospecting has generally failed to produce significant revenues and incentives for conservation. Nevertheless, policies based on similar hopes have been elaborated in subsequent iterations of the CBD (Corson and MacDonald 2012). Analogous reasoning appears in narratives that describe markets in carbon-sequestration services as a new source of development benefits as well as conservation gains and investor profits (UNDESA 2012; UNEP 2011).

In practice, market-based management of carbon does not mean a return to laissez-faire. Markets are supported by government actions; neoliberalization entails changes in the forms of state interventions, not their elimination (Peck 2008). As noted above, most environmental economists call for voluntary or public-sector regulations that curb pollution and destruction of ecosystems by tweaking markets to make it costly to pollute or destroy natural assets (Muradian and Rival 2012; TEEB 2010). This, along with the creation of tradable property rights to ecosystem functions, lays the epistemological and practical foundations for the conceptualization of 'ecosystem services' (ES) as a commodity that can be owned, bought, and sold. ES – carbon sequestration by forests, the buffering of tides and storms by wetlands, and the provision of habitats for desirable species – can then be traded under cap-and-trade regulatory systems or on a voluntary basis.[4]

Expectations for private-sector-led responses to global ecological crisis are seen most clearly in climate policy (Stern 2009). North-to-South aid flows have diminished while persistent indebtedness and poverty persist in much of the global South, and even while the human impacts of global warming are increasingly visible there. In this context, the World Bank has found a wide audience for its argument that private investment can finance forest conservation, climate-mitigation technologies, and low-carbon industrialization (Watson 2007; World Bank 2010). Support for this approach, however, falls far short of consensus.

By boosting investor expectations and by supporting the prices of the GHG offsets that it brokers, the World Bank seeks to catalyze international carbon trading. Most of the financing for Bank-managed carbon funds has been contributed by governments. Germany has been the largest donor to the Bank's Forest Carbon Partnership Facility (FCPF), followed by Norway, rich in oil revenues and keen to avert alternative policies that might stifle its lucrative fossil-fuel industries. The European Union has taken extraordinary measures to support its own, flagging, regional carbon-trading scheme (Methmann and Stephan, this volume). Thus, carbon 'markets', like most PES projects, are shaped and subsidized by public institutions (McAfee and Shapiro 2010). The World Bank's Carbon Finance Unit expects the private sector eventually to become the primary engine of a multi-billion-dollar market in credits for climate-mitigation and biodiversity conservation actions (World Bank 2013).

The Bank expects carbon-market revenues to be used to help finance REDD+.[5] In the Bank's vision, REDD+ revenues would be generated by public and private investments in forest conservation and forest 'enhancements', including tree plantations. Some of the profits from the sale of REDD+ credits are to be accrued by governments that demonstrate improved practice in limiting deforestation. Those governments, in turn, are expected to distribute REDD+ payments to landholders by means of PES and similar schemes, or to use REDD+ funds directly to strengthen protected-area zones and other conservation programs.

REDD+ is a multi-headed, multi-limbed creation with no single, standardized architecture, set of criteria, or financing mechanism. As of late 2013, at least 44 countries were pursuing REDD+ projects and preparatory processes in conjunction with the World Bank FCPF. Others are affiliated with UN-REDD+, established in 2008 as a joint project of the UN Food and Agriculture Organisation, the UN Development Programme, and UNEP and backed by a multi-donor trust fund. Still other projects designated as REDD+

by their sponsors are independent initiatives by governments, private companies, and nongovernment organizations (NGOs). Many governments, including those of the EU, Indonesia, and Guyana, have embraced REDD+, but others, such as Brazil's, have frequently argued against market-based REDD+ and favor a version of REDD+ supported by public financing.

Contradictions of carbon markets and REDD+

To many of its critics, however, REDD+ is little more than a scheme for pollution offsetting in the global North at the expense of the global South (Böhm and Dabhi 2009; McAfee 2012b; No REDD+ 2012). Some see REDD+ as a tool that enables states to recentralize their control over rural resources and territories (Phelps *et al.* 2010). Even before 2007, some rural-based organizations interpreted REDD+ as a mechanism that would facilitate new forms of land enclosures and evictions of forest dwellers and peasants. Meanwhile, among scientists and policymakers there is considerable doubt about whether carbon-offset trade, and REDD+ in particular, can achieve net reductions in GHG emissions or yield significant development benefits, much less both (Angelsen *et al.* 2012; Lohmann 2005; Pattanayak *et al.* 2010).

The case for carbon-market financing of poverty alleviation and sustainable development has several damning debilities. First, for-profit carbon-credit investments are linked tenuously, if at all, to the objective of slowing global warming. Consumers who wish to offset their individual carbon footprints might genuinely desire that their purchases of offsets will result in forest conservation or greener industrial technologies, but, as capitalist enterprises, the energy corporations, manufacturing firms, banks, hedge funds, and individual speculators who comprise the majority of carbon-credit buyers must make profit their priority. Their concerns about the actual, environmental effects of their carbon-market transactions are secondary at best.

Offset credits, like any marketed commodity, have little commercial value unless they are scarce relative to demand. But in the wake of the failure to strengthen the Kyoto Protocol and in the absence of rigorous state and regional limits on GHGs, carbon credits are anything but scarce. The value of the global carbon market in 2013 was less than half of its value in 2011 (Point Carbon 2014: 1), mainly because supply of credits has far exceeded demand. Point Carbon's review of the carbon markets reported that, while the ETS comprised about 95 percent of the value of all carbon-linked trade in 2013, '[o]versupply and consistently low prices threaten to render the EU ETS insignificant' as an incentive for reducing atmospheric pollution (Point Carbon 2014: 1). As the senior carbon analyst for Thompson Reuters observed, '[w]ithout ambitious climate targets there is no need for deep emission reductions and carbon prices will remain at low levels' (Point Carbon 2014: 1). In 2013, prices on unregulated voluntary carbon markets ranged between a mere US$1 to $8 a ton, while investments in new CDM projects have ceased almost entirely.[6] And, as a decade's experience has shown, carbon trading is particularly vulnerable to manipulation and fraud (FERN 2013; Macken 2011). As long as offsets are so easily and cheaply obtained there is scant economic incentive for enterprises to lower their GHG emissions.

Meanwhile, the attempt to standardize and fine-tune methods for certifying the ecological legitimacy of credit-earning investments has become a major undertaking, funded largely by public institutions and enrolling thousands of economists, business and legal consultants, ecologists, and more rarely, social scientists (e.g. Engel *et al.* 2012). Yet this effort is confounded by the irreducible complexity of actual ecosystems and their social settings. Ample literature documents the impossibility of measuring accurately the quantity of organisms saved or the quantity and continuity of GHGs sequestered or emissions averted by any particular land-use or industrial-technology intervention (e.g. Norgaard 2010). Estimates of net environmental loss or gain inevitably rely on far-reaching approximations, weakly founded claims about the commensurability of ecosystems and their attributes, untestable counterfactual scenarios, and assumptions about future human behaviors. Moreover, valuation of the resulting credits in monetary terms can never be bias-free: decisions about discount rates – the expected values of resources and ecosystems to future generations in different regions – are inevitably arbitrary in that they require value judgments by those who deduce the monetary worth, now and in the future, of species, landscapes, and natural processes (cf. Howarth and Norgaard 1993; e.g. Landis and Bernauer 2012).

Market-oriented PES schemes offer myriad examples of how living eco-social systems are recalcitrant to calculation (McAfee and Shapiro 2010; Muradian and Rival 2012). Problems arising from their ecological and social complexities include: 'uncertainty' about how much carbon is actually stored or how much GHG emissions are prevented; 'leakage', when destructive activities are shifted to an activity or place that is under less scrutiny; 'impermanence', concerning what will happen after conservation payments end, investment priorities shift, or when climate change transforms ecological relationships; and various forms of 'moral hazard': conflicts of interests among regulators, monitors, governments, landholders, and investors or opportunities for rent-seeking, outright fraud, and selective representation of data (Lohmann 2009; Muradian *et al.* 2013). These dilemmas are widely acknowledged by supporters of market-oriented PES and REDD+ (Chomitz 2006; Wunder 2013).

While most REDD+ and PES projects have been designed with reference to monetary values and narrow notions of economic efficiency, extensive literature on collective action and common property has shown that the choices and actions of land users are often better explained by cooperative norms, non-monetary values, and non-material incentives than by the expectation of competitive, individual, benefit-maximizing behavior that frames 'Western' models (Ostrom 1990). Recent reviews offer evidence that this is often true of PES project participants, whose ES payment incomes commonly fall short of their required costs in time and effort (Corbera *et al.* 2007; Kosoy *et al.* 2008; Van Hecken and Bastiaensen 2010).

Another deep flaw in the green-economy strategy of trading ecosystem services for development revenue is less often acknowledged. It is that the profit incentives necessary to spur substantial private investment in international climate mitigation depend upon continued inequality between lower-income and higher-income world regions (McAfee 2012a, 2012b). Revenue for conservation, green technology, and development in the global South is expected to come from energy, manufacturing, and transportation firms that purchase carbon credits to offset the damages caused by their polluting activities.[7] But businesses, brokers, and speculators in carbon or biodiversity markets will buy such permits only insofar as such investments offer a profit advantage over investments in other places and activities.

The predicted profit opportunities in North–South carbon trading are derived from the fact that offset credits can be obtained at less expense in economically poorer regions. In the language of orthodox economics, carbon credits in low-income countries are cheaper because opportunity costs – for example, potential income lost if land is used for carbon sinks instead of food or oil crops – are smaller where prices of labor, land, and other factors of production are lower and life expectancies are shorter. In other words, investment in greening in the global South is economically 'efficient' because nature and human lives are cheaper, in effect, and bargaining power is weaker in places where people are poor. Inequality is thus built into the framework and rationale for global carbon markets from the outset.[8]

To its advocates, the market-based efficiency derived from this wealth difference is the great virtue of transnational trade in carbon or biodiversity credits. Of course, the claim that 'free' trade results in the greatest gain for the least expenditure is at the heart of the neoclassical argument for the superiority of market-based management of nature and of everything else. But any trading of environmental goods and bads, whether in atmospheric pollution, biodiversity conservation, or other assets and damages, involves shifting the costs of conservation from one location, that is, one eco-social system, to another. Thus, what appears to be a 'triple-win' solution for business, nature, and 'humanity' has highly unequal human consequences and, therefore, deeply political implications.

Moreover, conventional economic logic dictates that the greatest conservation gains can be achieved by the most efficient allocation of available funds, whether the sources of those funds are public or private. Market efficiency, therefore, requires that payments for carbon sequestration or other ES be made to landholders whose decisions to conserve trees or fell them will be influenced by the payments. Payments to relatively wealthy 'stakeholders', such as prospective investors in logging or mining operations, large-scale ranching or agricultural plantations, shopping centers, or resorts, would typically be inefficient because the payments would need to be large enough to match the high opportunity costs of abstaining from such profitable activities. But these activities are major drivers of deforestation and wetlands loss and these actors often have decisive political power.

Middle-sector landholders, with more modest options for profit but with the capacity and intention to deforest, are seen as the most appropriate PES recipients (Wunder 2013). However, given the problems of uncertainty, leakage, and impermanence listed above, it is not clear whether market-oriented distribution of ES payments can achieve significant forest conservation in the medium or longer term. Some REDD+ project designers and investors are crafting 'jurisdictional' REDD+ methods meant to strengthen accountability and link multiple REDD+ projects at subnational and national scales (ROW 2013; VCS 2013). But these efforts, too, face daunting technical and political challenges: uncertainty about carbon cycles, conflicts of interests in project implementation and monitoring, and politically laden decisions about how baselines for the measurement of 'improved forest management' are being determined.

In any case, payments for ES to the very poor achieve relatively little in forest-conservation gains because such land users lack the necessary tenure rights, capital, or market access to be able to engage in significant deforestation with or without PES. Many REDD+ and proto-REDD+ projects have, nevertheless, targeted such 'marginalized' or indigenous communities. There is much debate in the PES literature about whether or not 'pro-poor' projects of this kind are providing net conservation gains or social benefits (Kronenberg and Hubacek 2013; Luttrell *et al.* 2013; Pattanayak *et al.* 2010). But, as the World Bank and other economists have pointed out, to the extent that poverty alleviation is a priority, market efficiency is likely to be compromised (Pagiola 2007).

In addition, farm and forest carbon-sequestration projects, like many conservation schemes, can have directly negative effects on poor land users. By the late 2000s, in the context of increased integration and financialization of the global economy and the closing of rural frontiers, world-market prices of food, fiber, and mineral commodities were already rising. Consequently, land prices soared in many regions. As ES offsets have become a new tropical export commodity, anticipation of profits from carbon- and biodiversity-market investments has added new economic value to rural landscapes. In many global-South regions, forests and swamps are being reconceptualized as carbon sinks, and farmlands repurposed as biofuel plantations (Akram-Lodhi 2012; Fairhead *et al.* 2012; Li 2010; Moore 2010).

This has accelerated the processes that critics call land grabbing – the illegal or unjust acquisition of land by the economically powerful – and green grabbing: expulsions of forest dwellers and small-scale farmers for ostensible environmental goals. Projects carried out under the rubric of PES and REDD+ appear to be contributing to this trend by displacing, or threatening to displace, peasant and indigenous communities (JPS 2014; Rocheleau 2014). Even where land users are not evicted, they may face reduced access to sites of cultural significance, passageways, and sources of food, forage, medicines, and shelter materials. In the context of resurging struggles over land and territorial sovereignty in many regions, markets in carbon and biodiversity offsets have inevitably political implications at local and national levels, regardless of their intended scientific neutrality and regardless of whether their sponsors are guided by conservation-efficiency or pro-poor priorities.

Opposition to green economy and REDD+

Climate policy also remains deeply contested at the international level. Despite appeals to the presumed common interests of humanity and to scientific expertise, green-economy proponents have not succeeded in separating environmental goals from ongoing tensions rooted in colonial legacies and continuing North–South inequalities. For three decades, the political dimensions of international trade and development negotiations have been partially blunted by the economic-liberal claims that growth is limitless and that market exchange per se benefits all parties. In contrast, environmental governance discourses have long been cast in terms of limits, tradeoffs, scarcity, and the need for constraints on economic development (Barbier 2011; Hardin 1968; Latour 2008). If there are supposed to be insufficient material resources and 'ecological space' for economic development as it has been known, then, leaders of global-South states contend, it is their peoples who will be required to sacrifice.

The North–South political dimensions of green economy were starkly evident at the 2012 Rio+20 Earth Summit. 'Green economy in the context of sustainable development and poverty eradication' had been endorsed by the UN General Assembly as a leading theme for Rio+20. However, members of the G-77 group voiced concerns that green-economy policies would reinforce existing North–South inequalities by displacing sustainable development as a global goal, saddling Southern states with new policy constraints, and erecting additional barriers to Southern exports. Several Latin American delegations decried green economy as 'environmental neocolonialism' and as a further abdication by developed countries of obligations to their former colonies (IISD Reporting Services 2012). Delegates from countries of the Bolivarian Alliance for Our Americas (ALBA) explicitly condemned green economy as a 'false and dangerous solution' (Bolivia n.d.; IISD Reporting Services 2012).[9]

As at most such multilateral conferences, much of the debate took place before and outside of the formal Rio+20 sessions (UNCSD 2012: 39). At business-oriented side events, participants cheered carbon markets and the green-economy theme (Suarez 2013). At another Rio venue, ecological economists debated the applicability of green-economy concepts with varying degrees of ambivalence (ISEE 2012). A headline in an NGO Rio+20 newspaper proclaimed 'The Green Economy: The New Enemy' (IPS 2012). Most participants in the Indigenous Peoples Global Conference on Rio+20 and the NGO-organized People's Summit for Social and Environmental Justice were strongly critical of market-centered environmentalism (People's Summit 2012).

The People's Summit declaration describes green economy as 'just another facet of the current financial phase of capitalism'. It calls instead for democratic governance, *buen vivir*, 'cooperatives and the solidarity economy, food sovereignty, a new paradigm of production, distribution and consumption, and a change in the energy mix, . . . expansion of the concept of work, the recognition of women's work, and a balance between production and reproduction' (People's Summit 2012; Utting 2012).[10] The statement by the Indigenous Peoples Global Conference on Rio+20 endorses 'collective control of productive resources' and rejects 'false solutions to climate change' including 'geoengineering, carbon markets, Clean Development Mechanism and REDD+' (People's Summit 2012).

The official Rio+20 Summit declaration, *The Future We Want*, is legally toothless, with many ambiguous assertions reflecting the contradictory agendas of various states, NGOs, and business lobbyists. Some of its contradictions are ironic: governments that condemned green economy as 'commodification of nature' were unwilling to support language critical of fossil fuels. It is, nevertheless, significant that the emphasis on green economy in the original draft was muted in the final declaration: green economy is described there as merely '*one* of the important tools' that '*could* provide options for policy making but should not be a rigid set of rules' (UNCSD 2012: 9, emphasis added). Because economic valuation and trading of ES, particularly carbon sequestration, have been central to most green-economy policy proposals, this revision of the original draft represents a symbolic setback, at least, for supporters of universalized, market-based management of the carbon economy.

A month later, peasant and indigenous leaders at a conference in Bukit Tinggi, Indonesia offered examples of 'how . . . REDD/REDD+, "climate smart agriculture" and agrofuels are destroying the coping mechanisms, resilience and autonomous adaptive capacities of small-scale food producers, and creating opportunities for corporations and traders to acquire rural peoples' lands, forests, coasts and other natural resources' (Guttal 2013: 743). Participants in the conference, organized by the multinational federation of food producers, La Via Campesina, agreed that

> IFIs and multilateral institutions facilitate land and natural resource grabbing by promoting extractive, destructive and economic growth-driven development models, discouraging states from legislating and regulating in favor of workers, small-scale producers and the environment, and collaborating with capitalist interests to design and push instruments of financialization such as the Green and Blue Economies, REDD+ and carbon trading
>
> (Guttal 2013: 744)

This view is not unanimous among indigenous and rural activists. For example, a controversial report by members of the UN Permanent Forum on Indigenous Issues (PFIP) gave qualified support to REDD+, concluding that REDD+ can have positive outcomes if indigenous rights are respected and if strong safeguards are enforced (Lang 2013). In more than 30 countries, rural and indigenous organizations have joined REDD+-linked and other carbon-offsetting projects. But 2008–2012 also saw the release of at least 32 declarations against REDD+ by social-movement organizations and more than 15 position papers critical of REDD+ by development NGOs and research organizations (ibid.).

While noting that indigenous groups have the right as sovereign peoples to choose whether or not to engage with REDD+, a leading indigenous-rights activist, Tauli-Corpuz, reported to the PFIP in 2010 that 'there is a strong preference that forests should not be used as offsets':

> There are many serious concerns about REDD because of how it was originally conceived and being shaped to be part of the [Kyoto] carbon trading mechanism. One concern is that instead of cutting back their GHG emissions in the home front, [industrialized countries] will just buy cheap forest carbon credits from tropical developing countries who are implementing REDD Plus. This reduces the pressure on Annex 1 countries to cut their own emissions [and] in the end, will not result into any substantial cuts in emissions reduction. If forest carbon becomes part of the carbon market, speculative or hedging activities (sub-prime carbon) will be facilitated.
>
> (Tauli-Corpuz and Baer 2010: 13)

Some indigenous groups have come to oppose carbon-offset projects that they initially supported. COONAPIP, the National Coordinating Body of Indigenous Peoples in Panama, withdrew from participation in a UN-REDD program in March 2013, alleging that its objective 'is to privatize the forests of Panama . . . and allow the State to cash in on carbon credits in utter contempt for the rights of Indigenous Peoples' (Kühne 2012). The program was reinstated that December after issues of indigenous rights were tentatively resolved to COONAPIP's satisfaction. After three years of negotiations, the largest indigenous organization in Peru complained to the FCPF about exclusion of indigenous communities from the REDD+-linked Forest Investment Plan (REDD Monitor 2013). After several years of involvement, the Civic Council of Popular and Indigenous Organisations of Honduras wrote to the FCPF in opposition to that country's REDD+ Readiness Preparation Proposal and to 'REDD+ in general', stating that 'the carbon market does not solve the problem of greenhouse gases because the industrialized countries and companies continue to emit CO_2 and the projects of death and destruction, plunder and extermination of Indigenous Peoples and commons of nature will continue' (AIDESEP 2013).

In Mexico, some indigenous communities are accepting REDD+ agreements while others are adamantly opposed. At least 13 local organizations in Chiapas and 27 in Brazil, mainly from Acre, have asked the state of California to drop plans to permit polluters to apply offsets from the states of Chiapas and Acre in order to comply with California's new GHG cap-and-trade program (MIU *et al.* 2013; REDD Monitor 2013). Ecuadorean NGOs have decried their government's participation in UN-REDD, contending that it violates a new constitutional provision forbidding the appropriation of environmental services. Grassroots and NGO opposition to REDD+ has been growing in Southeast Asia, particularly Indonesia (FOEI 2014). A coalition of African NGOs announced the formation of the No REDD+ in Africa Network in March 2013 (IEN 2013).

In some countries, the quest for REDD+-linked funding and land-tenure rights has spurred competition and conflict among government agencies and other political interests and economic actors. In Chiapas, Mexico, a much-heralded REDD+ payments scheme in the Lacandon rain forest was pointedly scrapped in 2012 after state elections brought a different party to power. Space here does not permit recounting of other examples: suffice it to say that the green-economy expectation that market-led, science-based management of climate crisis can transcend politics is being repeatedly refuted at multiple levels.

Green economy and its others

The meanings of 'green economy', 'commodification of nature', and even 'REDD+' vary among those who have embraced or denounced them. These differences aside, interpretations of green economy and carbon markets tend toward two poles. Some view markets in nature and its functions – alongside new technologies, accounting methodologies, and financial incentives to shift investment and consumption toward greener patterns – as key to averting climate catastrophe. Markets are seen as efficient, apolitical, and therefore fair, at least potentially. The opposing perspective interprets economic, ecological, and social crises as interrelated consequences of globalized capitalism and the deep inequalities it has created, sharpened by legacies of colonialism. From this perspective, carbon markets are necessarily and profoundly unjust.

In green-economy discourse, nature, society, and economy remain conceptually distinct, with the former two categories represented in terms of the latter. This approach differs from earlier versions of economic and development thinking in that it takes environmental problems seriously. However, it is consistent with neoliberalism in that it presumes that such solutions can and must be achieved within the framework of capitalism and economic growth, albeit with increased public investment and regulation, and in its assumption that carbon trading and other markets in nature can overcome ecological limits and resource scarcities.

Emerging critiques from the left and 'from below' question these assumptions. Many are convinced that carbon trading will yield far more 'losers' than 'winners'. Some contend that formerly colonized countries are owed compensation for past and continuing damages. Such allegations, voiced in calls for a New International Economic Order by the Non-Aligned Movement in the 1960s and 1970s, were revived at Rio+20. The concepts of ecological debt and unequal ecological exchange have added new dimensions to these longstanding demands (Wittman and Caron 2009).

The political dimensions of market-based conservation and climate policy are manifest in the form of widespread opposition to market-oriented climate and conservation schemes. In Africa, Asia, and Latin America, struggles against individual land titling, dams, wind farms, biofuels, and privatized conservation, and restrictive PES and REDD+ projects echo centuries of peasant resistance to dispossession (Peluso and Lund 2011; Rosset 2013; Sassen 2010). Indigenous peoples' organizations contend that green grabbing is already violating peoples' rights to territory, to free, prior and informed consent (FPIC), and to less tangible rights to cultural identities and ways of living. Coalitions of rural-based social movements, many of them rooted in local struggles, are calling for human and territorial rights and for compensation for indigenous and peasant contributions to biodiversity conservation and climate-change mitigation (Böhm and Dabhi 2009; Carbon Trade Watch 2013; Fairhead *et al.* 2012; LDPI 2012; No REDD+ 2012). Assertions of human rights to food and water, women's rights and needs, and rights of the poor, especially small-scale farmers, further expose the politics embedded in green-economy discourse (Guttal 2013).[11]

What these various demands and rights claims have in common is that they disrupt the totalizing vision implicit in most green-economy discourse: that of a market-world in which individual interest and the competitive quest for profitable advantage are the most objective and efficient determinants of how resources are used, and in which the pursuit of growth is the necessary basis of both greening and development. Critiques and propositions are emerging mainly from social movements and their intellectual allies, but also from some global-South states. They challenge the categories and assumptions behind carbon-offset trading, market environmentalism, and neoliberal governance more broadly. Moreover, academic critics of green economy are explicitly questioning the necessity of economic growth – green or otherwise – along a single pathway (Jessop 2012; Martínez-Alier *et al.* 2010).

Protagonists of today's rural social movements are attempting to create new forms of governance, new interpretations of rights, and new conceptualizations of food, energy, and territorial sovereignty (CSRG 2011; Edelman and James 2011; Escobar 2010; ISS 2014; Martínez-Alier 2011; Mueller and Bullard 2011; Shrivastava and Kothari 2012; Starr *et al.* 2011). Some espouse meanings of human–nature relationships that contrast with the narrowly conceived conceptualizations of property rights and territorial borders that underpin market-centric conservation strategies, including most versions of REDD+ (Baletti 2012; Bryan 2012; Icaza and Vázquez 2013; Rosset 2013). In the words of Nicaraguan rural leader Faustino Torres:

> In our spaces of collective construction we are currently engaged in the joint search for the combination of the peasant perspective with indigenous cosmovisions, the latter of which can be summed up in the phrase Buen Vivir (living well; at peace with ourselves and in harmony with each other and with the Mother Earth).
>
> (Torres 2013)

Implicit or expressed in such actions and discourses are ontologies of human–nature relationships at odds with the foundational concepts of modern, institutionalized science as well as with the premises of neoclassical economics and mainstream development theory, in which 'nature' and 'society' are unitary, distinct, and opposed. These understandings are informed by what Escobar (2010) has called 'relational ontologies': they are rooted in ideas about what is important and what is real and that are not universal, but instead are linked to specific ecological contexts, histories, social arrangements, and cosmologies.

The limits and contradictions of the market panacea for conservation and development are increasingly evident and acknowledged. Counter-discourses to green economy challenge the conventional consensus that economic growth, green or otherwise, is the *sine qua non* of human progress, widening the cracks in long-hegemonic ideologies and creating a discursive space in which alternate possibilities become imaginable. The far-reaching criticisms of the green-economy paradigm reflect the ways that 'environmental' concerns have become more openly political and central to the challenges facing both those who seek to save capitalism and those who desire alternatives or are already living them.

Today, the construction of environmental policy discourse in terms of scarcity and the implied need for limits on development begs the deeply political question: who, where, is being required to sacrifice? But if one assumes, instead of scarcity, a world of sufficiency, then the prospects are transformed. The concept of buen vivir conveys ideas of plentitude, even while recognizing the existence of biophysical limits to *how* people, communities, and countries can develop themselves. It connotes values and measures of well-being that are material as well as social but that cannot be reduced to monetary prices and might not be quantifiable or commensurable across places and cultures. Buen vivir rejects the construction of ecological limits as absolute scarcity, focusing less on the finitude of resources and carbon sinks than on the anti-entropic, life-giving relationship between human labor, water, soil, and sun, and the activities of other species.

Capitalist instabilities, rooted in overproduction and intensified by financialization, are raising the stakes in the contest between these models. In the context of a global 'triple crisis', transnational carbon-offset banking and trade is not only a response to the threat of calamitous climate change. It is also a product of competition for new sites and mechanisms of accumulation at a time of stifled economic growth. While potentially profitable carbon offsets attract new investors to capital's shrinking hinterlands, others there – states, communities and social movements of the poor – have also staked claims to these ecosystems and territories. Consequently, the issue of global carbon-market making is inevitably and increasingly political.

Notes

1 To an earlier version of REDD, REDD Plus adds the goals of conservation and sustainable management of forests and enhancement of forest carbon stocks in developing countries.

2   The majority of PES projects involve payments for carbon sequestration or for hydro-logical services financed at sub-regional and national scale. Biodiversity offsets and cred-its for wildlife conservation are also bought and sold internationally.

3   I=PAT is meant to show that total human impact on the environment (I) can be calcu-lated by multiplying population (P) by quantities indicating level of affluence (A) and technology (T).

4   Cap-and-trade systems for regulating GHG emissions and are designed to raise the cost of polluting without stifling economic growth. Governments or groups of governments set legal limits ('caps') on how much GHGs or other pollutants may be emitted and then dis-tribute 'allowances', such as carbon-offset credits, to the enterprises under regulation. Then, as the regulatory bodies gradually reduce the supply of offsets, the rising prices of these credits will supposedly persuade managers of polluting businesses to invest in technological innovations and greener practices in order to minimize their needs to purchase offsets.

5   In the early 2000s, climate negotiators had begun discussing a global project to achieve Reduced Emissions from Deforestation (RED). Building on forest-related provisions for GHG offsetting in the CDM and on the model of PES, the Bank introduced REDD+ at the UNFCCC climate conference at Bali, Indonesia in 2007. It elaborated REDD++, which includes the improvement of forests as well as their preservation, at the 2010 UNFCCC conference in Cancún, Mexico.

6   A post-Kyoto global climate regime with specific GHG-reduction targets and an expanded version of the CDM might encourage such investments by reviving the global price of credits for emissions-reduction or carbon-sink-preservation activities. Private-sector carbon traders have called for 'investment-grade climate policy' for just this reason.

7   Provisions in the EU's cap-and-trade scheme (ETS) that allowed firms to meet their emissions-reduction obligations by buying carbon credits linked to mitigation schemes in developing countries are being phased out. Provisions for offsets linked to REDD+ have been proposed as an addition to the US state of California's GHG cap-and-trade program. They are supported by industry groups and some conservation organizations but opposed by critics on the grounds that they are environmentally unjust and produce no net global reduction in GHG emissions nor contributions to California's economy.

8   The potential for carbon markets to reinforce or even increase inequalities is not con-fined to North–South trade. It can arise in offset trading within countries and regions, and even in PES and REDD+ schemes that are financed by public subsidies, as are the majority of PES and REDD+ schemes to date (Böhm *et al.* 2012; McAfee 2012b).

9   As of January 2014, ALBA, the Bolivarian Alliance comprised nine member states: Antigua and Barbuda, Bolivia, Cuba, Dominica, Ecuador, Nicaragua, St. Lucia, and St. Vincent and the Grenadines.

10   Buen vivir, sumak kawsay, and *lek'il kuxlejal* are among the terms being put forward by indigenous spokespeople and other critics of conventional, development to convey the idea that 'living well' – adequately in material terms and sustainably with fellow humans and other species – is a more worthy and achievable goal than the pursuit of economic growth and monetary income.

11   Categories of rights – human and territorial; the right to development; food sovereignty; rights to produce and to choose not to emigrate; rights to nature – have different histo-ries and implications. Some contend that conceptions of rights rooted in Western juridi-cal systems, such as land-tenure rights, inevitably introduce logics of commodification and reinforce power inequalities.

References

AIDESEP (2013) Carta No. 021. *Asociación Interetnica Dedesarrollo de la Selva Peruana*. Available from: https://www.climateinvestmentfunds.org/cif/sites/climateinvestmentfunds.org/files/FIP_Peru_IEDAPA_Spanish.pdf [accessed 13 February 2014].

Akram-Lodhi, A. H. (2012) Contextualising land grabbing: contemporary land deals, the global subsistence crisis and the world food system. *Canadian Journal of Development Studies/Revue canadienne d'études du développement*, 33, 119–142.

Angelsen, A., Brockhaus, M., Sunderlin, W. D. and Verchot, L. V., eds. (2012) *Analysing REDD+: challenges and choices*. Bogor: Center for International Forestry Research (CIFOR).

Baletti, B. (2012) Ordenamento Territorial: Neo-developmentalism and the struggle for territory in the lower Brazilian Amazon. *Journal of Peasant Studies*, 39, 573–598.

Barbier, E. (2011) The policy challenges for green economy and sustainable economic development. *Natural Resources Forum*, 35, 233–245.

Böhm, S. and Dabhi, S., eds. (2009) *Upsetting the Offset: the political economy of carbon markets*. London: MayFly Books.

Böhm, S., Misoczky, M. C. and Moog, S. (2012) Greening capitalism? A Marxist critique of carbon markets. *Organization Studies*, 33 (11), 1617–1638.

Bolivia, (n.d.) Proposal for RIO+20 by the Plurinational State of Bolivia. *Global Alliance for the Rights of Nature*. Available from: http://therightsofnature.org/proposal-for-rio20-by-plurinational-state-of-bolivia/ [accessed 15 July 2013].

Boulding, K. E. (1966) The economics of the coming spaceship earth. *Environmental Quality in a Growing Economy*, 2, 3–14.

Bruntland, G. H. (1987) *Our Common Future*. New York/Oxford: Oxford University Press.

Bryan, J. (2012) Rethinking territory: social justice and neoliberalism in Latin America's territorial turn. *Geography Compass*, 6, 215–226.

Carbon Trade Watch (2013) Protecting carbon to destroy forests: land enclosures and REDD+. *Carbon Trade Watch*.

Chichilnisky, G. and Heal, G. M., eds. (2000) *Environmental Markets: equity and efficiency, Economics for a sustainable earth series*. New York: Columbia University Press.

Chomitz, K. (2006) *At Loggerheads? Agricultural expansion, poverty reduction, and environment in the tropical forests*. Washington, DC: World Bank.

Coase, R. H. (1960) The problem of social cost. *Journal of Law and Economics*, 3, 1–44.

Corbera, E. (2012) Problematizing REDD+ as an experiment in payments for ecosystem services. *Current Opinion in Environmental Sustainability*, 4, 612–619.

Corbera, E., Kosoy, N. and Martínez Tuna, M. (2007) Equity implications of marketing ecosystem services in protected areas and rural communities: case studies from Meso-America. *Global Environmental Change*, 17, 365–380.

Corson, C. and MacDonald, K. I. (2012) Enclosing the global commons: the convention on biological diversity and green grabbing. *Journal of Peasant Studies*, 39, 263–283.

CSRG (2011) Statement by the Civil Society Reflection Group on Global Development Perspectives on Rio+20 and beyond. Available from: https://www.reflectiongroup.org/stuff/input-rio-2012 [accessed 27 February 2014].

Daly, H. E. and Farley, J. C. (2003) *Ecological Economics*. Washington, DC: Island Press.

Declaration (2012) Declaration of Chiapas in REDDellion: enough of REDD+ and the green economy. *Climate Connections*. Available from: http://climate-connections.org/2012/09/24/declaration-of-chiapas-in-reddellion-enough-of-redd-and-the-green-economy/ [accessed 19 July 2013].

Edelman, M. and James, C. (2011) Peasants' rights and the UN system: Quixotic struggle? Or emancipatory idea whose time has come? *Journal of Peasant Studies*, 38, 81–108.

Engel, S., Palmer, C., Taschini, L. and Urech, S. (2012) Conservation payments under uncertainty when nonuse benefits have market value. Grantham Research Institute on Climate Change and the Environment Working Paper No. 72. Available from: http://dx.doi.org/10.2139/ssrn.1973449 [accessed 26 February 2014].

Escobar, A. (2010) Latin America at a crossroads. *Cultural Studies*, 24, 1–65.

Fairhead, J., Leach, M. and Scoones, I. (2012) Green grabbing: a new appropriation of nature? *Journal of Peasant Studies*, 39, 237–261.

FERN (2013) *EU ETS myth busting: Why it can't be reformed and shouldn't be replicated*. FERN.

Ferraro, P. J. (2008) Asymmetric information and contract design for payments for environmental services. *Ecological Economics*, 65, 810–821.

FOEI (2014) Indonesia: REDD must cover the rights of indigenous peoples and communities on their territories. *Gathering of Peoples on Forests, Biodiversity, Community Rights & Indigenous Peoples*. Available from: http://ecologicalequity.wordpress.com/themes/stories-of-right-stories-of-might/redd-must-clearly-cover-the-rights-of-indigenous-peoples-and-communities-on-their-territories/ [accessed 26 February 2014].

GEC (2013) Who we are. *Green Economy Coalition*. Available from: http://www.greeneconomycoalition.org/about [accessed 19 July 2013].

Georgescu-Roegen, N. (1971) *The Entropy Law and the Economic Process*. Cambridge: Harvard University Press.

Gills, B. K. (2010) *Globalization in Crisis*. Oxford: Routledge.

Guttal, S. (2013) Porto Alegre, Nyéléni and Bukit Tinggi: evolving views of agrarian reform. *Journal of Peasant Studies*, 40, 739–746.

Hardin, G. (1968) The tragedy of the commons. *Science New Series*, 162.

Howarth, R. B. and Norgaard, R. B. (1993) Intergenerational transfers and the social discount rate. *Environmental and Resource Economics*, 3, 337–358.

Icaza, R. and Vázquez, R. (2013) Social struggles as epistemic struggles. *Development and Change*, 44, 683–704.

IISD Reporting Services (2012) Rio+20: Third PrepCom and the UN Conference on Sustainable Development (UNCSD): summary of the meeting. Available from: http://www.iisd.ca/uncsd/rio20/enb/ [accessed 19 July 2013].

IMF, International Monetary Fund (2012) IMF Survey: economy faces triple threat to sustainable future, says Lagarde. *IMF Survey Magazine*. Washington, DC: IMF.

IEN (2013) Africans unite against new form of colonialism. *Indigenous Environmental Network* . Available at: http://www.ienearth.org/africans-unite-against-new-form-of-colonialism/ [accessed 26 February 2014].

IPS, Inter Press Service (2012) Green economy, the new enemy. *TERRAVIVA Interpress Services RIO+20*, 21 June. Available at: http://www.ips.org/TV/rio20/green-economy-the-new-enemy/ [accessed 22 July 2014].

ISEE, International Society for Ecological Economics (2012) Conference. Available at: http://www.isecoeco.org/tag/isee2012/ [accessed April 15 2013].

ISS (2014) Food sovereignty: a critical dialogue. *Conference program*. The Hague: International Institute of Social Studies.

Jessop, B. (2012) Economic and ecological crises: green new deals and no-growth economies. *Development*, 55, 17–24.

JPS (2014) Special issue: resistance to land grabs. *Journal of Peasant Studies*.

Kosoy, N., Corbera, E. and Brown, K. (2008) Participation in payments for ecosystem services: case studies from the Lacandon rainforest, Mexico. *Geoforum*, 39, 2073–2083.

Kronenberg, J. and Hubacek, K. (2013) Could payments for ecosystem services create an "ecosystem service curse"? *Ecology and Society*, 18 (1), 10–22.

Kühne, K. (2012) REDD resistance around the world. Available from: redd-monitor.org [accessed 26 February 2014].

Landis, F. and Bernauer, T. (2012) Transfer payments in global climate policy. *Nature Climate Change*, 2, 628–633.

Lang, C. (2013) Joan Carling on REDD: NGOs are 'imposing their ideological views' on indigenous peoples. Available from: http://www.redd-monitor.org/2013/05/31/joan-carling-on-redd-ngos-are-imposing-their-ideological-views-on-indigenous-peoples/ [accessed 26 February 2014].

Latour, B. (2008) It's development, stupid! Or how to modernize modernization? Available from: http://www.espacestemps.net/articles/itrsquos-development-stupid-or-how-to-modernize-modernization/ [accessed 26 February 2014].

LDPI (2012) Land Deal Politics Initiative. *Contested Global Landscapes.* Available from: http://www.cornell-landproject.org/activities/2012-land-grabbing-conference/papers/ [accessed 19 August 2013].

Le Blanc, D. (2011) Special issue on green economy and sustainable development. *Natural Resources Forum,* 35, 151–154.

Li, T. M. (2010) Indigeneity, capitalism, and the management of dispossession. *Current Anthropology,* 51, 385–414.

Lohmann, L. (2005) Marketing and making carbon dumps: commodification, calculation and counterfactuals in climate change mitigation. *Science as Culture,* 14, 203–235.

Lohmann, L. (2009) Regulation as Corruption in the Carbon Offset Markets. *In:* S. Böhm and S. Dabhi, eds., *Upsetting the Offset: the political economy of carbon markets.* London: MayFly Books, 175–191.

Luks, F. (2010) Deconstructing Economic Interpretations of Sustainable Development: Limits, Scarcity and Abundance. *In:* L. Mehta, ed., *The Limits to Scarcity: contesting the politics of allocation.* London/Washington, DC: Earthscan, 93–108.

Luttrell, C., Loft, L., Fernanda Gebara, M., Kweka, D., Brockhaus, M., Angelsen, A. and Sunderlin, W. D. (2013) Who should benefit from REDD+? Rationales and realities. *Ecology and Society,* 18 (4), 52–70.

Macken, K. (2011) Strengthening Credibility in the EU ETS Following Security and Fraud Related Incidents. *In:* 9th International Conference on Environmental Compliance and Enforcement: *Conference Proceedings.* Whistler, British Columbia: INECE, 284–289.

Martínez-Alier, J. (2003) *The Environmentalism of the Poor: a study of ecological conflicts and valuation.* Cheltenham: Edward Elgar.

Martínez-Alier, J. (2011) The EROI of agriculture and its use by the Via Campesina. *Journal of Peasant Studies,* 38, 145–160.

Martínez-Alier, J., Pascual, U., Vivien, F.-D. and Zaccai, E. (2010) Sustainable de-growth: mapping the context, criticisms and future prospects of an emergent paradigm. *Ecological Economics,* 69, 1741–1747.

McAfee, K. (1999) Selling nature to save it? Biodiversity and green developmentalism. *Environment and Planning D: Society and Space,* 17, 133–154.

McAfee, K. (2012a) Nature in the market-world: ecosystem services and inequality. *Development,* 55, 25–33.

McAfee, K. (2012b) The contradictory logic of global ecosystem services markets. *Development and Change,* 43, 105–131.

McAfee, K. and Shapiro, E. N. (2010) Payments for ecosystem services in Mexico: nature, neoliberalism, social movements, and the state. *Annals of the Association of American Geographers,* 100, 579–599.

Mehta, L., ed. (2010) *The Limits to Scarcity: contesting the politics of allocation.* London/Washington, DC: Earthscan.

MIU, Aliança RECOs, FEPHAC, Centro de Concentração Indígena Yuna Baka Nai Bai, CIMI *et al.* (2013) Open letter to the government of California. Available from: http://

www.redd-monitor.org/wordpress/wp-content/uploads/2013/04/Open_Letter_Acre_english_portugese_spanish.pdf [accessed 16 June 2013]

Moore, J. W. (2010) The end of the road? Agricultural revolutions in the capitalist world-ecology, 1450–2010. *Journal of Agrarian Change*, 10, 389–413.

Mouffe, C. (2005) *On the Political (Thinking in Action)*. London: Routledge.

Mueller, T. and Bullard, N. (2011) Beyond the 'Green Economy': System Change, Not Climate Change? Global Movements for Climate Justice in a Fracturing World. *In:* UNRISD conference: *Green Economy and Sustainable Development: Bringing Back the Social Dimension*, Geneva.

Muradian, R. and Rival, L. M. (2012) *Governing the Provision of Ecosystem Services*. New York: Springer.

Muradian, R. *et al.* (2013) Payments for ecosystem services and the fatal attraction of win-win solutions. *Conservation Letters*, 6 (4), 274–279.

No REDD+ (2012) Exposing REDD: the false climate solution. Available from http://www.redd-monitor.org/2012/10/30/new-video-exposing-redd-the-false-climate-solution/ [accessed 26 February 2014].

Norgaard, R. B. (2010) Ecosystem services: from eye-opening metaphor to complexity blinder. *Ecological Economics*, 69, 1219–1227.

OECD, Organisation for Economic Co-operation and Development (2011) *Towards Green Growth*. Paris: OECD.

Ostrom, E. (1990) *Governing the Commons: the evolution of institutions for collective action.* (The Political Economy of Institutions and Decisions series). New York/Cambridge: Cambridge University Press.

Pagiola, S. (2007) *Guidelines for 'Pro-poor' Payments for Environmental Services*. World Bank. Available from: http://siteresources.worldbank.org/INTEEI/Resources/ProPoorPES-2col.pdf [accessed 26 February 2014].

Pattanayak, S. K., Wunder, S. and Ferraro, P. J. (2010) Show me the money: do payments supply environmental services in developing countries? *Review of Environmental Economics and Policy*, 4, 254–274.

Peck, J. (2008) Remaking laissez-faire. *Progress in Human Geography*, 32, 3–43.

Peluso, N. L. and Lund, C. (2011) New frontiers of land control: Introduction. *Journal of Peasant Studies*, 38, 667–681.

People's Summit (2012) Final declaration. Available from: http://rio20.net/en/propuestas/final-declaration-of-the-people's-summit-in-rio-20 [accessed 27 February 2014].

Phelps, J., Webb, E. L. and Agrawal, A. (2010) Does REDD+ threaten to recentralize forest governance? *Science*, 328, 312–313.

Point Carbon (2014) Global carbon market contracts by 38% in 2013 as prices and volumes. *Press release*. Thompson Reuters, 20 January.

REDD Monitor (2013) *We reject REDD+ in all its versions*. Letter from Chiapas, Mexico opposing REDD in California's Global Warming Solutions Act (AB 32). Available from: redd-monitor.org [accessed 26 February 2014].

Rocheleau, D. (2014) Green land-grabbing and resistance in Chiapas. Forthcoming in *Journal of Peasant Studies*.

Rosset, P. (2013) Re-thinking agrarian reform, land and territory in La Via Campesina. *Journal of Peasant Studies*, 40, 721–775.

ROW, The Red Offset Working Group (2013) California, Acre and Chiapas: partnering to reduce emissions from tropical deforestation. Available from: http://greentechleadership.org/documents/2013/07/row-final-recommendations-2.pdf [accessed 26 February 2014].

Sassen, S. (2010) A savage sorting of winners and losers: contemporary versions of primitive accumulation. *Globalizations*, 7, 23–50.

Shrivastava, A. and Kothari, A. (2012) *Churning the Earth: the making of global India*. New Delhi/New York: Penguin.

Starr, A., Martínez-Torres, M. E. and Rosset, P. (2011) Participatory democracy in action: practices of the Zapatistas and the Movimento Sem Terra. *Latin American Perspectives*, 38, 102–119.

Stern, N. H. (2009) *The Global Deal: climate change and the creation of a new era of progress and prosperity*. New York: PublicAffairs.

Storm, S. (2009) Capitalism and climate change: can the invisible hand adjust the natural thermostat? *Development and Change*, 40, 1011–1038.

Suarez, D. (2013) You Cannot Manage What You Do Not Measure: Natural Capital Accounting at Rio+20. *In*: AAG Annual Meeting: *Blue and Green Economies 1: Paradigm Shift or Hegemonic Realignment in Environmental Discourse?* Los Angeles.

Swallow, B., Leimona, B., Yatich, T., Verlarde, S. J. and Puttaswamaiah, S. (2007) The Conditions for Effective Mechanisms of Compensation and Rewards for Environmental Services: CES Scoping Study. *Issue Paper No.3, ICRAF Working Paper no. 38*. Nairobi: World Agroforestry Centre.

Swyngedouw, E. (2010) Apocalypse forever? Post-political populism and the spectre of climate change. *Theory Culture Society*, 27, 213–232.

Swyngedouw, E. (2011) Depoliticized environments: the end of nature, climate change and the post-political condition. *Royal Institute of Philosophy Supplements*, 69, 253–274.

Tauli-Corpuz, V. and Baer, L.-A. (2010) *The Copenhagen Results of the UNFCCC: implications for indigenous peoples' local adaptation and mitigation measures*. New York: United Nations.

TEEB (2010) *The Economics of Ecosystems and Biodiversity: Mainstreaming the Economics of Nature: A synthesis of the approach, conclusions and recommendations of TEEB*. Available from: http://www.teebweb.org/publication/mainstreaming-the-economics-of-nature-a-synthesis-of-the-approach-conclusions-and-recommendations-of-teeb/ [accessed 15 April 2013].

TNC (2013) Green Economy 101: Protecting Nature's Ability to Provide for Us. *The Nature Conservancy*. Available from: http://www.nature.org/ourinitiatives/habitats/riverslakes/howwework/green-economy-101.xml [accessed 21 July 2013].

Torres, F. (2013) Our evolving collective vision of agrarian reform and the defense of land and territory. *Journal of Peasant Studies*, 40, 762–767.

Turner, R. K., Pearce, D. W. and Bateman, I. (1994) *Environmental Economics: an elementary introduction*. Essex: Prentice Hall.

UNCSD, United Nations Conference on Sustainable Development (2012) The Future We Want.

UNDESA, United Nations Department of Economic and Social Affairs (2012) Guidebook to the Green Economy 1.

UNDESA, United Nations Department of Economic and Social Affairs (2013) Guidebook to the Green Economy 3.

UNEP (2011) Towards a green economy: pathways to sustainable development and poverty eradication. *United Nations Environment Programme*, Nairobi.

Utting, P. (2012) Green economy: the new enemy? *UNRISD. News & Views*. Available from: http://www.unrisd.org/80256B3C005BE6B5/search/CAC549BBACB76E25C 1257A380050CBE4?OpenDocument [accessed 4 August 2012].

Van Hecken, G. and Bastiaensen, J. (2010) Payments for ecosystem services in Nicaragua: do market-based approaches work? *Development and Change*, 41, 421–444.

Vatn, A. (2005) *Institutions and the Environment*. Cheltenham/Northampton: Edward Elgar.

VCS (2013) *Jurisdictional and Nested REDD+*. Verified Carbon Standard. Available from: http://www.v-c-s.org/JNRI [accessed 17 September 2013].

Wade, R. (1997) Greening the Bank: The Struggle over the Environment: 1970–1995. *In:* D. Kapur, J. P. Lewis and R. C. Webb, eds., *The World Bank: its first half century*. Washington, DC: Brookings Institution, 611–734.

Watson, R. (2007) How to Finance International Low Carbon Investment. *In: Beyond Stern: financing international investment in low carbon technologies*, Centre Briefing Note 22. London: Tyndall Centre for Climate Change Research. Available from: http://www.tyndall.ac.uk/content/beyond-stern-financing-international-investment-low-carbon-technologies-and-projects [accessed 22 July 2014].

Wittman, H. K. and Caron, C. (2009) Carbon offsets and inequality: social costs and co-benefits in Guatemala and Sri Lanka. *Society & Natural Resources*, 22, 710–726.

World Bank (2010) *World Development Report 2010: development and climate change*. Washington, DC: World Bank.

World Bank (2012) *Inclusive Green Growth: the pathway to sustainable development*. Washington, DC: World Bank.

World Bank (2103) *Mapping Carbon Pricing Initiatives: developments and prospects*. Washington, DC: World Bank.

Wunder, S. (2013) When payments for environmental services will work for conservation. *Conservation Letters*, 6 (4), 230–237.

Xenos, N. (1989) *Scarcity and Modernity*. London/New York: Routledge.

Žižek, S. (1999) Carl Schmitt in the Age of Post-Politics. *In:* C. Mouffe, ed., *The Challenge of Carl Schmitt*. London: Verso, 18–37.

13 Political sellout!

Carbon markets between depoliticising and repoliticising climate politics

Chris Methmann and Benjamin Stephan

Introduction

In the first half of 2013 the European Emissions Trading System (EU ETS) became once more an issue of European higher politics. In April, the European Parliament stopped a proposal by the EU Commission with a vote of 334 to 315 to postpone the auctioning of 900 million emissions permits in an effort to halt the dwindling prices of EU ETS allowances – a plan known as backloading. In a second attempt on 3 July, the parliament eventually accepted the Commission's proposal with a vote of 344 to 311 in favour. In both instances, many of the conservative and libertarian MEPs voted against the proposal, arguing that '[t]he EU ETS was established as a market-based mechanism and must continue to operate according to market principles. We are therefore concerned about the impact of Commission intervention to adapt the auction timetable in order to manipulate the carbon price' (A spokesman for conservative MEPs quoted in Harvey and Vaughan 2013).

Having made your way through this edited volume, you, the reader, will at best manage a weary smile in light of the ingeniousness of this quote. In contrast to what is being indicated by these MEPs, there is nothing natural or fundamentally principled about markets. Social scientists have long done away with this myth, demonstrating that markets are indeed deeply social and political institutions (most prominently Polanyi 2001). Nevertheless, the implied quasi-naturalness of the EU ETS epitomises what this chapter is about – how carbon markets can have depoliticising effects, and how they may be repoliticised.

The political is, for us, no distinct sphere of society as it is usually understood – the strategic manoeuvring in and around parliaments, governments or during international negotiations. Instead, it is a latent feature of all social life. If we understand the social as a formation of sedimented and hence naturalised routines and taken-for-granted knowledge, the political refers to those issues and areas where existing social structures are challenged and actors try to establish new routines and truths.[1] Our account of the political thus follows Laclau's and Mouffe's 'post-foundational' understanding (Marchart 2007). This wider conception of the political 'has to do with the establishment of that very social order which sets out a particular, historically specific account of what counts as politics and defines other areas of social life as *not* politics' (Edkins 1999: 2, emphasis in the original).

For example, while in earlier centuries gender relations were thought to be unpolitical, they have increasingly become contested and thus entered the sphere of the political. The social and the political must be reconceptualised as a dynamic continuum, with moving boundaries and contradictory dynamics.[2]

This conception allows us to overcome the usual distinction between states and markets and point to the political dimensions of establishing and operating carbon markets. Paraphrasing Andrew Barry, we understand carbon markets as 'political machines', a 'set of skills, techniques, practices and objects with which it is possible to evade and circumscribe politics' (Barry 2001: 7). Carbon markets remove climate policy from the sphere of the political and thus help to keep basic social structures – the use of fossil fuels, industrial logging and agriculture – untouched. Moreover, they restrict consideration of these issues to a small group of carbon professionals while excluding the wider public. However, as Andrew Barry points out, technology can also be 'politically productive' as it has the potential to 'open up new sites of political contestation' and 'raise questions about the properties and capacities of technical objects and devices' (Barry 2001: 208). Dynamics of de- and repoliticisation can hardly be separated. In this chapter, we start by investigating the depoliticising effects of carbon markets for global climate governance as well as the possibilities for repoliticising this established and very technocratic way of climate protection.

Carbon markets emerging as the dominant climate policy

Carbon markets have become the prevalent policy option to organise greenhouse gas reductions (see also Simons and Voß, this volume). We argue, first of all, that this development has already depoliticised climate politics by locking it into the cage of carbon markets. What started out in the 1960s and 1970s as a scholarly debate among North American economists on how to incorporate environmental externalities into economic decision making has become, according to most recent estimates, a market worth 38.4 billion euros in 2013 (based on the value of traded allowances and certificates (Point Carbon 2014)). Coase (1960) suggested addressing externalities through market-based bargaining between involved parties, which made some scholars develop proposals for permit trading schemes for air (Crocker 1968) and water pollution (Dales 2002). The newly created Environmental Protection Agency (EPA) then started to test and work with these scholars (Cook 1988). During these early years, permit and hence emissions trading came to be articulated in a manner crucial for its subsequent success: today it is taken for granted that emissions trading is the most efficient policy instrument to address greenhouse gas emissions, outpacing carbon taxes or command and control measures. But as Lane (2012) has shown, this is not a natural trait. It is the result of a longer process through which permit trading has been constructed as an efficient policy tool pitted against so-called 'command and control measures', which have been framed as inefficient. The construction of permit trading in this manner was crucial, as it successively allowed both environmentalists, who demanded to avert climate change, and corporations, who demanded cost-reduction strategies, to coalesce around the notion of emissions trading and carbon markets and support them.

The emissions trading idea and the coalition promoting it internationalised during the 1990s. Supported by a pro-carbon trading coalition consisting of North American environmental NGOs such as Environmental Defense, the World Resources Institute and the Center for Clean Air Policy, as well as a growing number of transnational corporations, the US government called for a creation of an international emissions trading system in 1996. In 1997, after initial reservations of the EU and a number of developing countries, international emissions trading, complemented by two offsetting mechanisms – the Clean Development Mechanism (CDM) and Joint Implementation (JI) – became part of the Kyoto Protocol. Surprisingly, the emissions trading idea did not die in 2001 when US President Bush announced that the USA – the biggest supporter of emissions trading – would not ratify the Kyoto Protocol. Instead, the carbon trading coalition won over the EU Commission, who started to consider a European emissions trading system. Supported by a number of European corporations and industry associations, the carbon trading coalition also co-opted major European environmental NGOs such as Greenpeace, which originally had been highly critical of emissions trading (Meckling 2011; Stephan 2011). The EU Parliament and Council adopted a Commission proposal to create a European emissions trading system in 2003. It started to operate in 2005 and is, to date, by far the largest system in the world, constituting about 88 per cent of the volume of globally traded carbon allowances and certificates in 2013 (Point Carbon 2014).

Currently, the carbon market seems to be in crisis. Prices for allowances and credits are collapsing – not just in the EU ETS. The inability of governments to agree to a continuation of, or a successor to, the Kyoto Protocol has killed the demand for the CDM's Certified Emissions Reductions. As a result, the value of the carbon market shrunk by 60 per cent between 2011, when it peaked at 96 billion euros, and 2013 (Point Carbon 2014). Furthermore, carbon markets have been shaken by a series of scandals – fraud, cyber criminality or value added tax (VAT) carousels to name just a few (see below). While some economists question that low prices represent a crisis, low prices at least do not provide the necessary incentive to invest in low-carbon technologies, as the recent German surge in lignite consumption demonstrates (Ottery 2014). Yet, despite these challenges, carbon markets – at least for now – still represent the dominant response to mitigate carbon emissions. And the coalition of carbon market supporters continues to grow: trading schemes are being considered or established in countries such as Brazil, China or Mexico. In addition the World Bank is promoting carbon markets through its Partnership for Market Readiness. When the European Parliament voted over the backloading of the ETS, a large coalition of corporate actors pushed for the Commission proposal in order to strengthen the EU ETS (Alstom 2013). It is thus fair to say that carbon trading has become by far the dominant solution to global climate politics with supporters ranging from the NGO and business world, to governments, scientists and elsewhere. Nonetheless, it remains to be seen to what extent the current expansion will be successful. As a result, other policy options, such as a carbon tax or stronger emission standards for coal-fired power plants are often neglected in favour of the cap-and-trade approach.

Unpolitical carbon

Social science research has shown that markets are complex social institutions based on myriad routines and conventions (Fligstein and Dauter 2007; Prudham 2009). However, these routines have to be established and become mutually accepted. In the case of carbon markets, the international agreements or the domestic legislations that have created the respective markets provide only some of these routines and conventions, and they do so only in a very broad way. Most of them have been worked out 'on the fly' once the market is being implemented (see, for example, Callon 2009). A negotiation and discussion process among policy makers, scientists and market participants is necessary to create a common framework of reference – the foundation for this new social institution. The interaction between natural science and market procedures is particularly interesting in this regard. Many core aspects of the carbon market are based on scientific findings. But whereas it is okay for science to deal with uncertainty and have unresolved discussions about particular processes, 'scientific debates . . . must be silenced so that ecological information can be intelligible in the logic of capital' (Robertson 2006: 368). As the following discussion of different aspects of carbon market creation show, even though these discussions seem technical in nature they can be quite political since they involve struggles about what is perceived to be the right way of doing things. For a market to be able to function, these struggles have to be pacified and one option has to emerge as the proper, mutually accepted way of doing things (Stephan 2012). This is not to say that this pacification would ever be complete. As we will show below, it is always already dislocated and open to contestation. Moreover, the emergence of competing ways of commodification is always possible. In this sense, the following three sections show examples of how issues have been depoliticised during the creation of markets, but we also highlight the wider depoliticising consequences for climate politics in general.

Making carbon

The first important step in the development of carbon markets is the measurement of greenhouse gas emissions. Measuring emissions in most cases is neither a simple nor straightforward endeavour. It is highly complex, involving significant levels of scientific uncertainty and disputes on how to do it properly. The struggles around these questions have to be settled, too, and one approach has to be accepted as appropriate for the market to function properly.

Let us take a look at the example of forest carbon emissions. It is not possible to measure forest carbon emissions directly. Instead scientists have to find a way to approximate it. To determine global estimates, scientists resort to biome averages, the amount of carbon stored per hectare for different forest types (tropical equatorial forest, tropical dry forest, etc.), in conjunction with data from satellite-based remote sensing indicating land cover changes. Together, these two factors are used to determine how much carbon dioxide is being emitted. Yet, biome averages lack the necessary accuracy. Gibbs *et al.* (2007: 6) have shown that the results using different datasets vary drastically. For example, in their analysis the highest estimate of the carbon stored in Brazil's forests is 51 per cent greater than the lowest estimate, while for Indonesia the difference is 149 per cent. Determining a forest's carbon content through field inventories is significantly more accurate. Here, scientists measure each tree within a specific plot, to determine its biomass, from which they then derive its carbon content (for exactly how they do this, see Lovell and MacKenzie, this volume). In addition, they take soil samples to determine the amount of carbon stored in the ground. As field inventories are very laborious they are only done in a small number of sample plots across a forest for which the data are being collected. Similar to the biome average approach, these results are then used in conjunction with remote sensing data to determine the emissions from a forest. This final estimate can vary depending on a number of different aspects, the procedure through which the ground measurements are being conducted, the way of sampling field inventory plots or the manner in which satellite images are being analysed. To get results consistent enough for a successful commodification of forest carbon, the actors involved have to agree on uniform procedures.

With its *Guidelines for National Greenhouse Gas Inventories* (Aalde *et al.* 2006) and its *Good Practice Guidance for Land-Use, Land-Use Change and Forestry* (Penman *et al.* 2003), the IPCC outlined a three-tiered approach for measuring and accounting forest carbon and provided default values (conversion rates, emission factors, etc.) where different options had been discussed within the scientific debate. In so doing, the IPCC has provided the necessary orientation and silenced scientific debates (at least for policy makers and carbon market actors). This is one of the reasons why integrating Reducing Emissions from Deforestation and Degradation (REDD+) into the carbon market currently seems to be a feasible option for many actors. This stands in stark contrast to the UNFCCC negotiations in 2000/2001. During these negotiations, avoiding deforestation was excluded as an eligible project type from the CDM because, inter alia, many actors did not perceive it to be possible to measure forest carbon accurately enough (see, for example, Fogel 2005). In this sense, actors promoting carbon measurement techniques have successfully depoliticised and pacified an intense political debate by creating a seemingly neutral and transparent way of measuring the carbon content of a particular area, activity or project. And this does not only apply to forest carbon emissions, but also to other forms of mitigation projects. For example, CDM guidelines prescribe how much carbon a particular project is supposed to mitigate.

However, this depoliticisation comes at the cost of excluding larger structural issues from the carbon content equation. Reducing activities and projects to their immediate payoff in terms of carbon emission reductions in the pursuit of feasibility reduces ecosystems to carbon sinks and excludes wider structural issues that are difficult to calculate. A crucial element not considered is the question of innovation (Lohmann 2009: 507). The structural impact of an emissions reduction project is not considered, although planting a forest or implementing solar power generators will differ regarding further CO_2 emissions beyond the single project. The latter, for instance, might stimulate innovation in solar power generation, which in turn spurs the implementation of this technology, while a forest that is planted may degrade later on. In a similar vein, carbon prices do not account for the long-term consequences of choosing a particular emission reduction strategy. It can, thus, increase path dependency of existing social and economic structures (Lohmann 2009: 506). For example, the CDM creates an incentive to focus on cost-efficient projects which immediately deliver large amounts of Certified Emissions Reductions (CERs) – the low-hanging fruits – even though much more fundamental changes would be necessary. Their initial costs are much higher though (Driesen 2007; Prins and Rayner 2007). Measuring carbon thus disregards the impact of individual projects on wider social structures. And this is mostly due to the fact that these effects can hardly be measured in terms of carbon emissions (Spash 2010: 176). Moreover, even if we would know the exact amount of carbon mitigated by a particular project, it would still be necessary that these effects are additional to what would happen otherwise – and this counterfactual reasoning (see below) makes the assessment even more difficult. As Kevin Anderson of the Tyndall Centre for Climate Change Research illustrates in a UK parliamentary hearing with regard to wind turbine projects:

> [T]hose wind turbines will give access to electricity that gives access to a television that gives access to adverts that sell small scooters and then some entrepreneur sets up a small petrol depot for the small scooters and another entrepreneur buys some wagons instead of using oxen and the whole thing builds up over the next 20 or 30 years, so it is the same thing. The additionality test would be, if you can imagine Marconi and the Wright brothers getting together to discuss where they will be in 2009, easyJet and the internet will be facilitating each other through internet booking. That is the level of . . . certainty you would have to have over that period. You cannot have that. Society is inherently complex.
>
> (cited in Gilbertson and Reyes 2009: 54)

In this sense, the measurement of carbon, which is a precondition for the implementation of carbon markets, is a highly artificial decision, which may depoliticise and hence solve a highly politicised political discourse (as in the case of forest emissions) while simultaneously depoliticising climate politics in general by excluding a vast array of important questions from the carbon equation.

However, at the same time, it also opens up possibilities for repoliticisation through contestation, as we will see below.

The low-hanging fruits and the limits of the bowl

Another example for the depoliticising effects lies in the accounting for different greenhouse gases within a wider scheme of carbon markets. For that end, the ton of carbon dioxide equivalents (tCO_2e) (Paterson and Stripple 2012: 571–573) as the common accounting metric for the carbon market, and the development of the related concept of Global Warming Potential (GWP) are an example for such struggles and their pacification. As different greenhouse gases are covered through the Kyoto Protocol, there is the need to commensurate reductions in different gases and make them fungible. Gases like methane or the very potent HFC–23 (a by-product in the production process of refrigerants) are converted to carbon dioxide equivalents. The GWP of these gases defines their conversion rates. Carbon dioxide is the reference gas, which has been assigned a GWP of one. The GWP of other gases depends on their radiative forcing relative to carbon dioxide over a 100-year period. Based on this, a refrigerant plant in China reducing one ton of HFC–23 in the context of a CDM project receives CERs in the amount of 11.700 tCO_2e (see also MacKenzie 2009: 444–447).

Even though the GWP now serves as a key foundation for carbon markets, it was originally proposed as merely a 'simple means of describing the relative abilities of emissions of each greenhouse gas to affect . . . the climate' (IPCC 1990: 58). The goal stated in the first IPCC report was to provide a tool that would allow the comparison of different policy options. The IPCC authors introduced the concept with a disclaimer to policy makers, arguing that there is 'no universal accepted methodology for combining all relevant factors into a single global warming potential for greenhouse gas emissions . . . However, because of the importance of greenhouse warming potentials a preliminary evaluation is made' (IPCC 1990: 58). Policy makers, however, ignored the deficiencies and uncertainties attached to the concept when they adopted a slightly updated version (see IPCC 1995: 22) of this preliminary concept as the basis for accounting carbon emissions into the Kyoto Protocol. The 100-year time frame for which the GWP is calculated is a rather arbitrary number. In its reports the IPCC offered calculations for a 20-, a 100- and a 500-year period (Shine 2009: 470). However, depending on the time frame chosen, the GWP and hence the relative value of different CDM projects differs (see also Mac-Kenzie 2009: 444–447).

Closely related to the GWP, the tCO_2e was invented in the wake of the Marrakech Accords – a full four years after the Kyoto Protocol. Its emergence was rather uncontroversial and there have not been any issues raised since. The GWP story is slightly different. Its use in the context of the Kyoto Protocol and in relation to carbon markets has been described as problematic by a variety of scholars (see, for example, O'Neill 2000; Shine 2009; Skodvin 1999). Nevertheless, the scientific discussion about the uncertainties and limitations has been silenced in the policy world, enabling the development of the current national carbon accounting approaches and the creation of the carbon market. The fact that alternative metrics were not yet available during the first half of the 1990s might be an explanation of why no broader debate about this issue evolved.

A number of examples from the history of the CDM, the most important carbon trading mechanism within the UNFCCC, demonstrates what this means in practice. The most prominent case is that of HFC–23, which thus far generated the bulk of CDM credits. The high GWP of HFC–23 makes it the perfect candidate for generating a lot of CERs. The revenue generated by these projects was actually so high that manufacturers began to produce additional refrigerants in order to be able to mitigate more HFC–23 emissions, which could then be sold as CERs through the CDM (Gilbertson and Reyes 2009: 56). A similar pattern can be observed in the production of N_2O emissions (Schneider *et al.* 2010). Here, the CDM generated the demand for products whose production emitted greenhouse gases, and so turned the CDM on its head. Again, the artificial pacification of debates about the commensuration of different greenhouse gases results in the creation of loopholes through which structural problems that generate carbon emissions are excluded, while mitigation projects focus on the 'low-hanging fruits' such as HFC–23, with perverse incentives and effects. As we will see below, this depoliticisation has already laid the foundation for a later repoliticisation of this practice.

Not only are other structural drivers of greenhouse gas emissions excluded from political action, but other non-carbon-related problems are often disregarded when it comes to including certain activities into carbon markets as well. The most recent example of this problem dates back to COP-18 in Durban in 2010 (UNFCCC 2011). Here, the UNFCCC Meeting of the Parties decided to make carbon capture and storage (CCS) projects eligible for carbon credits. It is highly debatable if CCS is really that climate friendly. What is more, it subsidises and legitimises a very controversial technology by neglecting controversies about the dangers of storing carbon under the ground, corporate power or the inflexibility and incompatibility of coal-fired power plants with renewable energy. The logic of the CDM, however, fences these debates exclusively under the umbrella of carbon emissions and so tends to exclude these questions from consideration. Another example would be hydroelectric dams, which have massive social consequences, such as displacement (Haya 2007). For example, in 2009 more than 900 hydroelectric projects in China received or sought approval from the CDM (Copeland Nagle 2009: 23) even though they might create even more carbon emissions by decomposition than those that are saved through electricity generation (Graham-Rowe 2005). Nonetheless, the logic of the carbon market is inattentive to these questions.

Finally, there is a whole range of CDM projects that, from a common sense perspective, would rarely appear as sustainable. For example, afforestation has been used to create vast non-native eucalyptus plantations, which quickly generate CDM credits through carbon sequestration, but drain local water resources (Carbon Market Watch 2013). Land grabbing, violation of indigenous land rights and displacement have been reported in some cases. The experiences with the afforestation and reforestation under the CDM thus create sceptical views on the desirability of REDD+, which might include forest management into the logic of carbon markets. Again, forests are treated as carbon storages, but not as ecosystems and people's homes.

Against this backdrop, it is quite ironic that the suspected source of the 'swine flu' virus H1N1 is an agricultural factory destined to become a CDM project (Point Carbon 2009). Although project officials claim that the planned waste management system would have mitigated the outbreak of the epidemic, it is quite telling that the CDM dwells on such projects that pursue a model of agriculture, which is far from being sustainable. In this sense, carbon markets often ignore the entanglement of the sources of global warming with other social and economic problems and grievances. In other words, not only do carbon markets such as the CDM focus on the low-hanging fruits, they also put these into a very narrow basket and exclude everything from consideration that lies beyond the limits of this basket. Carbon markets thus have a depoliticising effect, but they simultaneously open up new possibilities for repoliticisation. When things such as the swine flu arise, this also affects the debate about the CDM.

Of parallel dimensions

While the previous examples are struggles that had to be solved before the market could start to operate, our third case – business-as-usual scenarios or baselines – is an example where it is constantly necessary to establish a particular way of seeing or doing things as appropriate. A business-as-usual scenario is needed to assess the amount of emissions reduced through a CDM or any other carbon offset project. It is also key to determine the additionality of reduction projects, that is, if the project is different from what would happen otherwise. These scenarios contain a number of counterfactual assumptions, describing a world in the absence of the particular reduction measures. The emissions reductions provided through a new wind farm, for example, are determined by comparing the emissions-free electricity produced through the turbines against the emissions of the average electricity mix that is transported through the grid (defined through what has been called grid factors). The counterfactual assumption in this case is that, in the absence of the wind farm, the consumers of its electricity would just consume the same amount from the average electricity mix. These assumptions do not consider whether the electricity might not be consumed at all or if this electricity demand might cause changes in the electricity mix.

One does not need to take a postmodern view on the world to determine that there is not one, objective baseline to be identified. Even the rather positivist economists and lawyers involved in the carbon market business acknowledge this: the UNFCCC defines a baseline as a 'scenario that *reasonably* represents the anthropogenic emissions by sources of greenhouse gases that would occur in the absence of the proposed project activity' (UNFCCC 2006: 3/CMP.1, Annex, paragraph 44, emphasis added). 'These methodologies allow project participants to quantify the estimated emissions in the most plausible alternative scenario to implementation of the project activity' (Baker and McKenzie 2013). What represents a reasonable or plausible baseline is determined through a technocratic expert discourse. The CDM Executive Board represents the institutionalisation of this discourse. It approves or refuses project proposals, inter alia, on the basis of the (im)plausibility of their baselines. This, of course, constantly involves a significant degree of struggle (see also Lohmann 2005: 218, 228): different proposals to calculate, for example, grid factors are pitted against each other. In order for the CDM and the carbon market to function, these disputes have to be pacified. A project proponent is well advised to base its proposal on what represent the dominant positions in the discourse in order to increase the chances for approval.

Even if we move away from the project level approach of the CDM this problem remains. In the previously mentioned case of REDD+, negotiators have been debating whether historic deforestation rates, predictions about future deforestation rates or a combination thereof should constitute the basis for baselines. Assumptions about future deforestation rates are themselves based on predictions about population growth, economic growth and future infrastructure development. This example shows that the amount of counterfactual assumptions that go into a baseline vary depending on the type of activity: while the wind farm is a relatively straightforward example involving a limited amount of counterfactual assumptions, a REDD+ baseline is more complex based on a significantly higher amount of 'What if?' presumptions.

What is more – and this is crucial for the depoliticising effect – counterfactual logics naturalise and so legitimise the status quo. This becomes apparent in the application of so-called supercritical coal technology, which improves the efficiency of coal-fired power plants (Gilbertson and Reyes 2009: 57). The implementation of this technology has become an accepted criterion for becoming a CDM project. Now, the CDM subsidises the construction of coal-fired power plants, for example, in India. Obviously, the slight improvement in power plant efficiency blurs the overall fact that climate change is mostly caused by burning fossil fuels, and that a significant proportion of carbon stored in the earth's surface may not be burned. Moreover, this happens against the backdrop of an officially acknowledged business-as-usual scenario in which the construction of coal-fired power plants is taken for granted. This narrows the scope for alternative political action – banning coal power stations in general – and thus excludes these options from political consideration altogether.

Repoliticising carbon markets

Even though carbon markets dominate today's climate politics, their hegemony is by no means carved in stone. They might represent the dominant solution in the current social and discursive structure but, as this structure is contingent, characterised through a fundamental instability, there is always the possibility for its destabilisation and hence for change. Laclau and Mouffe, from whom we introduced the notion of the political earlier in this chapter, pay particular attention to the contingent character of social structures in their hegemony and discourse theory (Laclau 1990: 33). They introduce the notion of dislocation to capture these uncontrollable moments when new developments challenge the established discursive order, reveal its contingency and hence have the chance to destabilise and potentially change it (Howarth and Torfing 2005: 16). Unforeseen events, such as the discovery of anthropogenic climate change or the global financial crisis, are examples of such dislocations; they could not be accommodated within the hegemonic discursive structures at the time but revealed their deficiencies and hence their contingent character. Once dislocations have taken place, the gaps that have occurred have to be filled: societies will struggle over the most appropriate way of dealing with climate change or the financial crisis. Dislocative moments, however, do not only occur at the obvious level as the aforementioned examples indicate. We can also find dislocations at the micro level. The scientific disputes that have been pacified during the establishment of the carbon market might erupt again. Or new scientific findings might challenge established market routines. Furthermore, actors do not have to wait for dislocations to magically appear. If they are able to reveal the tensions and inconsistencies of hegemonic structures and articulate appealing alternatives, dislocative effects can occur, too. We will use the remainder of this section to first highlight the fragility of the institution of carbon markets, pointing to a number of smaller dislocations that have occurred. We then turn to the attempts by opponents of carbon markets to repoliticise the issue.

A fragile institution

One of the major challenges global carbon markets are currently struggling with is their collapsing prices, primarily caused through an overallocation of emissions allowances in emissions trading systems. That is to say, more allowances are issued than are needed to cover the actual emissions. This is a structural phenomenon that can be found in pretty much every emissions trading system. The Kyoto Protocol's international emissions trading system is troubled by what has been called Hot Air; the prices for European Unit Allowances (EUA) have collapsed, forcing the EU Commission to come forward with the backloading proposal mentioned in the introduction; and even the alleged success story of the Acid Rain Program began with an initial period of overallocation. It is estimated that, without any additional actions by the EU, an emissions surplus of 2.6 billion EUAs would accrue in the EU ETS by 2015 – more than the total amount of annually issued allowances (Neuhoff and Schopp 2013: 10). In addition, the World Bank estimates that the potential supply of offset credits generated through the CDM between 2013 and 2020 exceeds the potential demand through the different emissions trading systems by 20 per cent (World Bank 2013: 21).

The recent price development and the repeated overallocations present a dislocative moment to the current dominant position of carbon markets, namely that they are regarded as policy instruments that efficiently coordinate mitigation efforts and help to transform the economy into a low-carbon one. There is the fear, however, that without a certain price level the carbon market is unable to trigger the innovations and transformations that are necessary. There are estimates that currently a price of at least 30–35 euros would be necessary per ton of carbon in regions like Europe to trigger the necessary shifts to stay below the 2 °C threshold by 2100 (see, for example, Bowen 2011). In early 2013, however, EUAs traded for less than 3 euros while CERs traded well below 1 euro. These developments are not congruent with the expectations that have been put into carbon markets and hence do not fit the dominant discursive structures.

Carbon markets as a crime scene

In addition to the overallocation and price problem, carbon markets have been shaken by a series of fraud cases. Some of them have led to investigations by both Interpol and Europol (see, for example, Interpol Environmental Crime Programme 2013). European member states alone have lost an estimated 5 billion euros, due to what is known as VAT carousels or 'missing trader intra-community fraud' (Europol 2009). In the case of an emissions trading system, setting up a VAT carousel is comparatively easy as no physical goods have to be moved over borders. All can take place virtually within a few seconds. Furthermore, there have been a number of incidents of cyber criminality involving hacked trading accounts or phishing attacks to access registry accounts. In 2010, seven German companies fell victim to a phishing attack, losing over 3 million euros and causing a week-long suspension of carbon trading in Germany and other EU member states (Interpol Environmental Crime Programme 2013: 24). A year later, the cement manufacturer Holcim lost 1.6 million credits due to a hacker attack, worth at the time more than 23 million euros (Krukowska and Carr 2011). More recently there have been reports of carbon boiler rooms where private individuals are being offered carbon credits as investments. These are often sold to them at higher than the market price and cannot – contrary to what is claimed by the companies selling them – be easily resold with a profit.[3]

These instances of fraud might be interpreted by some as a rather positive sign of maturing carbon markets, which like other markets now also have become interesting for organised crime. However, there have also been fraud cases directly related to some of the fundamental structures of the markets. In the case of the CDM and other offset mechanisms, the claims by project proponents about measured emissions, or the details of the baselines that have been calculated, have to be verified by external bodies. Transnationally operating certification bodies such as the German TÜV Süd, the French Société Générale de Surveillance (SGS) or the Norwegian Det Norske Veritas (DNV) provide such services. They are the largest verifiers licensed by the UNFCCC to audit and verify CDM and JI projects. However, they are paid by the companies proposing the projects. Payment is only made if the registration and verification process for a project is complete. Hence, there are substantial incentives to rubber-stamp projects where possible. The CDM Executive Board temporarily suspended DNV in 2008, SGS in 2009 and TÜV Süd in 2010 because their verification procedures did not live up to the required scrutiny (Szabo 2010). These cases challenge the image of an efficiently functioning policy tool. Particularly in the cases where criminal activity is linked to specific characteristics of the carbon market, they have the potential to dislocate the current discursive structures. As Barry notes, technology such as the carbon market 'disentangles an object from its relations, translating the object into a new form. Yet in doing so, it may serve to create new forms of entanglement' (Barry 2001: 209). In this case, carbon becomes entangled with a whole new range of criminal or fraudulent behaviour, dislocating the claim of an efficient and technocratic way of regulating the atmosphere.

Science and uncertainty

A third aspect which has the potential to dislocate the current dominance of carbon markets is the scientific uncertainty related to the measurement and accounting procedures upon which carbon markets are based. As we pointed out in the first section of this chapter, the creation of carbon markets is a deeply depoliticising process because it depends on the assertion of particular measurement and accounting procedures as mutually acceptable. Despite remaining uncertainties, actors have to come to an agreement on what the proper way of going about things entails. Yet this consensus – necessary for the market to function – is not eternal. It is, rather, a temporary stabilisation and institution of discursive structures. New scientific findings, for example, have the ability to challenge and dislocate this consensus. As Barry puts it:

> [T]he work of scientists can have political rather than anti-political effects. For scientific work can identify the weaknesses in this vast exercise in routine monitoring and measurement. Potentially at least, far from restricting the space of contestation, further scientific calculations may serve to open it up.
>
> (Barry 2002: 274)

The scientific work on issues related to the GWP might create dislocative effects that lead to a repoliticisation of the issue. Due to the uncertainties, the GWPs have already had to be adjusted for a number of gases. Hence, the numbers from the second IPCC Assessment report, that – defined through the Kyoto Protocol – serve as the basis for the carbon market, have been outdated (for the differences in the GWP of key gases between the Second and the Fourth IPCC Assessment Report see Table 13.1). The revision of the GWP numbers through science thus far has not had any consequences for carbon markets. They have simply been ignored and the system sticks to the 1995 numbers. Scientists, however, also work on alternative metrics which they now assume to be better suited to compare different gases. As different measures entail a change in the relative weighting of the greenhouse effect of different gases, they also raise the interest of many non-scientists. Countries such as Brazil or New Zealand with a high share of methane emissions, for example, would have a much smaller overall greenhouse gas footprint if the Global Temperature Change Potential (GTP) introduced by Shine *et al.* (2005) and featured in the Fourth IPCC Assessment Report (IPCC 2007: 215–216) was used instead of the GWP (Shine 2009: 469–470). We have yet to see how these scientific developments affect carbon markets. While the dislocative moments emerging from the GWP debate themselves will not result in the end of carbon markets, they are likely to cause changes in the practices and routines underwriting the market.

Active Repoliticisation of Carbon Markets

The repoliticisation of carbon markets does not solely depend on the events and developments previously outlined. It can also deliberately be brought about by actors. Even though many NGOs and other actors, initially critical of carbon markets, have joined the carbon market bandwagon during the past 15 years, there are groups and networks of activists and scientists that still oppose carbon markets. They have tried – thus far with limited success – to repoliticise the issue. Their criticism, however, remains relatively marginalised in the mainstream debate. Within the UNFCCC, Bolivia is one of the few parties that repeatedly argues against carbon markets. Among the major environmental NGOs, Friends of the Earth is the only one currently opposing carbon markets (see, for example, Clifton 2009). In addition there are smaller groups. Some of them, for example, Carbon Trade Watch, have a global focus, while others address the issue on the local level – for example, targeting individual CDM projects.

Carbon market critics have tried to repoliticise the issue by questioning the environmental integrity and effectiveness of carbon markets. They also have raised equity and justice concerns and criticised the lack of transparency.

Table 13.1 Changes in selected Global Warming Potential estimates between the second and the fourth IPCC Assessment Reports

	Second Assessment Report	*Fourth Assessment Report*
Methane	21	25
Nitrous oxide	310	298
HFC 23	11.700	14.800
Sulphur hexafluoride SF6	23.900	22.800

(IPCC 1995: 22; IPCC 2007: 212–213)

Carbon offsetting through the CDM in the developing world has been described as carbon colonialism (Bachram 2004). It is seen as a mechanism based on a system that has largely been devised in the developed world, by which industrialised countries' companies profit from offset projects they conduct in the developing world. These projects lock land and resources without adequately incorporating local communities into the decision-making process.

Critics, of course, also exploit the dislocative moments we have pointed out in the previous paragraphs, which have formed independent of their work. Just before the consultations about the possibilities of backloading started within the EU Parliament, Carbon Trade Watch, FERN, the Corporate Europe Observatory (CEO) and 41 other civil society groups issued a call to 'scrap the EU ETS' altogether (Scrap the EU ETS 2013). Instead of wasting its time with an emissions trading system, they argue that the EU should stop supporting fossil fuel subsidies and move towards a complete fossil fuel phase out.

Conclusion

For concerned environmentalists, the recent debate surrounding the aforementioned EU ruling on backloading should be deeply unsettling. As usual, the legislative procedure was accompanied by strong corporate lobby campaigns. However, the difference this time was that large corporations – among them oil multinationals such as BP and Statoil – were in favour of the proposed regulation (Alstom 2013). Among the environmental movements in the EU, this has caused (or supported) considerable suspicion regarding the EU ETS. If even those companies that generate large amounts of greenhouse gases and should, in theory, 'suffer' from such a regulation are in favour of it, what does this tell us about carbon markets in general? This chapter has sought to provide a tentative answer for this question. We have argued that carbon markets have deeply depoliticising effects. They narrow the scope of climate politics to those options that leave the fundamental causes of global warming – fossil fuel dependence, industrial agriculture, individual mobility – untouched. And they legitimise contentious policies, such as hydroelectric dams or CCS by labelling them one-sidedly as climate friendly and thus desirable. What is more, today carbon markets appear often as the most viable and efficient policy instruments and so become synonymous with climate policy in general, thus further ruling out other alternatives. However, this depoliticising tendency is not without tensions and contradictions. A number of scandals have shaken the discursive ground on which carbon markets rest. They have shown that the system is not without failure. And this provides a huge opportunity for environmentalists to put their concern to action. If they are willing to fight, these dislocations allow for delegitimising carbon markets as the dominant policy option in climate politics. And a different approach is possible: unable to get it through congress, US president Obama turned away from the option of a national carbon market and asked the EPA for stronger environmental standards by decree. This is but one example that shows there are alternatives, and it is viable to implement them. To look back to the first substantive chapter in this book (Lane, this volume) one alternative is to economically reconstruct the understanding of the environment that made emissions trading possible in the first place. Among the things that still stand in the way of these alternatives: carbon markets.

Acknowledgements

The authors would like to thank Richard Lane, Arno Simons, Johannes Stripple and Ian Rinehart for helpful comments on an earlier draft of this chapter.

Notes

1 In this sense, our account of the social is truly post-structuralist, combining a structuralist understanding of society with the political contestation and instability inherent in the prefix 'post-'.
2 For an assessment of the political in the broader field of environmental politics see Kenis and Lievens (2014).
3 For detailed information on a variety of cases see http://www.redd-monitor.org/tag/boiler-room/

References

Aalde, H. *et al.* (2006) Forest Land. *In:* S. Eggleston, L. Buendia, K. Miwa, T. Ngara and K. Tanabe, eds., *IPCC Guidelines for National Greenhouse Gas Inventories – Volume IV: Agriculture, Forestry and Other Land Use.* Geneva: IPCC.

Alstom (2013) Alstom one of 38 companies calling for EU ETS backloading to encourage low carbon investment. Available from: http://www.alstom.com/press-centre/2013/4/alstom-one-of-40-companies-calling-for-eu-ets-bacloading-to-encourage-low-carbon-investment/ [accessed 5 March 2014].

Bachram, H. (2004) Climate fraud and carbon colonialism: the new trade in greenhouse gases. *Capitalism Nature Socialism,* 15 (4), 5–20.

Baker & McKenzie (2013) *CDM Rulebook: establishing a baseline.* Available from: http://www.cdmrulebook.org/85 [accessed 5 March 2014].

Barry, A. (2001) *Political Machines: governing a technological society.* London/New York: Athlone Press.

Barry, A. (2002) The anti-political economy. *Economy and Society,* 31 (2), 268–284.

Bowen, A. (2011) *The Case for Carbon Pricing.* London: Grantham Research Institute on Climate Change and the Environment.

Callon, M. (2009) Civilizing markets: carbon trading between in vitro and in vivo experiments. *Accounting, Organizations and Society,* 34 (3–4), 535–548.

Carbon Market Watch (2013) Forestry/land-use projects in the CDM. Available from: http://carbonmarketwatch.org/category/sustainable-development/forestry-land-use-projects/ [accessed 5 March 2014].

Clifton, S.-J. (2009) *A Dangerous Obsession: the evidence against carbon trading and for real solutions to avoid a climate crunch – A Research Report.* London: Friends of the Earth UK.

Coase, R. (1960) The problem of social cost. *Journal of Law and Economics,* 3, 1–44.

Cook, B. J. (1988) *Bureaucratic Politics and Regulatory Reform: the EPA and emissions trading.* New York: Greenwood Press.

Copeland Nagle, J. (2009) Discounting China's CDM dams. *Loyola International Law Review*, 7 (1), 9–26.

Crocker, T. D. (1968) Some economics of air pollution control. *Natural Resources Journal*, 8 (2), 236.

Dales, J. H. (2002) *Pollution, Property and Prices: An Essay in Policy-making and Economics*. (New Horizons in Environmental Economics series). Cheltenham: Edward Elgar Publishing.

Driesen, D. M. (2007) *Sustainable Development and Market Liberalism's Shotgun Wedding: emissions trading under the Kyoto Protocol*. Syracuse: Syracuse University.

Edkins, J. (1999) *Poststructuralism & International Relations: bringing the political back in*. Boulder/London: Lynne Rienner.

Europol (2009) Carbon credit fraud causes more than 5 billion euros damage for European taxpayer. Available from: https://www.europol.europa.eu/content/press/carbon-credit-fraud-causes-more-5-billion-euros-damage-european-taxpayer-1265 [accessed 5 March 2014].

Fligstein, N. and Dauter, L. (2007) The sociology of markets. *Annual Review of Sociology*, 33, 105–128.

Fogel, C. (2005) Biotic carbon sequestration and the Kyoto Protocol: the construction of global knowledge by the Intergovernmental Panel on Climate Change. *International Environmental Agreements: Politics, Law and Economics*, 5 (2), 191–210.

Gibbs, H. K., Brown, S., Niles, J. O. and Foley, J. A.. (2007) Monitoring and estimating tropical forest carbon stocks: making REDD a reality. *Environmental Research Letters*, 2 (4), 1–13.

Gilbertson, T. and Reyes, O. (2009) *Carbon Trading: how it works and why it fails*. Uppsala: Dag Hammarskjöld Foundation.

Graham-Rowe, D. (2005) Hydroelectric power's dirty secret revealed. *New Scientist*, 4, 2488.

Harvey, F. and Vaughan, A. (2013) MEPs reject proposed reform of emissions trading scheme. Available from: http://www.guardian.co.uk/environment/2013/apr/16/meps-reject-reform-emissions-trading?INTCMP=SRCH [accessed 5 March 2014].

Haya, B. (2007) *Failed Mechanisms: how the CDM is subsidizing hydro developers and harming the Kyoto Protocol*. Berkeley: International Rivers.

Howarth, D. R. and Torfing, J. (2005) *Discourse Theory in European Politics: identity, policy, and governance*. New York: Palgrave Macmillan.

Interpol Environmental Crime Programme (2013) *Guide to Carbon Trading Crime*. Lyon: Interpol.

IPCC (1990) *First Assessment Report*. Geneva: IPCC, WMO, UNEP.

IPCC (1995) *Second Assessment Report – Working Group One 'The Science of Climate Change'*. Geneva: IPCC.

IPCC (2007) *Climate Change 2007. Synthesis Report*. Cambridge: Cambridge University Press.

Kenis, A. and Lievens, M. (2014) Searching for 'the political' in environmental politics. *Environmental Politics*, 1–18.

Krukowska, E. and Carr, M. (2011) Organized crime blamed for roiling $110 billion carbon market. Available from: http://www.bloomberg.com/news/2011-01-31/organized-crime-may-have-stolen-carbon-permits-amid-weak-security-eu-says.html [accessed 5 March 2014].

Laclau, E. (1990) *New Reflections on the Revolution of our Time*. London: Verso.

Lane, R. (2012) The promiscuous history of market efficiency: the development of early emissions trading systems. *Environmental Politics*, 21 (4), 583–603.

Lohmann, L. (2005) Marketing and making carbon dumps: commodification, calculation and counterfactuals in climate change mitigation. *Science as Culture*, 14 (3), 203–235.

Lohmann, L. (2009) Toward a different debate in environmental accounting: the cases of carbon and cost-benefit. *Accounting, Organizations and Society*, 34 (3–4), 499–534.

MacKenzie, D. (2009) Making things the same: gases, emission rights and the politics of carbon markets. *Accounting, Organizations and Society*, 34 (3–4), 440–455.

Marchart, O. (2007) *Post-foundational Political Thought: political difference in Nancy, Lefort, Badiou and Laclau*. Edinburgh: Edinburgh University Press.

Meckling, J. (2011) *Carbon Coalitions: business, climate politics, and the rise of emissions trading*. Cambridge: MIT Press.

Neuhoff, K. and Schopp, A. (2013) Europäischer Emissionshandel: Durch Backloading Zeit für Strukturreform gewinnen. *DIW-Wochenbericht*, 80 (11), 3–11.

O'Neill, B. C. (2000) The jury is still out on Global Warming Potentials. *Climatic Change*, 44 (4), 427–443.

Ottery, C. (2014) German brown-coal burning rises amidst calls for reform of ETS. Available from: http://www.greenpeace.org.uk/newsdesk/energy/analysis/call-ets-reform-drive-out-german-lignite [accessed 5 March 2014].

Paterson, M. and Stripple, J. (2012) Virtuous carbon. *Environmental Politics*, 21 (4), 563–582.

Penman, J. et al., eds. (2003) *Good Practice Guidance for Land Use, Land-Use Change and Forestry*. Geneva: Intergovernmental Panel on Climate Change.

Point Carbon (2009) Suspected epicentre of swine flu is CDM project. Available from: http://www.pointcarbon.com/news/1.1107815 [accessed 5 March 2014].

Point Carbon (2014) Global carbon market contracts by 38% in 2013 as prices and volumes. Available from: https://www.pointcarbon.com/aboutus/pressroom/pressreleases/1.3780672 [accessed 5 March 2014].

Polanyi, K. (2001) *The Great Transformation*. London: Beacon.

Prins, G. and Rayner, S. (2007) Time to ditch Kyoto. *Nature*, 449 (7165), 973–975.

Prudham, S. (2009) Commodification. *In*: N. Castree, D. Demeritt and D. Liverman, eds., *A Companion to Environmental Geography*. Malden: Wiley-Blackwell.

Robertson, M. M. (2006) The nature that capital can see: science, state, and market in the commodification of ecosystem services. *Environment and Planning D: Society and Space*, 24 (3), 367–387.

Schneider, L., Lazarus, M. and Kollmuss, A. (2010) *Industrial N_2O Projects under the CDM: adipic acid – a case of carbon leakage?* Stockholm: SEI.

Scrap the EU ETS (2013) Civil society groups urge European Parliament to forget about backloading and call for a phase-out of fossil-fuels. Brussels: Fern, Corporate Europe Observatory and Carbon Trade Watch.

Shine, K. P. (2009) The Global Warming Potential: the need for an interdisciplinary retrial. *Climatic Change*, 96 (4), 467–472.

Shine, K. P., Fuglestvedt, J. S., Hailemariam, K. and Stuber, N. (2005) Alternatives to the Global Warming Potential for comparing climate impacts of emissions of greenhouse gases. *Climatic Change*, 68 (3), 281–302.

Skodvin, T. (1999) Making climate change negotiable: the development of the Global Warming Potential index. *Cicero Working Paper*, (9).

Spash, C. L. (2010) The brave new world of carbon trading. *New Political Economy*, 15 (2), 169–195.

Stephan, B. (2011) The power in carbon: a neo-Gramscian explanation for the EU's adoption of emissions trading. *Global Transformations towards a Low Carbon Society*, 4, 1–20.

Stephan, B. (2012) Bringing discourse to the market: the commodification of avoided deforestation. *Environmental Politics*, 21 (4), 621–639.

Szabo, M. (2010) U.N. panel suspends two more carbon emissions auditors. Available from: http://uk.reuters.com/article/2010/03/26/us-carbon-un-suspensions-idUKTR E62P5E420100326 [accessed 5 March 2014].

UNFCCC (2006) *FCCC/KP/CMP/2005/8/Add.1 – Report of the Conference of the Parties serving as the meeting of the Parties to the Kyoto Protocol on its first session, held at Montreal from 28 November to 10 December 2005.* Bonn: UNFCCC.

UNFCCC (2011) *Modalities and procedures for carbon dioxide capture and storage in geological formations as clean development mechanism project activities. Proposal by the President.* Durban: UNFCCC.

World Bank (2013) *Mapping Carbon Pricing Initiatives.* Washington, DC: World Bank.

Index